The Handbook of MAINTENANCE MANAGEMENT

Joel Levitt

Industrial Press Inc.

Library of Congress Cataloging-in-Publication Data

Levitt, Joel, 1952 -
    The Handbook of Maintenance Management / by Joel Levitt - 1st ed.
    488 p. 15.6 x 23.5 cm
    Includes glossary and index.
    ISBN 0-8311-3075-X

TS
192
.L467
1997

INDUSTRIAL PRESS INC.
989 Avenue of the Americas
New York, NY 10018

FIRST EDITION

10 9 8 7 6 5 4

Dedication

This book is dedicated to the four most important women in my life.

Firstly I would like to thank my wife, Barbara. She has taught me steadfastness in her commitment to me. Thank you Barbara and I love you.

Secondly I would like to thank my two sisters. It is hard to imagine a better pair of sisters. Marjory has been for me my whole life (and even before). I really got her presence and love. She has taught me insight into people and leadership. Thank you Marjory and I love you. JoAnn has always given me perspective about the events in my life. I count on her ability to see clearly. Thank you JoAnn and I love you.

Lastly and by no means leastly there's my mother Sophie. She is a remarkable woman who raised three kids in love and is now teaching her grandchildren about unconditional love. Thank you for my life and your continued involvement in my families' lives. I love you.

Joel Levitt
April 1997

Contents

How to Use this Book   vii

1. Where We Are Today   1
2. Patterns in Maintenance   10
3. Strategic Assessment of Maintenance Options   20

PATTERNS

4. Selling Maintenance Improvements to the Whole Organization   27
5. How Assets Deteriorate   37
6. Estimating Maintenance Budget for Buildings and Equipment   42
7. Evaluating Maintenance—Two Questionnaires   48
8. Benchmarking Maintenance   66
9. How to Evaluate Worker Productivity   77
10. Maintenance Budgeting   87
11. Life Cycle Costing   96

STRATEGIES

12. The Science of Customer Service   106
13. Reengineering Maintenance   112
14. Contracting, Outsourcing   121
15. In-Sourcing   129
16. Guaranteed Maintainability   136
17. Maintenance Quality Improvement   139
18. ISO 9000 and Its Relation to Maintenance   146
19. Techniques for Continuous Improvement in Maintenance   153

20. Improving Maintenance Reliability—RCM 160
21. Accounting Issues of Maintenance 164
22. Maintenance Information Flow 175
23. Capturing Maintenance Information 186
24. What is PM? 209
25. How to Install and Run a PM System 215
26. PM Task List Development 231
27. Predictive Maintenance 247

SUPPORT FOR MAINTENANCE STRATEGIES

28. Projects that Put You in the Driver's Seat 261
29. Use of Statistics in Maintenance 273
30. Planning 278
31. Project Management 282
32. Estimating Job Duration 287
33. Maintenance Scheduling 292
34. CMMS—Computerized Maintenance
    Management Systems 298
35. Maintenance Parts and Part Vendors 325
36. Maintenance Stockroom and Inventory Control 336
37. The Internet and Maintenance 346
38. How Maintenance Interfaces to Other Departments 354
39. Elements of Maintenance Leadership 363
40. Craft Training of Maintenance Workers 375
41. Special Issues of Factory Maintenance 394
42. Special Issues of Fleet Maintenance 401
43. Special Issues of Building Maintenance 413
44. Special Issues of Field Service 437

APPENDIX

Resources 449
Glossary 463

How to Use this Book

This book is intended to be used in different ways by different groups. It was designed as a complete survey of the field for students or maintenance professionals, as an introduction to maintenance for non-maintenance people, as a review of the most advanced thinking in maintenance management, as a manual for cost reduction, as a primer for the stockroom, and as an element of training regime for new supervisors, managers, or planners.

I would like to present a curriculum—or road map—for some of the more frequent groups of readers. I encourage all levels of the maintenance profession to study different aspects of maintenance. Some of the best ideas come from people on the outside of the core management group applying lessons learned in other arenas.

New Manager: A new manager needs increased analytical skills, increased involvement with the maintenance budgeting process, and the ability to see the big picture. He or she needs to be able to lead the department with both a steady hand and an open door. The following chapters are heavy on the analysis aspects of the book and lighter in some of the other areas: 1–9, 11–23, 32, 38, 39–41, 43, and 44, depending on your business.

New Supervisor: The supervisor might be moving to management for the first time. He or she needs to know what management is about, and usually quickly. The supervisor is usually responsible for the PM effort, so that is covered in detail. The issues of worker productivity, customer service, and improvement of maintenance are core to this job along with the scheduling and planning functions. Read Chapters 1–5, 10, 13, 17, 20, 22, 24–28, 30–40, 41–44, depending on your business.

Update Skills to Current Ideas in the Field: This presupposes a long-time maintenance professional who wants to upgrade skills with the advances in the field over the last decade. These chapters skip maintenance basics and concentrate on new technologies, techniques and thinking: 1–3, 13–20, 31, 33–37.

Non-maintenance Person Review of Field: There is a decided lack of basic maintenance knowledge throughout our organizations. The field of maintenance is hard to approach because without a background, the language is strange. People seem to care about incomprehensible issues (as any technical field looks to outsiders). This choice of chapters introduces the fundamentals of maintenance. It will allow the outsider to join the conversation. Good readings for members of interdepartmental teams in quality, problem solving, RCM, etc. Read Chapters 1–5, 7, 8, 11, 12, 17, 18, 22–27, 35, 38, 40, 41 to 44, depending on your business.

Installing a CMMS: A computer system is a tremendous change for a maintenance department. This selection of chapters introduces many of the issues that an installation team is likely to come in contact with. Featured are the sections on the information flow through maintenance, the whole PM story, and the importance of the parts side of the business. Read Chapters 1–4, 8, 12, 23–31, 34, 35, 38, 41 to 44, depending on your business.

Cost Reduction Road Map: Long-term cost reduction requires knowledge of causes and effects in maintenance. Without that knowledge, one cost will decrease and another will pop up. The following chapters feature many techniques and approaches to cut costs: 1–4, 7–18, 20, 21, 23, 25–28, 30–31, 34, 35–36, 37–40, 41 to 44, depending on your business.

Storeroom Primer: The storeroom is a primary service provider to the maintenance department. They can make or break a maintenance effort. We encourage storeroom personnel and maintenance personnel to get to know each other's turf. Start by reading Chapters 1–8, 11, 20, 23, 25, 28, 34–37, and 41 to 44, depending on your business.

1
Where We Are Today

In traditional organizations, maintenance is a department responsible for the function of maintenance. As a department, it was supposed to defend itself from incursion by operators, housekeeping, drivers, users, and engineering, while servicing all maintenance requests. Its function was either to preserve an asset or to preserve the ability of the asset to safely and economically produce something. The differences can be seen when examining the mission of an art museum's staff in relation to a piece of art (preserve the asset itself) and the mission of an art museum's staff toward the boiler (economical and safe heating of water).

The distinction between function and department is real. As managers of a function we are more like owners of a business than members of a department. The customer does not care about what you can produce or how you do it. They are only interested in their needs. A smart organization identifies the customer's needs, determines metrics (benchmarks) that measure the satisfaction of the needs and then builds a maintenance function to service the needs with the least amount of wasted effort.

Maintenance, as a function, has logical processes:

Users of all types
Administration
Production   Part vendors
Warehousing   Engineering
Satisfied Customers   Planners
(need is fulfilled)
(awaiting the next
need)

☞   service request ➤ planning ➤➤   execution ➤➤ QC ➤➤ User
        service desk function   Dispatch function   Satisfaction
                                       Outside shops
                                       Outsourcing Contractors
                                       Tradespeople

Unsatisfied customers   ——PROCESSES ➤➤
(need something)   (we mobilize and deliver the goods)

1

The new paradigm of business is to focus on streamlining, reducing the inputs, and making the process (whether you produce cars, educate children, or move freight) more responsive and efficient. As a result, the mission of maintenance must conform with the continuous improvement of all processes in the organization. The new mission includes the idea that maintenance should work endlessly to reduce, and where possible, eliminate the need for maintenance.

To constantly improve the process of maintenance service, we need the best possible resources deployed in the most efficient ways. All possible inside and outside maintenance providers must be considered. Don't fall into the trap of limiting yourself to traditional providers or traditional techniques of doing business. For example, successful maintenance departments build long-term partnership-type relationships with outside vendors that would have been unthinkable a few years ago. They rely on these vendors to be experts in their fields. When chosen carefully, treated well, and allowed to make a profit, these vendors multiply your ability to respond quickly and inexpensively to your customer's needs.

The maintenance process starts with the customer's communication with the maintenance function. It also starts when a PM task list is due or when an inspector sees a problem and submits it for future corrective action. The process continues with the planning function, material ordering, prioritization, scheduling, and execution. The process is also responsible for communicating with the customer that the job is complete (or why it is not, etc.).

A bad process depends on the individual heroism of maintenance management, support staff, and technicians to serve the customer. Individual heroism—such as digging through the dumpster and re-machining a discarded part to fit—might occasionally be the only way to serve the customer. Good service with a defective process is the exception and the result of special effort. This heroism is seldom rewarded or even acknowledged. In fact, it may anger people. Take a look at what can happen to a simple job; also look at the amount of time spent on value-added tasks compared to non-value-added tasks.

A Simple Job that Went Bad!

A call comes in that the press is making screeching noises. The call is logged to the call log-in sheet, and then entered to

CMMS to create a work request. A new work order is created and reviewed by the manager.

The new work order is handed off to and reviewed by the maintenance supervisor and by the planner. The planner goes out to the job to see what has to be done. The planner pulls files on the asset to locate vendors and old job steps. The planner, production manager, maintenance manager, and production controller have a meeting and agree to take the downtime to repair the machine, but determine that the repair should be planned and staged to run in one day as the machine is needed for a critical run. The risk or consequence of an unplanned outage drives the decision.

The planner, using her experience coupled with the history in the maintenance files, calls vendors looking for a specific part for the obsolete press and finds one after much searching. The planner writes up the requisition for a part not in stock with results of her research.

The planner sends the requisition via inter-office mail to the purchasing department across town with next-day delivery; the requisition sits one day in the in-bin of the PA who gives it to a buyer. On the next day, the buyer starts making calls from scratch because policy forbids a single source without a letter of justification from engineering; in fact, the notes on the vendor are covered by a staple. The buyer is not personally familiar with the part and calls engineering for clarification of the vendor's questions. As this is happening, the planner calls the stock room to see if computer stock levels are accurate on parts showed in stock.

Engineering has to research the part and gets back to the buyer who has to reach the vendor again to find out that he didn't have what they needed. Engineering does not know about the contents of the maintenance files. Since the part cannot be found, the buyer asks the engineer to design a work-around for the part, which the engineer does on a hot ASAP basis since downtime is potentially involved.

The work-around takes three days at top speed. Two of those days are spent getting internal engineering approval for the work-around. The buyer orders new parts to replace the old unavailable part. The buyer informs the planner of his success by E-mail, and says that everything will be in next Monday. He does not mention the work-around.

The stock clerk calls back in one day with information that 6 of the 7 parts are actually in stock. The stock room enters a requisition for the missing part. The missing part is ordered. Delivery is within 48 hours.

The planner asks the supervisor if the special tool bought last year will be available for use next week. The supervisor doesn't return the call for three days. After hearing back from the buyer, the stock room, and the supervisor, the planner completes job package, reserves the parts in the stock room, and informs the supervisor that the job should be ready to fly next week when the last of the materials get in. All parts come in on Monday as promised, the supervisor reviews the work plan and assigns key skills to start the next morning.

The press is taken out of service and disassembled, and parts are taken from the reserve cage to the work area. Mechanics notice that the parts don't reflect the work plan and call the planner. Work stops. The planner can't figure what the parts are for and calls the buyer. The buyer, in a team meeting which cannot be disturbed, calls back right before going home and explains that engineering designed a dandy work-around.

The planner—in a panic—realizes that the other parts will probably be wrong using the work-around and calls engineering for drawings of the work-around. The engineer had gone home after a long day. The planner calls the original vendor and, armed with a special verbal authorization via car phone from the plant manager, orders the part to be taken to the airport for the next flight out service. Extra charges are only $250, which she considered a bargain.

The planner gets the maintenance crew back on the job doing everything but the section that requires the special part, and re-plans the job on the fly. The planner leaves for the 90-minute drive to the airport, meets the airplane, gets the part, drives back to the plant by 2:30 AM. Third-shift mechanics tighten the last nuts at 6:45 AM and start up the machine for the incoming day shift to run the job.

The planner returns to work by 9 AM, haggard but satisfied to face irate calls from the engineer, buyer, accountant, and not a word from production. And so, a new day dawns in the maintenance department.

On the other hand, a good process provides good service with only occasional acts of heroism. High quality lies at the core of the process. Improvements in the quality of maintenance delivered envelope the entire process from improved supplier relationships, to improved work order handling, to better understanding of true customer needs today and tomorrow.

Why Manage the Process?

What is the mission of your maintenance effort? It is clear, as you look at the housing stock, public buildings, old factories, and chemical plants, or some fleets, that the mission of some organizations is to spend nothing (or as little as possible) on upkeep. Spending nothing is a strategy that has certain payoffs and certain consequences. The payoff is lower costs for a period of time. The consequence of not spending could be catastrophic! The properties deteriorate at accelerated rates, trucks are unsafe, morale is poor at the factory, and you will have to fight OSHA, EPA, DOT and have a hellish existence. This is something to think deeply about and discuss with your advisors. Spending nothing on maintenance, in the words of Dr. Mark Goldstein, could subject "the senior member of the team to federal code violations which could result in a prison sentence."

Our assets age and deteriorate whether or not we manage the repair and maintenance effort. We can manage (try to control what happens and plan) or just let things happen (repair what breaks after it breaks). Webster's unabridged dictionary lists the definition of "to manage" as "to have charge of, to direct, to conduct, to administer." The word is from the French word manége which refers to the training of a horse. In maintenance, we mean manage as doing what you say when you say it, taking corrective action before a major breakdown, keeping to estimates, keeping to budget, and having a tight, low-cost operation where the customers are satisfied.

In the truck/bus world, this debate (to manage or not to manage maintenance) has raged since the beginning of the automotive era. In October 1926, for example, an article entitled "Maintenance costs cut through regular inspections" appeared in *Bus Transportation*. This article argued the advantages of periodic inspections versus breakdown mode maintenance. Earlier references include articles in the magazine of the Society of Automotive Engineers like "Care and Maintenance of Motor Trucks" in the April 1921 issue.

Each maintenance department has to deal with this issue in its own unique way, and the ideal level of management for each group of assets changes over time. As with all change, the people that make up the department will usually resist changing the level of management in any direction.

There are excellent reasons to manage the process of maintenance, including many of the following.

1. Managing maintenance will reduce long-term costs. The lowest cost overall long-term cost structure is the well-managed one!

2. A managed process of delivering timely maintenance can assure capacity. In a factory, fleet, or revenue-producing facility, the process can be designed to reduce downtime. Reduced downtime allows the organization to provide its product or service to its customers with fewer assets, thereby lowering costs.

3. Cost control. A well-managed plant, building, or fleet's cost of operation will vary in a controlled way (rather than random breakdowns totally driving yearly costs). Failure of major components will be published and budgeted for, well in advance of the event.

4. Maintenance management will help you meet a competitive challenge. You could be in a very competitive segment of the market. One of the edges that separates you from the competition is your effective management of maintenance.

5. Maintenance management will help preserve your physical assets. Management systems detect when something is going to fail (before it does). Good management initiates remedial action before deterioration goes too far. Breakdowns are always the most expensive and the most disruptive type of repairs. Effective management will extend the life of the asset. Proper cleaning techniques and processes will also help preserve the cleaned surfaces for longer life.

6. Maintenance management can increase the level of perceived service to the user. In buildings, a responsive department that stays ahead of deterioration and keeps the place looking clean will have less trouble keeping and attracting good tenants. In a production plant, the increased level of service can enhance the feeling of team effectiveness when requests are serviced in a short time. Good management provides a better level of service to fleet users or outside customers. Service is always defined according to the needs of the user, i.e., on-time, reliable, undamaged, clean, etc.

7. Maintenance management promotes fire and health safety. Frequently your users need to be reminded about what practices to follow for their own safety. Your management program inspects for hazards that could save a life or avoid a lawsuit. In addition to maintenance efforts, the housekeeping program will also help fire prevention and safety.

8. In the building maintenance field, maintenance management will help increase cash flow. Withholding rent to stimulate the manager/landlord to take care of some problem is not always allowed by the law but it is a time-honored tactic of tenants. Well-organized and documented maintenance reduces the need for the tenant to resort to such means and provides an excellent defense in court. Well-managed maintenance is more predictable. There will be far fewer budget busting catastrophes. Cash outflow becomes more reliable.

9. Employee needs. A managed department is a better place to work. The operators have well-maintained and safe equipment to operate. The mechanics have a productive and managed atmosphere that rewards a longer term view of the operation.

10. Maintenance management can improve your quality of life. While irate users and late night calls are part of the business, good management practices will reduce them to lower levels. Dealing with problems (rather than letting them sneak up on you) can reduce your stress level. You provide an important service. Whatever segment of the market you serve, your quality of life and your good feelings about yourself can be enhanced by serving it well.

Good maintenance practice can also impact: code violation fines, insurance costs and availability, liability costs, vandalism, and bad public relations (and neighbor action against your company).

The Multiplier Effect of Saving Money

Saving money from each good maintenance decision provides a small trickle of funds back to the organization. The trickle might be $10 a day from a relamping program to $1000's from an improved specification. These trickles come together as the manager continues his/her improvement effort. Soon the trickles become small creeks of cash flow, then babbling brooks, and finally rivers of extra cash flow.

There are tremendous amounts of money available from investments in maintenance and energy improvements. Few organizations are set up to allow re-investment of savings from prior successful improvements. Consider fighting for a percentage of the returns from your successful programs as investments in new improvements.

In most businesses that operate for profit, only a few dollars per hundred are left over for the shareholders; for every dollar of revenue, expenses soak up as much as 95¢. That is a ratio of revenue to profit of 1:20. At that rate, every dollar of maintenance savings is equal to $20 of extra revenue. A $1500/month maintenance savings is equal to a $30,000/month extra sales (if you just look at profit dollars).

The "Whack the Gopher" Game
of Maintenance Management

There is a game in the arcades where you get a mallet and face a group of holes. Every few seconds a gopher pops out of one of the holes. As you move to whack the gopher another one sticks his head up. Maintenance cost reduction efforts look a lot like the gopher game. Short-sighted "savings" in some areas can cause dramatic increases in other areas. When looking for savings by improving your maintenance savvy, remember the whole cost structure of maintenance expenses:

maintenance labor (your staff or a contractor), maintenance parts and supplies, housekeeping labor and supplies, energy (electric, gas, oil), water and sewer, ownership costs, depreciation (how long will asset last without a major rehab), capital replacement costs (also related to above but apply to replacement of equipment), regulatory penalties, EPA clean air/water, downtime costs, insurance, workmen's compensation, and liability costs.

Each maintenance area has specialized costs related to it:

In factories: OSHA, scrap, production speed, out-of-spec parts (seconds), loss of good will.

In field service: loss of customer, travel costs.

In fleets: traffic tickets, late penalties, uncomfortable customers, reduced trade-in, loss of good will.

In buildings: vacancy costs, rent withheld pending repair costs, reduced comfort, reduced value.

2
Patterns in Maintenance

Patterns in maintenance are the sum of the circumstances that define the organization. Patterns are the reflection of the organization's situation in the world, the product or service, the personalities of the key players, the financial situation, and other factors. When we live in a pattern long enough, it looks like the way the world is rather than just an individual pattern. It is essential to see the pattern as a pattern and not as some objective reality.

Patterns themselves are not good or bad. Instead of good or bad, a pattern might be said to serve or not serve the long-term interests of the organization. Also, the implication or consequence of a particular pattern might not have been what the bosses, owners, supervisors, mechanics, politicians, or other stakeholders wanted. There may be a negative consequence when an organization cuts back on maintenance or refuses to spend adequate funds to preserve an asset. Consider (see *Why Buildings Fall Down*, in the Appendix) the Mianus River bridge on the Connecticut Turnpike that collapsed on June 27, 1983.

To understand the whole story, we have to go back 15 years before the failure. In 1968 the design of the span was deemed unsafe and ceased being used. It was faulted for not having structural redundancy. Without redundancy, if any element of the structure failed, the consequence would be catastrophic. The lack of redundancy of the structures of the Mianus River bridge was known to the highway department and presumably communicated to others in the government during budget discussions.

Meanwhile, the maintenance budget for the turnpike continued to be squeezed by the Connecticut legislature. These dollars that were freed up allowed the legislature to fund more politically pressing needs. One consequence of the squeeze was to eliminate a second snooper truck for close-hand inspections of the structural elements of the bridges. The lack of a back-up snooper truck (the one they had was out of service for 11 months) for close-up inspections meant that the inspections were done with binoculars, from the ground, 75' away.

Corrosion caused failure of one of the bolts (all of the engineers had differing detailed explanations of what happened). Two cars and two trucks plunged into the river. The continued pattern of deferring maintenance, not allowing funds for back-up equipment, and spending elsewhere led to the consequence that we all read about in the paper.

The industry or business that your organization is involved in helps shape the dominant pattern, but does not dictate the pattern. Two different buildings within the same agency might have different patterns because the personalities and backgrounds of the key players are different.

Consider the case of two very large defense department printing plants with similar equipment, funding, output, customers, size, and the same mission. The only difference is one is run by an Army general and the other is run by an Air Force general (for those of you loyal to one of the services, both maintenance systems were excellent, and very different). The maintenance systems are so completely different that they cannot communicate with each other. The basic language and attitudes are different. Personality and background can be a major component of the dominant pattern.

In another case, the reason for the difference seems to be more complicated. The following case compares two school systems in different parts of North America. School district A is large (171 schools) and barely adequately funded. While they are tight financially like any other large school system, there is enough money to keep ahead of the deterioration. They are an old line craft shop where even lightbulb changes are being done by electricians. The dominant pattern is "it can't be done—we'll never change them" (whoever "them" is). Any idea to help the situation is greeted with negativity. The only way to transform an organization with this attitude is with explosive revolutionary change. Everyone sees what is going on around them in industry and government and they are scared. Too scared to change.

School district B is about half the size (80 schools). A large $300,000,000 bond referendum partially earmarked to rehabilitate the aging school buildings has failed. At first there was shell shock because the bond was part of everyone's plans for the future. You could feel the disappointment because an old cooling plant or a roof had to be maintained for a few more years. Twelve to eighteen months passed, and a new can-do attitude emerged from the ashes of the failed referendum. The attitude that emerged was: let's get together and get the job done with what we have. Let's use our people as efficiently as possible. They voluntarily reorganized themselves by geography to reduce travel time. Other changes included transferring appropriate workload to the chief custodian, rethinking truck inventory, faxing work orders directly to the craftspeople's houses, and more. They were motivated by the challenge rather than paralyzed with fear for the future.

An organization's reaction to the dominant pattern is called the strategy. The most effective strategy is ruled by the dominant pattern. The same strategy in both school systems or both printing plants would cause different results. The underlying engineering and deterioration in two schools or printing plants might even be the same. Since the rest of the elements that make up the pattern are different, the strategy cannot be the same. It is when maintenance investment is allowed to fall below the deterioration rate that the pattern's consequences become a problem.

When considering the pattern, consider what group or entity has control of the shop's resources. In some shops, the blaring of the paging system clearly schedules the mechanical and electrical staff. In others, an engineer seems to be in control.

Who is in Control?

1. Mechanics are in substantial control of themselves. They review what is going on and assign themselves.

2. Equipment operators collar maintenance people as they walk around. A maintenance worker is not allowed to "just say no." This masquerades under the guise of providing great customer service.

3. Maintenance supervisors or managers are actually running the show. They can stand the heat when they assign people to activities that are going to be important to the whole operation rather than ones that are currently important to the requestor (even if the requestor is the University president).

4. Operations, production, building, fleet, and distribution manager has the control of the shop. Your great plans to PM the stuff on the third floor goes out the window at a perceived or real need.

5. Most breakdown dominated shops are truly run by the broken machines. The breakdowns scream so loud that supervisors and managers can only respond. Their plans and schemes will just have to wait.

Some of these patterns are tough for maintenance effectiveness. Others force maintenance professionals to adopt an excellent maintenance attitude (even if they didn't want to). The patterns discussed below are just examples with cute names. Your organization has a pattern which is far more complex than any mentioned here. These patterns can occur on an organization-wide basis, a plant or location basis, or on a department basis. It also could be argued that people have patterns.

Cheap: Cheap is distinct from scarcity because cheap is a choice where funds could be made available. Private organizations, where the profit flows into a real person's pocket, might have cheap as the pattern. A shelving manufacturer had "cheap" down pat. He had 10 junky presses in the hope of having one or two run. No money was invested. Organizations also might want to allocate their funds in areas other than asset upkeep. This would be common when a plant is being sold. Most buyers look at cash flow, product line, and only minimally at deferred maintenance or the quality of the equipment. You will not be paid much more (if anything) if your equipment is 100% maintained as opposed to 75% maintained. The extra 25% maintenance is lost money for the seller.

For sale: When an organization is for sale it works on cosmetic maintenance that will appeal to the buyer's eye. In real estate it is called improving the curb appeal. In vehicles you clean the vehicle and replace the obvious wear items (like brake pedals, mats, etc.) to improve the appeal. In one factory they painted all of the equipment. Deep substantive maintenance is avoided unless absolutely necessary.

Gung-ho growth: Some companies/agencies grow at tremendous rates. Think of Compaq in the 1980's or Netscape in the 1990's. Think about a new governmental agency just created. These organizations present unique challenges to their new maintenance departments. This pattern presents different problems compared to the same-sized organization that is well established. One of the new transmission plants in Ohio followed this pattern. Over 50% of their workload was getting the bugs worked out rather than actual maintenance. It would take a few years before they had a handle on their maintenance needs.

Guru-of-the-month club: There are all kinds of business "gurus" that will tell you how to transform your organization. If you're curious about this, go to the business section of any bookstore. Instead of really doing the hard work of substantive transformation, some organizations enlist the latest, most popular business guru. They will have meetings and workshops for all hands. Endless energy will be spent, and some good might come. Top management then loses interest or nerve and moves on to the next guru. This pattern is disruptive to maintenance because it violates the constancy of purpose needed to gain ascendancy over deterioration.

No shutdown: Think about the air traffic computer center or the utilities in an ICU (Intensive Care Unit of a hospital). The pattern is driven by no shutdown—ever. The costs might be peoples' lives or millions of dollars. Maintenance dollars appear out of the woodwork if an operating room is disrupted or an ICU goes down. There are severe liability and marketing consequences to ignoring this pattern.

Scarcity: This is present where the funding is only a percentage of what is needed to preserve the asset. It is seen in governmental agencies, small businesses, failing organizations. Scarcity creates strange decisions such as not having enough money to change all the bearings, so you just change the really bad one. Too much scarcity for too long creates higher than normal maintenance costs, downtime, and quality problems because of short-term decisions designed to keep cash outflow to a minimum. Scarcity is problematic when deferred maintenance starts to build up, which could crash down at any time and bury your team.

In 1995/6, the budget impasse in Congress caused agencies to operate on 75% of the last year's budget. A year or two like that in a well-maintained building would not usually be catastrophic. A few more years would be catastrophic for the physical plants involved. Without some extra resources there is no good solution (the only solutions don't involve traditional maintenance such as a sale of the asset and leaseback, transfer the product line to a vendor or competitor for private labeling, etc.). Some organizations go through this pattern periodically. This becomes the *cost crunch* pattern. When you have a tough quarter, when the price of oil or pulp is in the basement, there is a tendency to pull in. Even well-thought-out maintenance investments will be deferred.

Short sightedness: This is seen in organizations with bonuses tied tightly to quarterly performance. Some organizations allow maintenance to have a percentage of revenue dollars. In a pulp mill, when the price of pulp was high, shipments were also high. The mill ran around the clock and all year round. The maintenance department had a percentage of the large pot and was floating in money. The problem was that no downtime was allowed to spend all of the budget. When the price of pulp dropped, there was plenty of downtime allowed but no budget money. They were not allowed to save money from flush times for projects to be done during scarce times. Short sightedness dominated the maintenance in this plant. Short sightedness is also seen where the people in control of maintenance have no maintenance knowledge and serve other masters. A good example is a city stadium like Veteran's Stadium in Philadelphia, where the sports teams Phillies (baseball) and Eagles (football) play. Good business practices dictate that a percentage of the rent paid to the city should have gone to upkeep. Instead most of the rent went for immediate consumption of value to the political bosses (keeping rec centers and libraries open, certainly an admirable use of funds). The result is that 15 years later the stadium needs 32 million dollars in repairs to keep it safe, where a few million spread over 10 years would have kept it in shape (based on a March 1, 1992 article in the *Philadelphia Inquirer*).

Throw it out: This is the strategy of assets being used and discarded. Rent it and return it. Many organizations sadly view their workers in this light. Nothing is fixed beyond a band-aid type repair. The auto rental business is based on owning billions worth of assets and not doing very much maintenance (allowing the warranties and car dealerships to perform and fund the maintenance needed). When the warranty runs out (and sometimes before) they sell the asset and replace it. This keeps the need to organize and manage maintenance to a minimum, and serves the customers who want late-model cars.

Virtual corporation: This is related to the throw-it-out philosophy. In some situations this philosophy works very well. Many movie studios own nothing. They rent, borrow, build their assets for as long as the project is going and then dispose of everything. They build a virtual company to produce the movie and then disappear when the movie is complete. There is not much maintenance in a virtual organization at the organizational level. You do see maintenance by the subcontractors (the firm that rents the cameras does do maintenance on them and includes that cost into the rental fee).

Cash cow: When a popular and mature product has paid for all of the facilities and development costs it becomes a cash cow. A cash cow brings in money without much effort from management. No one knows when a cash cow will shrivel up and die. The attitude is to put as little money into the production machinery as possible beyond what is necessary for basic uptime. This is related to the next strategy, tomorrow you may die.

Tomorrow you may die: We all know people who conduct their lives as if they will die the next day. They take no long-term positions. A good example of this attitude can be found among certain politicians who may lose the next election. Another example involves a major beverage bottler who rotated the best and brightest of the young potential managerial candidates through several key jobs. This is an imported concept and works very well in Japanese plants. Before you could be an executive you got to run a bottling plant, a distribution fleet, and a warehouse. You spent 1 or 2 years as manager of each area. The problem was that the people had to make an impression (being a shining star is not necessary or desired in a Japanese plant) immediately without necessarily knowing or understanding each area. The result is that these managers disrupt maintenance and know they will not have to clean up the mess. All the energy seeks things that can be impacted in the short term. It is similar to short sightedness as a pattern, with one major difference—these people know what they are doing (making a future mess) and do it anyway. A good long-term job is irrelevant to an outgoing beverage heir apparent or lame duck politician.

Speed to market: This pattern is characterized by a whole organization focused on getting the products from concept to the racks at the store. This attitude has a great impact on maintenance. Organizations driven by speed to market tend to make do with what they have and use it in ways it has never been used before. It might take too long for the official tooling so they adapt something to work. There is an infectious can-do attitude. It is tough on the maintenance function, but is personally rewarding.

Partnership: This is a powerful model where you enroll your vendors (and also inside departments) in solving the engineering problems. The bearing vendor is your expert and applications engineer for all bearing problems. The lubricant vendor plays the same function for lubricant issues. The advantages of forming partnerships is having a depth of expertise that most organizations cannot afford. The disadvantage is getting ripped off by poorly chosen partners.

Wheeler dealer: A wheeler dealer has no interest in the deal already conquered or the shopping center already bought, but looks for other deals to do. As a consequence, cash is held up to fund deals rather than cleaning up the deals already made. To the wheeler dealer, cash is king because it is the seed corn that deals grow from. It is much easier for the wheeler dealer to get other people to come along with a deal when they can show some of their own cash invested. The problem for maintenance is that the cash might come from deferred maintenance.

Rolls Royce: This kind of organization has the best of the best at almost any cost. They have the latest buildings, vehicles, equipment. Everything is well maintained. Usually the source of money is very high margins, public money, or great personal wealth. It is great fun to work there but it can spoil you for the real maintenance world if you have to change jobs.

Core values: Many organizations go through retreats, seminars, and groups to uncover their core values. They then work to capitalize on their core values in the marketplace (even public entities go through this process). This is great if maintenance is included; if not, it can be a disaster for the maintenance effort.

Safety: A safety-driven organization tends to have a well-thought-out maintenance effort. Maintenance and safety are so closely related that to be good at one you have to be good at the other.

Serious analysis: When an organization is run by analysis, then all the efforts are analyzed—often to death. There is a school that looks for hard numbers for everything that the company wants to do. This is not bad, and many times it is great for a maintenance manager that speaks that language. It becomes a problem when analysis becomes a substitute for actions.

Customer service: Anyone who has visited a bookstore and looked through the business section has seen some of the hundreds of books on improving customer service. The topic is hot in the 1990's. This is because there are choices, and the average consumer will go with the organization who surprises him/her with great service. In this model, maintenance does pretty well if it can be shown that customers will suffer—even better if all of the behind-the-scenes departments begin to be thought of as internal customers, then maintenance gets some very positive attention.

Tinker's paradise: In a tinker's paradise everything is made in-house from raw materials. I'm sure they would prefer starting from iron ore. The maintenance manager is highly skilled and hands-on. Sometimes when a builder of equipment or a truck mechanic makes good and has a company, it becomes a tinker's paradise. This becomes a company identity from the owner who is a tinkerer. Maintenance departments pride themselves on putting nothing outside and being able to do anything.

3
Strategic Assessment of Maintenance Options

Maintenance is war. The enemies are breakdown, deterioration, and the consequences of all types of unplanned events. The soldiers are the maintenance department. The civilians we protect are production, office workers, drivers, and all the other users of our organization's assets.

What is the pattern?	Define the strategies	What logistical support does the strategy need
assets	CMMS	analysis capability
mechanics	PM	computerization
layout	RCM	training
morale	PCR	time to think
economics	TPM	engineering
skill sets	PdM	purchasing
business	TPM	accounting support
culture		

Patterns

Military students study historical battles with an eye toward identifying the pattern of conditions which dominated the outcome. The conditions could include the topography, size of the enemy force, morale, motivation (are they defending their homes and families), skills and experience of the enemy, the ordnance available, the opposing general, and hundreds of other factors. Any one (or a combination) of factors could have had a decisive contribution to the outcome.

20

The tools available to identify the dominant pattern are economic modeling, component life analysis, failure curve fitting, equipment life cycle review, surveys, and exception reporting.

To choose the best strategy, the maintenance leadership must recognize different patterns. These patterns include accepting the weaknesses and strengths of the current plant, crew, management team, attitude, equipment age, purchase policies, and business conditions. Pattern matching proceeds without the assumption that you can change the department. It proceeds from where you are today. Recognition that your department is purely knee-jerk reactive might indicate that the best strategy is to build your department into the best fire fighters in your industry and support what is already strong in your culture.

Some organizations have responded to the increasingly competitive environment (the pattern) with a strategy of cutting costs. Lawrence Aubry of NSC Technologies talks about the folly of that approach (*Maintenance Technology*, June 1995): "Cutting costs while using the same ineffective process is a symptom of a lack of strategy, not a strategy itself. Strategy is understanding customers in enough detail to identify them, the problems you solve for them, and how you do it so that they are bound to you rather than someone else."

A strategy is supported by a collection of tactics. Each of these tactics is best given the right situation. The most advanced technology-based inspection program will fall flat on its face where there is no time or will power to shut down equipment that is not broken. A basic TLC (tighten–lube–clean) system might give the best outcome. Different equipment and differing service requirements respond to different tactics. An isolated machine with low downtime cost and exposure might be allowed to break (ignored until breakdown), while the same type of machine in a mission critical application might be intensively looked after in the same facility.

Many firms install a CMMS to catapult themselves from a strategy called bust'n'fix (ignore it until it busts) to one called proactive maintenance (where the maintenance action takes place before the breakdown). When they make the installation with inadequate logistical support, they shouldn't wonder why they still have so many unplanned events even though they are filling work orders and doing PM's as specified. The problem lies in the strategy not being supported.

We have many strategies and weapons at our disposal—some new and some old, some complex and some simple, some effective in one theater of operations and some better in another. Each strategy works best with the support of the correct weapons. The Nazi blitzkrieg depended on air support to soften and intimidate the ground resistance. Without that air power, the strategy would have failed.

Each weapons system has characteristic logistical support needed for optimum effectiveness. In a war, troops need food, bullets, medical support, bathrooms, sleep, and many other types of support. Support for troops can be manipulated within narrow boundaries. Troops can fight hungry, tired, or hurt. Their effectiveness is compromised as more and more support is missing.

Maintenance logistical support includes people, parts, systems, space, tools, good advise, company backing, training resources, access to equipment, access to information, and time. Each strategy has characteristic resources needed. In many ways, the choice of strategy depends on the level of logistical support that the maintenance effort can expect from the organization. To start a PM system off without trained people with the time and tools to win their battle (find and correct deterioration before breakdown) is like issuing guns without issuing bullets.

Frailty

In any war, human frailty is one of the greatest allies of the enemy (more soldiers died from jeep accidents in the Korean conflict than from enemy fire). In maintenance, dangerous human frailties include ignorance, not paying attention, fatigue, laziness, arrogance, somnambulism (sleep walking!), stupidity, cheapness (low bid), institutional rigidity, and corporate gluttony.

Alternative Strategies to Management of Maintenance

We live in a society where almost any service is available for a price. If we want to concentrate on our main businesses and not on maintenance, we have that alternative. We no longer have the option to ignore the proper management of assets. The costs are

Some Maintenance Strategies in Common Use*

Strategies

Outsourcing and contract maintenance	let someone else have the problem, just write a check
TPM—total productive maintenance	operator dominated maintenance (in-sourcing)
Challenge	make every request for resources an inquisition, always justify any use of resources
Full service leasing	separate use from ownership, just use the asset
Business solutions (including selling the building or product line)	stick to core competencies
Quality is #1	use quality components; the cost of the part is only a small part of the picture
Life cycle cost	look at total cost of ownership and use
Throw it away, disposable assets	use it then discard it; don't fix, just replace
Partnerships with vendors	lean on your vendors, and form a virtual company

Sampling of tactics

Automate	use machines and computers to do some of the work
Bust'n'fix	let it break
Bust'n'fix with management hysteria	let it break with a line mechanic right there to fix it
Design for maintainability	learn from the past to get it right the first time
TLC—tighten, lubricate, and clean	PM without the inspection (without brainpower)
Inspect/correct	PM with inspection and corrective maintenance
PCR planned component replacement	replace components before significant numbers of failures
RCM re-engineering and reliability centered maintenance	analyze and improve designs
High-tech communications	use all the technology to help the customer get what they want (E-mail, fax, page, cell phone, on-line work request)

Some Maintenance Strategies in Common Use (cont.)

Sampling of Tactics **(cont.)**

Replace problem "children"	replace old, outmoded, maintenance-intensive equipment
PdM—predictive maintenance	high-tech inspections
In-source work	find another internal group to perform maintenance tasks
Warrantee	buy for the good warrantee and retire equipment at end
Periodic shutdown	shut the line down, replace everything that moves
Condition-based maintenance	inspect and repair on a reading or condition
Certified users	intensively train the users in how to use the equipment, create a certification
Tools	upgrade the tools of maintenance, let mechanics pick!
Standardize	use the same equipment parts, even if you standardize on a heavier or bigger one

*Some items partially adapted from *Uptime* by John Campbell, published by Productivity Press.

far too high. From a cost effectiveness point of view, there are few alternatives; you can manage your maintenance or do one of the following.

1. Outsource your maintenance function. This is an increasingly attractive option for some sectors of industry. The building management field has used this strategy to good effect. The contractor can then make a profit by managing your maintenance. Your business might be diaper manufacturing or printing, while their strength is maintenance. In some situations, an outsider can run your maintenance effort better than you, make a profit, and still save you money. This is most common in building maintenance. A bank might hire a contractor to take care of a remote branch to save money and improve service.

2. Spin off maintenance as a separate profit center. Hire someone to run the separate company and give them an incentive

Process of Improving Delivery of Maintenance

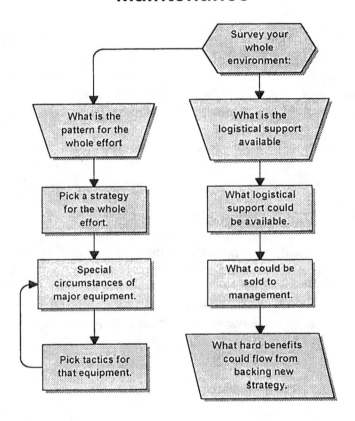

based on the new company profit earned. Because of the advantages of this method, many organizations set up maintenance subsidiaries. These subsidiaries are especially useful if the main organization is in some other business such as education, wholesale building products, beverage bottling, etc. This is a popular option in the fleet industry. The maintenance shop actually goes outside for extra business and can produce a substantial profit. To insulate itself from cutbacks, one large urban hospital maintenance department now successfully services local office buildings for profit.

3. Sell your equipment or trucks to the operators and make them business people. They own their own business. They become responsible for the management of their own small (1 unit) maintenance department. Owner-operator fleets are very popular among common carriers and in the textile industry.

4. In the fleet world, there is an additional option: turn your maintenance "problem" with all of the trucks over to a full-service leasing company that owns and maintains the equipment. The Ryder's and Leaseway's (or about 15,000 smaller lessors) would be happy to make a profit by owning and managing your fleet. They cannot only lease equipment but they will lease drivers or trips. The leasing companies can design a service that will expand to fit your peak periods and contract when you are slow. Until recently, even large fleets (such as Sear's Signal Delivery fleet) was operated under a full-service lease. Multi-function fleets such as airports and local governments can be handled by some of the specialized full-service lessors.

4
Selling Maintenance Improvements to the Whole Organization

The language of management is money. Few managements have the expertise, interest, or time to be concerned with the details of maintenance. Many maintenance professionals make a fundamental mistake and approach management as higher level maintenance people and talk maintenance to them.

We are in an extremely competitive battle for the organization's investment dollars. Investments in maintenance can earn big returns. In addition to rigorous financial justification, we must sell our strong suits, which are cost avoidance, reduced liability, improved customer satisfaction, and reduced downtime. While these areas are important, we must overcome the attitude that maintenance doesn't add value to the product and that it is a pure expense. The shift takes place when you realize that all organizations make a product or provide a service such as educating kids, picking up tons of garbage and moving it to the dump, converting pulp to paper cups, or whatever.

Maintenance can be seen as an independent company. Maintenance's product is the capacity in the machines, buildings, and vehicles to provide the services. We must shift our thinking and then shift the thinking of our organization. This requires a battle plan.

Prove Your Case—Steps to Take

1. Start off any justification with some background about the maintenance effort because few managers ever think about maintenance (when nothing is currently broken). Bring up the following in this section of the presentation.

27

a. If it is appropriate, calculate the average age and the total number of assets under your stewardship.

b. Estimate the *replacement* value (not book value) of the asset and the percentage of maintenance dollars to asset value.

c. The total value of the maintenance inventory and the number of SKU's. Also: number of vendors, number of material requisitions or PO's per year, and the amount of money your department returns to the community to local vendors and contractors.

d. The amount of uptime or downtime if that is relevant to your business (or capacity).

e. Current hours available after system imposed time. Quantity of current requests flowing into maintenance from users including projects, installations, other events. Also, a list of all of the user departments served.

f. Other things to mention are the number of completed work orders, number of completed PM tasks, number of maintenance workers per $1,000,000 of assets or sales over the last few years.

2. Through accounting analysis, prove the costs of the current operation. This would be expressed in dollars per year or cost per unit output (dollars per patient bed-day, or ton of steel, ton-mile). If the analysis is for a new unit, then check the cost of the old unit. Look at all cost areas for this proof.

3. Through sound economic modeling, prove the cost of the new operation. This would also be expressed in dollars per year or cost per unit output. Look at all cost areas. If the analysis is for a new unit, then check the cost of the new unit.

4. Show explicitly the return in financial terms and secondarily in human terms. Expect the financial issues to dominate the conversation even if they are the least important issues to you.

5. Whenever and wherever, include other departments in your presentation. This includes the production department, cost accounting, quality. Let each tell their story of the impacts of breakdowns or other events.

6. Xerox Corp. publishes a presentation training kit called *Training By Objectives*. It states that when making a presentation, know four things: always know who will be in your audience; know why your presentation is important to them; know what your boss wants to accomplish from the presentation; finally, know explicitly what you would like to get from the presentation.

7. Jay Butler, in *Maintenance Management*, has a list of action items for different groups within your organization. I call these groups stakeholders. Everyone touched by maintenance has a stake in how you do business. No one has a stake in the future way of doing business until they are sold, persuaded, and enrolled in joining the bandwagon. Some of the non-financial arguments are spelled out later in the chapter. Some of the highlights from Jay's lists are as follows.

   a. *Management:* neatly prepared materials, make an appointment, forward an outline of the presentation to your boss before the appointment, memo department heads not invited to appointment what you are doing and that you will update them afterwards, always talk cost reduction effort, make sure at least two members of your team actively participate in the session.

   b. *Users/customers:* solicit their help in preparation, assign part of the presentation to one of their team members, get away from their phone system, use their reports, stress improved response or improved service.

   c. *Maintenance support and technicians:* what will the project mean to us, why it will help the customers, what are the steps, this is not a layoff/downsizing or reengineering program, couple the presentation to an improvement for the maintenance team (such as building a break room, improving some specialized tools, etc.).

Techniques to Evaluate Competing Investments

Few organizations use all the methods to evaluate competing investments. Determine what techniques are used in your organization, learn them and use them.

Return on investment (ROI) is the most commonly used measure for investments. ROI is expressed as percentage of return you earn per year. If the yearly income varies, you can evaluate each year separately or average the years together (see ARR).

ROI of common investments:

Savings account	3.25%
Money market	5%
Mutual fund	12%
Small corporate investment	50%
Capital improvements	30%

Formula: ROI (return on investment) = yearly income / total investment.

Example: Replacing old-style fluorescent fixtures with new technology and electronic ballasts has an investment in a school of $300,000 or $150 per fixture. The reduction in energy, ballast replacement, and lamp replacement will yield a savings of $75,000 per year. The ROI calculation: 25% = 75,000 / 300,000. 25% is below the school's standard ROI of 33%

The school could not justify the investment until they contacted their utility who offered a rebate of $45 per fixture or

Rebate 2000 × $45 = $90,000 recalculation with the rebate results in a new ROI calculation:
$210,000 (new investment) = $300,000 (old investment) - $90,000(rebate)
33% = $75,000 / $210,000.

The school was able to make this investment.

Average rate of return (ARR) is the same as return on investment (ROI) over the entire life of an investment. The ROI will vary from year to year. Once you add all the return together and all of the investments together you can determine the ARR.

ARR Formula: ARR = average yearly income after tax/investment over life.

Example: Springfield Controls purchased a small public warehouse 50,000 sq. ft with material handling equipment for a total investment of $1.4 million. It was entirely paid for with internal funds. The *average* net income (after all expenses and taxes) over the years of the analysis is $210,000.

ARR = $210,000(average income) / $1,400,000 (total investment)
ARR = 15%

It's interesting to note that by borrowing you can sometimes significantly improve the ARR (or ROI). Why is this true? Consider the impact of borrowing funds where the rate you pay is below your ARR or ROI requirement. The organization will earn a return on the borrowed money equal to the spread between the ARR and the loan interest rate. The U.S. government supports this type of decision by making the interest deductible from the organization's income tax bill! As companies found out in the late 1980's and early 1990's, there is significant risk in excessive debt because you have to make payments even if sales go down and profits evaporate.

Payback method is the second most common method of evaluating investments. It involves determining the number of years (or months) it will take to pay off your investment based on the investment's return. The payback method is frequently used along with ROI and is the reciprocal.

Formula: payback in years = total investment / yearly income from investment.

Organizations are vitally interested in how soon their money will come back. In the relamping example above, the payback improves from four years to three years with the rebate.

Other Methods of Evaluating Investments

There are many other ways of evaluating competing investments. Some methods are used to pinpoint certain aspects of the investment (such as cash flow analysis or first year performance). For increased rigor, firms will run several types of analyses on the same investment to complete the picture.

Cash Flow Analysis: Most organizations use cash flow analysis to plan their overall investment program. A few will look at the cash flow from an individual maintenance improvement investment unless it is large in relation to the cash flow of the organization. The idea is to plot the movement of cash into and out of a project or an investment. Some investments (installing a PM system) require constant monthly outflows of cash for a long period of time (1–2 years) before providing returns. Other investments (such as re-powering a line haul tractor to improve efficiency) require significant cash on day one and provide immediate return. Some analyses are calculated on monthly rather than yearly cash flow. Cash flow analysis is a powerful tool to coordinate several investment projects. You can alter the timing between projects to minimize the cash out and maximize the overall return on investment.

year 1	year 2	year 3
CF=Outflow-Inflows	CF=Outflow-Inflows	CF=Outflow-Inflows

First Year Performance: This method looks at the ROI for the first year only for the competing investments. If two investments have similar average returns, the one with better first year performance may be the better overall investment. Some organizations that stress cash management (or are short on cash) look very hard at how the investment will act in its early stages.

Present Value/Present Worth: Present value/present worth techniques are by far the most complete (and complex) because they take into account when investments occur and when income is received. The basic theory is that money's value decreases over time—receiving $50,000 today has more value than receiving the same money five years from now.

Most organizations have internal standards of the value of money over time. The money used for your maintenance improvement could have been invested in a low-risk T-bill or mutual fund and earn a return. Other organizations use their borrowing rate as the factor in present value calculations. Present value formulas discount the return from your improvement by the internal rate of return assigned by the finance department. Any yield after that discount is over and above the cost of the money.

In the late 1970's, the prime interest rate was around 20%. This meant that competing internal investments had to provide very high returns to compete with market-type investments. During that period comparatively few maintenance investments could be made. In the early 1990's the prime fell to 6% which enabled all kinds of investments.

Simplified Formula for Net Present Value:

$$NPV = \text{Sum of } t{=}0 \text{ to } n \text{ (Net Return in year } t/(1 + k)^t\text{), where}$$

$$
\begin{aligned}
k &= \text{your internal rate of return} \\
n &= \text{total number of years} \\
t &= \text{the number of the year} \\
^t &= \text{to the power of } t.
\end{aligned}
$$

Notice that the denominator gets smaller quickly when k is larger. That means if k is larger then that investment requires quicker payback to be justifiable. Also the returns in the farther future are worth less and less today.

Example: A 5-year investment in a PM system with a 10% and a 20% rate of return
$(k = 0, 0.1$ and $k = 0.2) \, t/(1 + k)^t$

Year	[Investment]/return	$k=0\%$	$k=10\%$	$k=20\%$
1	[50,000]	[50000]	[50000]*	[50000]*
2	[30,000]	[30000]	[24793]	[20883]
3	40,000	40000	30053	23148
4	45,000	45000	30735	21701
5	45,000	45000	27941	18084
Total		**50,000**	**13935**	**[7950]**

*In this case, the initial $50,000 investment was made on day 1 of the project, and the $30,000 was invested at the end of the second year.

The conclusion is that an investment evaluated on a flat basis (without the time value of money) shows an average rate of return of 32% (($130,000/5 years)/ $80,000 investment). The payback is under 4 years. When you add in the time value of money the investment does not pay when competing with 20% investments (loss of $7950), it does pay when compared to 10% investments (extra return of $13,935).

In the next case, the effect on the bottom line is the critical factor. *Maintenance investments flow directly to the bottom line.* In non-profits, the language is different but the effect is the same: same students transported for lower dollars, same governmental offices for less cost per square foot, same streets plowed for less dollar.

Another issue is the excitement to the management of the various alternative investments and the way they are presented. A half-million dollars in sales seems much more exciting than $25,000 to $50,000 of maintenance savings (especially if the proof takes more than 3 sentences). The sales department may also have provided a 'nice' presentation and invited the president to a meeting with the prospective customers (he's now already involved).

Case Study: Effects on the Bottom Line

Tony Cavanaugh is the president of Springfield Manufacturing. He's been in the steel fabrication business since 1969 and has seen a lot of changes. Tony prides himself on the fact that Springfield is a profitable fabricator with the net about 7% of sales.

Barbara Strathmore, Springfield's sales manager, has proposed an expansion into some new areas. This expansion would require $94,500 in new equipment. The expansion would bring in an estimated $450,000 in new sales revenue per year.

Tom Duvane, maintenance manager, has proposed investment in a new computerized PM system. Calculations show the return was going to come from reduced parts in stock, increased up-time and reductions to maintenance costs. Tom also said that the system would allow his existing staff support more equipment. The investment is $75,000 with returns of $25,000 in the year one and $50,000 in the second and subsequent years (at present utilization figures).

Tony is not inclined to make both investments in the same year. Which one is better, and which one should be done first (even if they were both the same)?

Sales department investment:

Return per year (extra revenue × net profit percentage): $450,000 × 0.07 = $31,500
3 year payback = $94,500 investment / $31,500 average return
Payback (years to pay off investment; total investment/ return): 3 years.

Maintenance department investment:

2 year payback = $75,000 investment / $37,500 average return
Payback: *2 years.*

Questions to consider: How is Tom Duvane, maintenance manager, going to compete with the investment opportunities offered by the sales department and other departments?

Frequently there are overshadowing business reasons for a decision. In this case, what if the major customer requested the new capability? In another case, what if the market was getting more price competitive?

Non-financial Selling Points

Good maintenance management will help support the mission of the organization. The following table shows how good maintenance supports organizational objectives (partially adapted from the work of Darrell Travis, a project manager at Tompkins Associates).

Non-financial Selling Points for Improvements in Mainte-

Organization's Mission	How maintenance can support mission
Timely delivery, timely response to customer	Increased equipment availability, increased uptime, increased predictability
Expansion of market share	Increased output from existing asset base, lower cost per unit
Cost reduction	Greater asset utilization, reduced stockroom investment, avoid unnecessary capital replacements, decrease maintenance costs
Reduce number of people	Increase effective work time through better planning, increase cross skilling, reduce lost time
Better quality, ISO 9000 certification	Increase reliability and repeatability, support for documented procedures for ISO-9000, proof of meeting quality specification of customers
Better decision making	Better and more accurate information for life cycle analysis, ease the use of activity based costing, improved information on the performance of different equipment
Improved safety and regulatory compliance	PM will allow tracking and enforce completion of tasks relating better compliance with (in U.S.A.) FDA, USDA, DOT, EPA, etc.

Stakeholders

Each group in your organization has a stake in how you conduct maintenance. Each group is a constituency or a stakeholder. Adapt your argument to your audience. Use the language (and issues) of your organization to sell a PM program. In every organization there are issues which are more important than any others. You want to sell your improved maintenance management investments using these issues.

Organizational benefits of improvements to maintenance process:

1. Reduce the size and scale of repairs.
2. Reduce downtime.
3. Increase accountability for all cash spent.
4. Reduce number of repairs.
5. Increase equipment's useful life.
6. Increase safety of operator, maintenance mechanic, and the public.
7. Increase quality of output.
8. Reduce overtime for responding to emergency breakdown.
9. Increase equipment availability.
10. Decrease potential exposure to liability.
11. Reduce stand-by units required.
12. Ensure parts are used for authorized purposes.
13. Increase control over parts, reduce inventory level.
14. Decrease unit production cost.
15. Improve information available for equipment specification.
16. Lower overall maintenance costs through better use of labor and materials.
17. Lower cost/unit (cost per ton of coal, cost per widget, cost per student).
18. Improve identification of problem areas to know where to focus attention.
19. The frequency of user detected failures will decrease; decreased user problems translates to increased satisfaction.
20. Early detection avoids core damage.
21. Equipment has a breakdown curve: once over the threshold, failures increase rapidly and unpredictably. Working lower on the curve adds predictability.
22. Your inspectors are your eyes and ears into the condition of your equipment, facilities, and fleet. You can use their information on decisions to change your equipment make-up, change specification, or increase availability.
23. Predictability shifts the maintenance workload from emergency fire fighting due to random failures to a more orderly scheduled maintenance system.
24. Reduce energy costs.

Choose which items are most important to different stakeholders. Include these items in your "pitch" to the whole organization.

5

How Assets Deteriorate

One of the major elements of the pattern is dictated by the way assets fail. Eventually, equipment, fleet vehicles, and buildings require maintenance, and all fail in fairly characteristic ways. Our maintenance approach and support systems (such as stores, computer support, engineering support, etc.) need to be sophisticated enough to detect which critical wear curve is most likely to be most typical of the asset's deterioration. Once the curve is selected we must locate where on the critical wear curve we are and react accordingly. One complication is that every component system on each asset is on its own deterioration/failure curve. The electronics, belts, motors, gears, and sensors are all deteriorating in different patterns.

Random: The probability of failure in any period is the same.

> *Example:* The windshield of a vehicle will fail when it gets hit by a pebble. The probability of a failure is unrelated to life span. The windshield does not wear out in the traditional sense. This curve is common in electronics and in systems that become obsolete before they wear out.

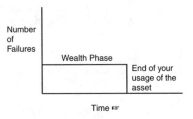

Infant mortality: The probability of failure starts high, then drops to an even or random level.

> *Example:* Many electronics systems fail most frequently during initial burn-in. After an initial period, the probability of failure from period to period doesn't significantly change. Most complex systems of any type have

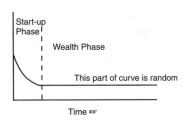

37

high initial failure rates due to warranty-type problems.

Increasing: The probability of failure slowly increases over time or utilization.

    Example: Consider the jaws of an aggregate crusher. These are massive blocks of manganese steel that get worn out by the rock. They wear in a predictable way and the probability of failure increases gradually throughout their life.

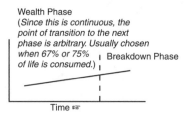

Increasing then stable: The probability of failure rapidly increases then levels off.

    Example: An electric heating element in a hot water heater. The probability of failure increases as the unit is turned on and then stabilizes to a random level.

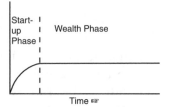

Ending mortality: The probability of failure is random until the end of the life cycle then it increases rapidly.

    Example: This failure mode is characterized by mechanical systems that wear until they reach a point and then they are at significant risk of failure. Failure modes related to corrosion usually go along until the amount of metal left is marginal to support the structure. Failure rates dramatically increase when this level of deterioration is achieved.

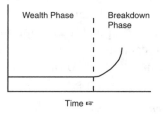

Bathtub: This curve is the combination of the infant mortality and the ending mortality curves. Probability starts high, then levels off, then starts to rise again.

    Example: Trucks initially have high failure rates due to defects in labor and parts and intrinsic design flaws. They then fall into a flat section of the curve until one of the critical systems experiences critical wear. After critical wear the whole reliability of the vehicle drops and the number of maintenance incidents increases until complete failure. This is a common curve for systems dominated by wear-out.

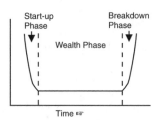

Proposition: The best strategy for dealing with maintenance is closely related to the type of failure and deterioration curve that dominates the asset deterioration. Each phase can be described, and countermeasures can be designed to either extend (as in the wealth phase) or minimize (as in the start-up or breakdown phases).

1. *Start-up phase* (Infantile mortality). This is represented most strongly by the infant mortality and bathtub curves. It includes failures of materials, workmanship, installation, and/or operator training on new equipment. Frequently the costs are partially covered by warranty. Unless you have significant experience with this make and model asset there is a lack of historical data. The failures are very hard to predict or plan for, and it is difficult to know which parts to stock. This period could last from a day or less to several years on a complex system. A new punch press might take a few weeks to get through this cycle, while an automobile assembly line might take 12 months or more to completely shake down. Be vigilant in monitoring misapplication (the wrong machine for the job), inadequate engineering, and manufacturer deficiencies.

Countermeasures: Start-up cycle (designed to shepherd the asset through this cycle and into the next)

Enough time to test run equipment

Enough time and resources to properly install

Operator training and participation in startup

Operator certification

Operator and maintenance department input into choosing machine

Maintenance and operator inputs to machine design to ensure maintainability

Good vendor relations so they will communicate problems other users have

Good vendor relations so you will be introduced to the engineers behind the scenes

Maintenance person training in the equipment (and periodic retraining when project wears on)

Maintenance person training in startup

Latent defect analysis (run the machine over-speed and see what fails and reengineer it)

RCM analysis to design PM tasks and reengineering tasks

Rebuild or reengineer to your own higher standard

Formal procedures for startup (possibility of videotape?)

2. *Wealth phase.* All curves have a wealth phase (except where the asset is not strong enough for the job, then it goes directly from startup problems to breakdown cycle). The bathtub, infant mortality, random, ending mortality have a well-defined middle phase. This cycle is where the organization makes money on the useful output of the machine, building, or other asset. This can also be called the use cycle. The goal of PM is to keep the equipment in this cycle or detect when it might make the transition to the breakdown cycle. After detecting a problem with the machine or asset, a quality-oriented maintenance shop will do everything possible to repair the problem. After proper startup the failures in this cycle should be minimal. Operator mistakes, sabotage, and material defects tend to show up in this cycle if the PM system is effective. The wealth cycle can last from several years to 100 years or more on certain types of equipment. The wealth cycle on a high-speed press might be 5 years, while the same cycle might span 50 years for a low-speed punch press in light service.

Countermeasures: Wealth cycle (designed to keep the asset in this cycle)

PM system

TLC—tighten, lubricate, clean

Operator certification

Periodic operator refresher courses

Close watching during labor strife

Audit maintenance procedures and check assumptions on a periodic basis

Autonomous maintenance standards

Quality audits

Quality control charts initiate maintenance service when control limits cannot be held

Membership in user or trade groups concerned with this asset

3. *Breakdown phase.* This is best represented in the bathtub and ending mortality curves. The increasing curve also has a breakdown phase but it is harder to see when it starts. This is the cycle that organizations find themselves in when they do not follow good PM practices. Breakdown phase is characterized by wear-out failures, breakdowns, corrosion failures, fatigue, downtime, and general headaches. This is a very exciting environment because you never know what is going to break, blow out, smash up, or cause general mayhem. Some organizations manage life cycle three very well and make money by having extra machines, low-quality requirements, and toleration for headaches. Parts usage changes as you move more deeply into life cycle three. The parts tend to be bigger, more expensive, and harder to get. The goal of most maintenance operations is to identify when an asset is slipping into the breakdown phase and fix the problem. Fixing the problem will result in the asset moving back to the wealth phase.

Countermeasures: Breakdown cycle

PM system (with emphasis on component replacement)

Maintenance improvement

Reliability engineering

Maintenance engineering

Feedback failure history to PM task lists

Great fire fighting capability

Superior major repair capabilities

Great relationships with contractors who have superior repair and rebuild capabilities

6
Estimating Maintenance Budget for Buildings and Equipment

One of the quickest ways to determine the pattern in an organization is to see how much money is available for maintenance in relation to the size and use of the asset. Some organizations squeeze maintenance, others (fewer and fewer) overspend in the maintenance area. The idea behind these formulas is that the maintenance investment required by an asset is related to its size, replacement value, and usage. Without this investment, the asset will *inevitably* deteriorate.

Many organizations squeeze the maintenance department's budget and reduce the dollars for maintenance every year. Every reduction below the base need level that is not covered by increases in efficiency, productivity, or by maintenance improvements will result in *deterioration*. This formula includes all direct maintenance labor, materials, and overhead. It does not include janitorial services. This quick estimate might be off by 30% and still will provide useful information.

There are only a few ways to deal with significant shortfalls of maintenance dollars. The most common way is to let the asset deteriorate. Walk around most major cities, an older steel mill, or most smaller scrap metal hauler operations and you can see the deterioration option at work. The other two strategies are ultra-high productivity of the maintenance workers where they can do more maintenance per dollar than most departments. The last strategy is a department that has an effective maintenance improvement or RCM program running to reduce overall amount of maintenance needed by the equipment and facility.

Elements that determine the estimate of the maintenance budget:

1. Size of asset
2. Type of construction, materials, and workmanship
3. Quality of design
4. Age of building/equipment/vehicle
5. History
6. How asset is used
7. Location
8. Knowledge and dedication of staff
9. Availability of spare parts and quality contractors
10. Type, knowledge, expectations of users
11. Laws, building codes, statutes, regulations that affect the business
12. Taxes
13. Amount of change in your organization (or your user's organization)
14. Competition
15. Amount of deferred maintenance
16. Hours of use a day
17. Production level or speed of operation

The formula for maintenance budget estimate:

$$((RCB \times MR) + (RCSE \times EMR)) \times CR = MB.$$

RCB = Replacement cost of building or facility. This number can be determined from the R. S. Means (address for R. S. Means can be found in the Resource section) costing information or experience. The basis includes all standard HVAC, plumbing, and electrical equipment. Replacement costs can be obtained from square footage estimates.

MR = Maintenance ratio. This number is the percentage of the asset value which must be reinvested to ward off deterioration. The range for standard type buildings is 1% to 3% per year. The exact amount depends on the use of the building, part of the country, and type of construction. Keep in mind a 2% MR would mean that the entire building would be replaced on a 50-year cycle.

RCSE = Replacement cost of production and mobile equipment. This would include production equipment, process boilers, mobile equipment, fleet trucks, etc. In factories, process plants, and other equipment-intensive applications, the equipment factor far outweighs the cost of building maintenance.

EMR = Equipment maintenance factor is equivalent to the MR but for production equipment. The equipment maintenance factor for all but the most unusual equipment is 7% to 15% per year of replacement value. Note that service companies provide full labor and parts service contracts on electronic equipment for 12% of replacement cost (which includes profit).

If you have several big categories of assets, then you might consider having different factors for the different classes. A paper mill might have the paper machine in one group and the delivery fleet in another. In the article "Another Way to Measure Costs" (1972, *Factory* magazine), the factor was calculated from 1% (labor intensive)–12%(capital intensive) range.

CR = Construction ratio is the ratio of time and materials spent doing renovation and new construction to the total time spent on all maintenance activity. It must be added to above. Construction or the installation and/or building of new equipment is not a maintenance function (not required to preserve the asset). It is frequently the responsibility of the maintenance department. This includes office construction, machine building, new installations, and truck body mounting.

MB = Maintenance budget with labor, materials, fringe benefits and overhead but without janitorial costs.

MB/Sq foot	= Maintenance costs per square foot of building or facility
MB/Rev	= Maintenance budget per revenue dollar
MB/Out	= Maintenance costs per unit output (such as linear foot of fabric, yards of refuse, patient-days, or ton-miles).

In some analysis of maintenance costs, both types of assets are lumped together in a replacement asset values (or RAV). The two maintenance ratios are then also lumped together.

Shopping Center Case Study:

Formula:$((RCB \times MR) + (RCSE \times EMR)) \times CR = MB$

750,000 sq foot shopping center with a co-gen facility

RCB: Cost about \$75/sq foot building value = \$56,250,000

MR: based on location and construction= 2%

$RCB \times MR$=\$1,125,000

RCSE: Cost for co-gen facility= \$2,600,000
EMR: based on similar facilities= 6%
$RCSE \times EMR$ = \$156,000

CR: very small amount of construction by maintenance dept(only 3%)= 1.03

$((\$56,250,000 \times 2\%) + (\$2,600,000 \times 6\%)) \times 1.03 = \$1,319,430$
$((RCB \times MR) + (RCSE \times EMR)) \times CR = MB$

MB/Sq foot = Maintenance cost per square foot

\$1,319,430/ 750,000 sq ft = \$1.759/sq ft.

This number is relevant only in relation to the factors that impact the budget from the first page: type of construction; quality of design; age, history, location of facility; knowledge and dedication of your staff; availability of spare parts and quality contractors; type, knowledge, expectations, and quality of tenants or users; building codes and statutes that affect business; taxes; amount of change in your organization (or your user's organization); competition; amount of deferred maintenance.

Details can be derived from the general budget estimate

From the overall maintenance budget we can determine the detailed items such as labor, materials, and overhead. We apply historic percentages to the overall budget and find the detailed amounts and crewing.

100% budget = Labor % + Materials %

The labor % is traditionally in the range of 30% to 50%. Historical rule of thumb (from *How to Manage Maintenance*, see Appendix) is labor + fringes equal ⅓ of total maintenance costs.

\$1,319,430 = 50% labor + 50% materials
\$1,319,430 = \$659,715 labor + \$659,715 materials

To determine the number of maintenance workers needed, we develop the cost of a fully burdened maintenance hour and multiply that by the likely hours per year plus overtime. That yields the cost of a person-year, which is divided into the available funds to yield crew size. The cost of a fully burdened maintenance hour might be:

Direct wages + Fringes + Overhead = Charge per hour
$17/hr wages + $5/hr fringes + $8/hr overhead = $30/hr

Direct wage = Cost paid per hour in the community of the facility being analyzed.

Fringes = include health and welfare, employer FICA, unemployment compensation, workman's compensation, but not vacation or holidays which are included in the 2080 hours below.

Overhead = Costs of supervision, clerical support, maintenance shop space costs, computer system, supplies, uniforms, large tool depreciation, etc.

O.T. rate = The overtime rate consists of $1.5 \times$ wage rate ($25.50/hr) plus the fringes of $5/hr for a total of $30.50. No extra overhead is required to work moderate levels of overtime.

Cost/yr= (Chg/hr × 2080 hr/yr) +(OT hrs × OT Rate/hr)
$65,450/yr = ($30/hr × 2080 hr)+(OT = 100 hrs × $30.50)

# of Craftspeople = Labor budget / yearly cost of craftsperson
10 = $659,715 / $65,450

The next calculation works backward to see what staff and support might be appropriate for a 10-person department. Extending the $8/hr allocated for overhead for the 10 craftspersons develops overhead (for staff support, supervisor, office expenses such as computers, fax machines, phones, etc.) of:

$166,400 = 10 craftspersons × 2080/ hr/yr × $8/hr.

The rough overall maintenance budget is:

1. Direct labor	$	384,315
2. Fringe benefits for above	$	109,000
3. Overhead	$	166,400
4. Materials, outside vendors	$	659,715
Total budget	$1,319,430	

Factory Example

General Formula:
$$((RCB \times MR) + (RCSE \times EMR)) \times CR = MB$$

Building worth (RCB)	\$10,000,000 MR = 1.5%
Equipment replacement value (RCSE)	\$20,000,000 EMR = 7%
Construction, installation, project (CR)	20%

$$((10M \times 1.5\%) + (20M \times 7\%)) \times 1.2 \quad = \$1,860,000 \text{ total maintenance budget}$$

If we use the same figures as in the above example, then 50% of the total would be labor, which yields: \$930,000 for labor and at \$65,450 per person year we would have 14 direct maintenance people. We would also generate about \$232,960 for overhead (14 people $\times$ \$8/hr $\times$ 2080 hrs per person = \$232,960) to run the department.

In "Maintenance as a Corporate Strategy," Andrew Ginder asserts that world class manufacturers spend about 2% of replacement asset value (RAV) on maintenance. He lays out a persuasive argument that when preventable maintenance is eliminated what is left approximates 2% of the RAV. (Formulas are partially adapted from the work of Dr. Bernard Lewis of Bernard Lewis and Assoc., Potomac, MD.)

7
Evaluating Maintenance— Two Questionnaires

The first questionnaire is detailed and requires knowledge about how maintenance is conducted. Parts of it can also be given to stockroom people, long-term contractors, and others with detailed knowledge of one or more areas of maintenance. The questionnaire is broken into nine critical areas:

A. Initiation and Authorization of Work
B. Systems and Procedures
C. Computerized Maintenance Management Systems
D. Preventive Predictive Condition Based Maintenance
E. Planning, Scheduling, and Followup
F. Purchasing, Parts, and Stores
G. Budgeting, Backlog, Maintenance Ratios, and Work Measurement
H. Guaranteed Maintainability

Specialized areas based on type of asset:
I. Training, Hiring, and Employee Development
J. Fleet Maintenance Questions
K. Building Maintenance Questions
L. Factory and Process Industry Maintenance questions

These questions can be used in several ways:

Give it to your staff and pose the question, where are we today?

Use it to start the discussion of what is good maintenance practice.

If you have periodic meetings, one section can be administered, scored, and discussed at each session.

48

Maintenance Questionnaire

Note: The numbers that appear before some of the questions (such as 2-4) denote the relative importance of that question. The importance number applies to all questions below until a new importance number appears. The lower the number, the higher the priority. Ones are the highest priority, two's lower, etc.

A. Initiation and Authorization of Work

1-1. A written work order on a printed (or computer generated) form is used for all jobs.

2-2. A written work order in evidence prior to starting a job for all work except genuine emergency repairs.

3. Regular meetings between management, users, and maintenance to set priorities.

4. A reasonable "date wanted, time wanted" space on each work order with restrictions against ASAP, RUSH, HOT.

5. All work done classified for repair reason, such as corrective, routine, breakdown, PM, diagnostics, construction, PCR (Planned Component Replacement), modernization.

One of the ways organizations have trouble is in formalizing the interface to maintenance. It is felt that giving up the ability to collar a maintenance worker on demand is also giving up good customer service. In the beginning in the growth of a department a maintenance person is hired. They get their work by pager, telephone message, or by being collared as they walk to a job. People grab them as they walk around to look at this "little problem." Since the maintenance person is so close to the user, there is usually no need for formal meetings, reports, or approvals. Since everything was done immediately there were no restrictions on rush jobs. If the maintenance person fell behind then every one knew, and either they waited or (in the case of a great emergency) a contractor was called in to take up the slack.

As time passes and the department grows, the maintenance

management people lose the direct connection with the customer's needs. In fact, as the whole organization grows, the various users themselves lose track of the greater needs of the organization. Maintenance becomes servants of people and their personal needs. These jobs might not have been contemplated at budget time. The result is a runaway situation that gets resolved by political and power considerations and not maintenance priority.

The second issue is the number of transactions. Without work orders, some percentage of the problems would fall through the cracks. Even small operations benefit with written work orders to avoid some of the mistakes of memory. The work order has significant other advantages for all sized departments in the areas of cost collection, proof of safety, job control, etc.

In a large bank office building and computer center, without a work order system, the dispatcher handled 75–175 requests a day. They estimated that as many as 5% of the work requested somehow got lost between the requestor and the mechanic. While 5% doesn't seem like a lot, it means that 4–6 people were ignored and have to call back another day. In about 10 cases a week, the VP of facilities was called by the user to make sure the job was done.

B. Systems and Procedures

1-1. Is there an absolute commitment on a long-term basis to improve the quality and reduce the cost of maintenance among all top management and maintenance management people.

2-2. All maintainable units (buildings, equipment, trucks, etc.) have unique ID numbers.

3. All work orders are costed for labor, parts, and other costs.

4. Are there up-to-date MSDS sheets, lock-out procedures, hazardous waste policies.

5. Are systems in place that detect craftsperson-induced problems. Is rework less than 3%

6. Do your systems show when overtime is below or above certain setpoints, such as 3%-9%

Systems and procedures tie together all of the activities of maintenance. Before any system can be effective management must support it. Also any system must tie the asset to the repair act and cost all activity. Maintenance has become the center of the effects of many new federal and state regulations. Many of the toxic waste rules fall heavy on maintenance (in some businesses, all of the toxic materials used are in the maintenance department). Implementing lock out/tag out and confined space regulations is almost completely a maintenance responsibility. One of the most important reporting functions of a maintenance system (either computerized or not) is to detect worker-induced problems (called iatrogenic failures). These workers may need some kind of intervention such as training, coaching, or discipline.

C. CMMS (Computerized Maintenance Management Systems) Information Systems

1-1. Do mechanics and supervisors have the training, knowledge, positive attitude, and access into the Maintenance Information System to investigate a problem.

2. Is garbage and faked data kept out of the system, and is the data coming out of the system commonly held by management and the workers to be accurate and useful.

3. Can the system give answers to most questions without the services of a programmer.

4. Does the system help in the day-to-day running of maintenance delivery.

2-5. Is repair history from the date in service immediately available with enough accuracy to detect repeat repairs, trends, and new problems.

6. Can the system isolate the "bad actors" using the Pareto principle to identify the problem tenants, problem machines, craftspeople, or parts.

3-7. Is the CMMS system supported by either a responsive vendor or a responsive data processing department.

Computers are becoming a ubiquitous tool in maintenance departments. In a typical maintenance management course, 75% or more of the attendees have some kind CMMS. The maintenance application is a very complex one with no external rules about how different issues should be handled. As a result, computerization has a spotty record in the field. Success has more to do with the attitude of the organization than it has to do with the quality of the software or the installation team.

The first issue in computerization is training, knowledge, attitude, and access for all people involved. In the most useful systems, the mechanics directly interact with the files. They could be looking up a unit history or entering their hours from the job they completed. For the reports and screens to be useful, garbage should be kept out of the database. All people involved should insist on accuracy and care in data entry.

Usefulness of a system means answers to questions quick enough to impact the decision. In some cases that could mean being able to print the last 5 repairs, while you wait, before starting another repair. In other cases it might require detailed statistical analysis of mean times between failure for which you can wait a day or more.

One trap that people fall into is thinking about maintenance as if it were an accounting problem. From a computerization point of view, maintenance is partially an accounting problem and partially an engineering problem. Since most programmers in the field are from accounting backgrounds, there is an overly financial orientation to the system (for example, keeping only the numerical data and having limited space to record work order comments or other non-numerical information) and roll-offs of data after the end of the fiscal year. On capital equipment, maintenance data should be kept until the asset is retired. Some summarization of minor jobs would be allowed, but the major repairs should be left intact for future analysis.

D. Preventive/Predictive/Conditioned Based Maintenance
1-1. Does top management support the PM system with their attention, money, and authorizations for downtime as required.

2. When deficiencies are found by inspection, are they written up and completed in a reasonable time.

2-3. Do repeated or expensive failures automatically trigger an investigation to find the root cause and correct it.

4. Was there an economic analysis of each item on the task list proving ROI (Return on Investment).

3-5. Is Reliability Centered Maintenance (RCM) considered when equipment failure could cause injuries or the equipment is critical or has high downtime costs.

6. Are units outside of the PM system because they are in very bad shape and fixing them up is not contemplated.

7. Does the actual failure history impact the frequency, depth, and items on the task list.

Many people choose the maintenance field because of the variety. They are attracted to the excitement of the breakdown and the ability to distinguish themselves "under fire." PM is boring. PM goes against human nature because human nature avoids action before it is absolutely necessary.

One of the keys to successful PM is vocal top management support. Another key is recognition to the PM workers when machines run well.

Part of a complete PM effort is inspection, which uncovers occasional deficiencies. When these deficiencies are uncovered they must be completed in a reasonable time. This is a necessary key to a successful PM program.

Much of breakdown maintenance is driven by the reflex to repair what breaks. PM is more reflective. There is time for root cause analysis or economic analysis of tasks or membership in PM. There can also be a periodic review of the failure history to see if the tasks are well directed.

E. Planning, Scheduling, and Followup
1-1. On longer jobs and project work, is the scope of work pinned down before a tool is lifted.

2. Are the five elements of planning (labor, materials, tools, machine access, permissions) considered before a job starts.

3. Is the maintenance schedule reviewed and updated at least weekly and signed-off on by the production manager and other interested parties.

2-4. Is PM or scheduled work routinely displaced by jobs with preferential treatment (lower priority but need to be done) such as a new electrical outlet for a boss.

5. Are maintenance work orders prioritized in a rational way to ensure that the most critical units are worked on first.

6. Does the mechanic show up to do a repair or PM on time (per schedule) 95% of the time, and is the unit returned to service as planned on-time 95% of the time, or are jobs routinely completed on time, on budget, as planned.

4-7. Is there a day's work for each craftsperson planned and written on a public schedule (that everyone can see) at least a day ahead of time for 75% or more of the craftspeople.

The process of running a great maintenance department begins with priorities being set in conjunction with the user. Proper planning will ensure that high-priority jobs (PM is one) are not displaced. It is well recognized that proper planning and scheduling will save significant execution time. For proper planning to take place, the scope of work should be pinned down and the five elements of planning considered before starting the job.

With these planning elements in place, the positive results of planning can be felt. Planning also means that mechanics show up when promised and machines are returned to service when promised. The schedule determines when a job should be done. The scheduling function also stages materials, tools, and sub-projects. In a scheduled environment craftspeople can know their next jobs ahead of time.

F. Purchasing, Parts, and Stores

1-1. Is the parts room well laid out with adequate space, good lighting, locations for parts for rebuild and returns, controlled access, shelving, bins, drawer units, computer access.

2. Can a maintenance craftsperson get the right part in less than 10 minutes 95% of the time or find out immediately that it is not a stocked item.

2-3. Are all parts used for maintenance purposes tied to an asset number through the work order.

4. Is there an ongoing effort to use vendors as long-term partners for stocking, engineering, and problem solving.

5. Is there a periodic review of high turnover, high cost items for possible savings.

6. Is there a review of expensive slow-moving items to see if they belong in stock at all.

7. Are reviews of the high use, high cost parts used to initiate a maintenance improvement study.

3-8. Is there adequate information system support to help easy look-up and investigation into unknown parts. Is there a cross reference to indicate all of the units that a specific part can be used on and all of the vendors of a specific part.

9. Are established reorder points, order quantity, safety stocks based on logical reasoning and not emotion. Is the type of buy (casual, phone bid, formal bid, engineered bid) factored into EOQ, ROP.

10. Is there an annual physical inventory and review of all parts for obsolescence, spoilage, quantity-on-hand, reorder point, lead time, and safety stock.

Parts consist of 40% or more of all maintenance costs. Yet in many departments the parts can be seen stuffed in their broken rotted boxes under stairwells or behind the boiler. To get maintenance under control, get the parts under control.

Frequently there is an adversarial relationship between stores, purchasing, and maintenance. In some cases, organization rules or laws interfere with good maintenance purchasing practices. Good maintenance practices include buying the best part from the best source. The best source might be the one that delivers when you need the part, not just on the days the truck is in the area.

Much of the inventory can be analyzed using traditional inventory ratios and formulas (such as E.O.Q, safety stock, min, max, etc.). Some items for critical units are stocked because the lead time on the part would create unacceptable downtime.

G. Budgeting, Backlog, Maintenance Ratios, and Work Measurement

1-1. Is the budgeting process real (developed against the maintenance demand created by buildings and facilities) or just some calculation against last year with some padding (to be cut away by top management).

2. Is the current budgeting process likely to serve the best overall long-term interests of the organization. Is it purely a nearsighted, short-term process.

3. Is there a five- or ten-year strategic maintenance plan which is discussed at budget time by both maintenance and top managers.

2-4. Is the relationship between the size of operation (such as number of rental units, tons of steel rolled, square footage, miles covered, hospital beds, students, etc.) and maintenance staffing well understood and used for budgeting.

5. Is there recognition that good maintenance practice has a major impact on other budget line items such as energy, capital replacement, regulatory compliance.

3-5. Are budget savings available for reinvestment into other cost reduction programs.

6. Do major pieces of equipment have a repair/replace budget.

7. Is the budget performance kept up-to-date with adjustments as necessary.

4-8. Is the ratio between scheduled and non-scheduled work tracked and trended for at least 3 years.

There were significant cuts in maintenance budgets throughout the 1990's. These cuts were designed to squeeze the fat out of maintenance. There is also the philosophy that only activity that directly adds value to the product should be funded. The budget is the battleground for these issues. The primary question is: is the budget a rational process likely to result in the best outcome for the organization?

To correctly and effectively control maintenance, the budget process needs increasing sophistication. Areas like energy and capital replacement have major impacts in tradeoffs with maintenance dollars.

The other issue is the long-term nature of maintenance. You might be able to defer opening a sales office or not. A roof, however, deteriorates at a regular rate and there is an optimum (lowest total cost, lowest disruption) replacement interval. Maintenance should maintain five- and ten-year projections for capital replacements.

The best maintenance departments measure themselves against the past and against the best in the business (when they can find the numbers). *The most important ratios are the ones that measure you against your primary mission.* If that is providing chilled water to the computer center, then that is the best measure of your success.

H. Guaranteed Maintainability

1-1. Are drawings and specifications on all new processes and buildings reviewed by the maintenance department early enough in the process so that changes can be made without adverse impact to the whole project.

2-2. Are specifications for buildings and new technologies on machines and building systems discussed with the maintenance department to see if they are in line with existing skills, parts stocked, training and test equipment and tools.

3. Are failure and cost histories available to designers or equipment buyers to make better decisions. Do they use it.

Avoid maintenance by proper design. Good design can have a great impact on future maintenance costs. The maintenance group has to be brought in the process of design for maintainability.

For an excellent example of this in action, look at the design of the typical fast food chain restaurant. They have refined the design to increase sales and reduce costs for three decades. Failure histories and cleaning costs have driven design improvements. Factors include making use of the existing skills, tools, and inventory. While different vendors might make better products, there is a cost to change from the existing vendor to a new one. If a change is indicated then time and resources for training is indicated.

I. Training, Hiring, and Employee Development
1-1. Is 1–5% of a technician's direct hours spent in craft, multicraft, or other related training. Is a determination made of benefit or return on investment from training.

2. Is cross training a goal for the department to allow craftspeople to do the "whole" job and provide scheduling flexibility.

3. Is there a consistent process to identify and hire the best maintenance workers, with the final decision being made by the maintenance department.

4. Has there been a recent assessment as to the technology and skill level required to maintain the machinery, equipment, and buildings with a comparison to the average skill level of the employees in the maintenance department.

3-5. Is training for the maintenance department part of all new equipment acquisition contracts. Are arrangements made for retraining in 3–5 years, when it is likely to be useful.

The only constant in the maintenance world is change. Hiring the best workers you can identify—and training them—is the best path to satisfied customers. In some cases, the way to determine the need for training is to look at the equipment and see what skill sets are needed to support them.

J. Fleet Maintenance Questions

1. Flat rates for all recurring jobs written on work order (called Repair Order or R/O).

2. Historical performance standards for common repairs (actual time).

3. Periodic issuance of Earned Hr. or other productivity report.

4. Use shop schedule to predict scheduled overtime and use of outside shops.

5. A good idea of the effect in hr./month of changes in fleet size or mix.

6. Repair history readily available to identify costs, frequencies, and component systems since equipment was purchased.

7. Comparison data of repairs to like units in like service.

8. Repair budgets for major units.

9. All parts are charged to units.

10. Parts catalog which includes cross reference.

11. Benchmark of how often units are down awaiting parts (a bad thing!) is trended.

12. Stores system identifies make, model where part is used.

13. Maintenance department head reports to V.P. of transportation or operations or plant manager.

K. Building Maintenance Questions

1. Regular meetings between management, tenants/users, and maintenance to set priorities.

2. All buildings and equipment have unique ID numbers. In addition, are systems identified (such as electrical distribution system, or waste water system).

3. Are maintenance routes established, do the route people repair a wide variety of minor problems and are they equipped and trained to do so.

4. Is the maintenance schedule reviewed and updated at least weekly and signed-off on by the building manager, major tenant or other interested parties.

5. Is there a five- or ten-year maintenance plan for roofs, facades, heaters, and other long-lived systems which is reviewed and updated at budget time by both maintenance and top managers.

6. Is the relationship between the size of operation (such as number of rental units, square footage, hospital beds, students, etc.) and maintenance staffing well understood and used for budgeting.

L. Factory and Process Industry Maintenance Questions

1. Regular meetings between production, production control, and maintenance to set priorities.

2. When statistical process control limits are exceeded (process variation), a maintenance request is initiated.

3. Are the operators trained to do routine maintenance (clean, lube, tighten).

4. Is there a certified operator training program to avoid common problems and assure upgraded skills.

5. Is there ongoing training of maintenance operators to improve their operation of the equipment, help them correct minor faults, and improve their powers of observation (to improve the quality and accuracy of the maintenance request).

6. Are high-technology inspections used such as vibration analysis, ultrasonics, etc., particularly where the interval between the observed deficiency and the breakdown is short.

7. Are the high-tech inspections integrated into and driven by the master PM schedule.

8. Are the task lists divided by skill level into interruptive or non-interruptive tasks.

9. Are maintenance work orders prioritized in a rational way to ensure that the most critical units are worked upon first.

10. Is PM scheduled by a user department such as production control, or is there a direct linkage between the master production schedule and the PM schedule.

11. Does your accounting system have capital spares category for expensive, critical, long lead time "insurance policy," parts.

12. Does the storeroom get PM lists, repair jobs in advance to kit up the parts in advance.

13. Is the relationship between the amount of output (production quantity such as cases processed or tonnage shipped) and maintenance staffing well understood and used for budgeting.

14. Is uptime tracked and are trends calculated along with downtime reason.

15. Is there a well-known cost of downtime by machine, process, or facility.

16. Is a Pareto analysis of downtime done and acted upon. Is a Pareto analysis done on failure modes of equipment.

17. Is there a maintenance improvement fund or a maintenance R & D account to charge experiments to improve maintainability or line up-time.

18. Is there recognition that good maintenance practice has a major impact on other budget items such as production, quality costs, energy, regulatory compliance.

19. Is the ratio of maintenance dollars to overall revenue dollars (or shipped product such as maintenance hours per tin of steel) tracked and trends kept for at least 3 years.

20. Is training for continuous improvement of skills, with the goal being to improve the cost per part, part of the mission of the department.

21. Is the PM system driven by measures of equipment usage such as machine hour, energy, cycles, pieces, tons, etc.

Short Questionnaire for Top Management and Other Users

Most maintenance departments are insulated from the rest of the organization. They attend to their problems, have their successes and failures, and don't reflect much on how they could be or what could have been. Much of the knowledge for a real maintenance renaissance is already inside the department. Some of these positive changes come about when the department undergoes a period of introspection. These two related questionnaires have been used by thousands of maintenance professionals in hundreds of organizations to start the process of introspection that leads to positive change.

This second questionnaire is designed to get the views about both the maintenance department and the maintenance function from top managers. It would be suitable for a director of administration, VP production, production manager, operations manager, VP real estate, or building manager. It can be completed in about 10 minutes. The scale is usually 1–10, with 1 being very very weak and 10 being very very strong.

Top Management's View of Maintenance Fitness Questionnaire

1. Are maintenance managers part of the top level strategic planning for the future of the facility, product, or organization. Are maintenance managers taken seriously.

2. Is there an absolute commitment on a long-term basis to improve the quality and reduce the cost of maintenance among top management, and are funds available for investment when properly justified.

3. Is the maintenance manager knowledgeable enough to put together a rigorous economic justification for a maintenance investment.

4. Is the budgeting process likely to serve the best overall long-term interests of the organization as far as avoiding the decay of assets.

5. Is there a five- or ten-year strategic maintenance plan including projected capital replacements which is reviewed and updated annually by both maintenance, operations, and top managers.

6. Is there a small maintenance improvement fund or a maintenance R&D account for experiments to improve maintainability.

7. Are the energy and capital replacement line items in the budget looked at along with maintenance line items at one time (since they are interdependent). Can an investment in maintenance be offset by savings in energy or extended asset life.

8. Are there regular meetings between users and maintenance to set priorities.

9. Do you think that maintenance workers and supervisors have the training, knowledge, positive attitude, and access into the CMMS (Computerized Maintenance Information System) to investigate a problem.

10. From what you've heard and seen, is garbage and faked data kept out of the CMMS.

11. Is the data coming out of the CMMS commonly held by both you and the workers to be accurate and useful.

12. In your experience, is the CMMS used to isolate the "bad actors" using exception reporting to identify the problem machines, craftspeople, or parts.

13. Are the printed reports set up to give you only the level of detail that you can reasonably use, with more detail available on demand. Are one-page management reports showing critical operating ratios with trends available to the management outside of maintenance.

14. Have you heard of repeated or expensive failures automatically triggering an investigation to find the root cause and correct it.

15. Are maintenance jobs routinely completed on time, on budget, as planned.

16. Is the storeroom well laid out in a location convenient to the maintenance action with adequate space, controlled access, shelving, drawer units.

17. Is there an ongoing effort to use vendors as long-term partners for stocking, training, engineering, and problem solving.

18. Is it okay with you if an outside maintenance part vendor who offers significant value added, as in the question above, charges a higher price for some parts.

19. Is 1–5% of a technician's direct hours spent in craft, multicraft, or other related training.

20. Are drawings and specifications on all new machines, processes, and buildings reviewed by the maintenance department early enough in the process so that improvements can be made without adverse impact to the whole project.

21. Do architects and designers or equipment buyers make use of existing failure and cost histories to make better design decisions.

22. Is training for the maintenance department part of all new equipment acquisition contracts. Is there thought about the issue of re-training 1, 3, and 5 years later when it is more likely that the machine will need service.

Bonus Questions
23. What is your overall rating of the maintenance department.

24. How familiar are you with the maintenance problems, department, and strategy.

8

Benchmarking Maintenance

There are hundreds of ways to measure or benchmark maintenance. Benchmarks are a type of measurement that have become popular in the business literature. Traditionally there has been a focus on the cost aspect to the exclusion of other factors. This is an issue because "You will readily see that too much emphasis on cost reduction will result in unacceptable performance in maintaining buildings and equipment at the desired level" (from *How to Manage Maintenance*).

The original definition of benchmark pertained to surveyor's marks that identified the altitude and coordinates of a location (reference from OED). These benchmarks were characterized by being public where they could be seen and referred to in deeds and other documents, and permanent so that years later the same benchmarks could be looked at to resolve disputes.

The difference between maintenance measurement and maintenance benchmarking is in the publication of the results to motivate the teams to change. Maintenance measures have always been used to privately evaluate a manager or department. A benchmark is widely publicized to the rank and file to be of maximal benefit. Our use of benchmark is a series of measures that will help you know if the maintenance function is improving, sliding, or stagnant.

There has been significant work in benchmarking different business functions in organizations. Leibfried and McNair's significant book *Benchmarking* shows some of the opportunities and pitfalls of formal benchmarking. Lawrence Aubry of NSC Technologies sums up the major pitfall eloquently in his article "Maintenance is a Process" (*Maintenance Technology*, June 1995): "But leadership cannot be achieved by continuous focus on a rear view mirror." To be a true leader you must go beyond what someone has done, into wholly new territory. In any case, the benchmarks will show you if you have arrived.

1. Internal benchmarks are based on prior periods in that location. An internal benchmark might be downtime hours due to maintenance problems. This benchmark would be tracked monthly (or even weekly) over a long period of time. Continuous improvement could then be measured against prior periods. This benchmark shows you if you are improving, but not where you are. The benchmark is problematic if the equipment is constantly changing, or is intrinsically unreliable.

2. Best-in-class is a benchmark of the best in your industry. Some large organizations will take the best plant of a certain type (if they have many similar plants). Other organizations will review trade literature or initiate a study to determine the best plant in the business and compare themselves to that plant. A benchmark might be the number and severity of customer complaints per month.

Compare yourself to the best plant of your type in the industry. Continuous improvement would measure your progress to catch up to them and eventually surpass them. It is very sobering to see just how good the best in class is compared to you.

One northeastern city wanted to see how they compared in getting garbage trucks repaired and back on the street. Their own measurement system showed significant improvement (at the start they had 46% availability, and after 2 years of improvement they had 61% availability). They called around and found another city that averaged 80% availability with a lower maintenance cost per vehicle!

3. Best-in-the-world is the ultimate comparison between functions. A best-in-the-world benchmark might evaluate your telephone answering function and compare it against the best in the world (such as Federal Express or Lands' End). In fact, Xerox did just that. When they wanted to learn about powerful telephone techniques, they studied Lands' End and L.L. Bean—not other copier companies.

The comparison organization might well be in a vastly different type of business (or even in government or education). Continuous improvement puts your organization against the best there is. Your achievements against the best-in-the-world's benchmark are tracked and reported upon. It is very sobering to see how far ahead some organizations have taken a technology or an idea.

David Peterson, General Manager of CSI Inc. in Knoxville, has studied world class maintenance organizations for years. He says that maintenance must increase its knowledge in business metrics to show top management what is going on in language that top management can understand. He has designed several measures which are noted in the discussion below as David Peterson measures.

One of the major problems with benchmarks is that so many of the most important maintenance issues cannot be directly measured.

Measurable	Difficult to measure
Labor hours	*Productivity*

Labor hours and ratios relating to hours are easy to gather and develop. It is very difficult to evaluate the true productivity of a maintenance work group. This includes evaluation of how much good work was done on units needing service.

Maintenance cost per unit output	*Unfunded maintenance liability*

Maintenance cost per ton of steel, per hospital bed-year, or other unit is easy to calculate. The ratios are meaningful over a long period and when operational conditions don't change too much. Cost per unit can be kept artificially low at the expense of increased unfunded maintenance liability (deterioration). How do you measure the amount of deterioration this year compared to last year?

Callbacks	*Knowledge, skills, attitude*

Callbacks (or rework) are trackable with some effort. A fleet might target vehicles that return within 4000 miles and have a problem with the same system. There are many reasons for re-work from inadequate skill, bad attitude, improper tools, cheap/bad parts to inadequate conditions.

It is difficult to measure the skills, knowledge, or attitude of the maintenance people. When someone is trained to service PLC's and their control programs, where does the asset show up on the measurable books?

Grievances   *Morale*

Grievances are countable. The morale is not. Grievances are too late in the process to be reliable measures of morale (morale can plummet well before grievances increase). The dedication of the maintenance department to show up during the snows, holidays, nights when everyone else is safely in bed is never measured or commented upon.

Maintenance Benchmarks

Maintenance benchmarks are divided into measures for costs, parts, work, and customer service.

Costs

Maintenance cost index: This is the most common measure of traditional maintenance departments. It plots the total maintenance cost for the last few years, perhaps by quarter. It is still useful to show what is being spent. You combine that information with what else you know, and it becomes another data point. In times of lower change and known inflation, the maintenance index trend would be very useful. Today there are so many variables that the trend is al-most meaningless. It is still kept by many companies.

Maintenance cost to budget (by line): The most common bench-mark is how are we doing in relation to how we said we would do. Variance reports show where problems might be developing.

Maintenance costs per square foot: In an office, shopping cen-ter, or apartment building, the maintenance cost varies with size rather than other measures. It is tough to relate maintenance costs to dollars sold in a department store or dollars of stock sold at a stock broker's office. In these cases the maintenance costs per size is the best item to track.

Maintenance cost per unit output (tons of steel, yards of garbage, patient-days, ton-miles): Assets that are used to make a product or provide a service require maintenance to be kept in a preserved state. This usage creates a need for an investment for every unit of usage. Your unit of usage could be making bricks, processing policies, or delivering blood. That activity can be related to the use of the asset (such as a building, truck, kiln) needed to provide that product or service.

In many industries, the maintenance cost per car assembled, hundred packages delivered, or thousand barrels of beer brewed is well known and understood. This is a David Peterson measure—maintenance contribution to cost per unit of production. (He also has a related metric called maintenance cost effectiveness which compares actual cost per unit output to calculated cost per unit.)

Maintenance cost per revenue dollar: Directly related to cost per unit is maintenance dollar per revenue dollar. This relates maintenance costs to total useful output, which is revenue. Most departments can show improvements in this area. This is a David Peterson measure.

Ratio of maintenance costs to asset value/Ratio of maintenance people to asset value: Consult the chapter on estimating maintenance budgets for details. This ratio can give a gross analysis if adequate maintenance is being invested. It is a very useful ratio if your organization is buying buildings and other assets and you want to keep ahead of problems. Another way of looking at this is to see how many people it takes to maintain $100 million of this kind of asset. This is a David Peterson measure.

Ratio of redundant equipment: How much equipment was added to ensure availability? Compare redundant equipment value to total asset base. This is a David Peterson metric. He says: "world class companies average 4% with newer plants approaching zero."

Ratio of maintenance hours to total plant hours/Ratio of maintenance payroll to total plant payroll: This is a traditional ratio which was more useful when processes and products were more static. In today's environment, the ratio is less of a useful guide because product mix, products, processes are changing daily. The average of all of the changes would probably result in increasing maintenance costs because the automation of production is proceeding more quickly than the automation of maintenance.

Contractor ratio: What is the ratio of contractor hours or dollars to in-house work? The design of your budget should predict the amount of contractor work in a given year. The ratio is only important as it relates to your prediction. If you predicted 10% and the last few months have been coming in at 50%, there had better be a bunch of construction or projects going on that were approved after your budget.

Parts

Maintenance labor to parts: This is a useful ratio when added to other knowledge because it provides input into formulas to estimate budgets for new buildings, fleet expansions, or plants.

Inventory turns: While maintenance inventory is different than a retail inventory, the analysis of turns is useful when insurance policy stock is removed from consideration. After that removal, the turns should be approximately reflective of an industrial distributer.

Purchase to issue ratio: The purchase to issue ratio is an advanced indicator of inventory accumulation or depletion. If you are trying to reduce your inventory, then you must run this ratio below 1.

Inventory level per mechanic/Inventory level per asset value: There is an optimum level of inventory (for a conventionally configured stores situation) to keep people busy and assets of a certain type humming along. Looking at similar industries such as utilities can uncover best practices within an industry. Of course the best in the world might use JIT inventory, strategic partnerships, consignment stock to radically improve the ratio.

Work Ratios

There are many measures that deal with the ratios of various types of work. How-you-spend-your-time benchmarks (this is defined more precisely in the chapter on work orders in the section "Reason for Repair") are essential to see how the mix of work is showing improvement.

Planned maintenance hours	*% Planned hours*

Planned hours from all sources should exceed 80% of the worked hours.

Emergency hours	*% Emergency hours*	*unscheduled*
DIN (Do It Now) hours	*% DIN*	*unscheduled*
Short repair hours	*% Short repairs*	*scheduled*
CM (Corrective Maintenance) hours	*% CM*	*scheduled*
Preventive Maintenance hours (PM)	*% PM*	*scheduled*

The first important breakdown is planned (PM+CM+Short repair+Project) to unplanned (DIN+EM). This ratio shows how much your facility is ahead of the breakdown curve or how much you are dominated by unscheduled events. The trends of these numbers gives you a feel for whether there is improvement.

Personal service work	*% PS*
Project/Capitalization work	*% Project/Capitalization*

A second measure to look at is how much non-maintenance work is done (this would include PS and Capital). There might be money saving opportunities in reviewing the details if the ratios look too large.

Total backlog by craft (in hours, weeks per person):
Many experts believe that managing the backlog (work immediately available to be done including pending) might be one of the most important jobs of maintenance leadership. The amount of backlog should not fall too low (1 week per person) or

too high (3+ weeks). *How to Manage Maintenance* says that low or no backlog indicates over-manning, and more than 10 days backlog indicates overtime is needed.

If people see the backlog running out, they tend to slow down to avoid layoff. If the backlog is too large then user's routine work doesn't get done quickly or reliably. Increased backlog is one reason to authorize contracting or overtime.

One calculation issue is whether to use true time available or 8 hours. Calculations show an 8-hour day is reduced by 1 hour and 20 minutes from meals and actual breaks and additional 30 minutes from meetings and other information exchange. A real workday might be closer to 6.0–6.5 hours. When a 10-person crew has a 490-hr backlog, we would calculate they have about 8 days (using 6 hr/day) not 6 days (using 8 hr).

Overtime	*% Overtime*
Unscheduled overtime *(or scheduled overtime)*	*% Unscheduled overtime*

Overtime is an interesting indicator because in most maintenance situations some natural overtime (3–9%) indicates that you are properly crewed. Natural means that people are not slowing down to create overtime. If there is no overtime, the temptation is to think that there are too many maintenance people for the workload. Of course, this does not include organizations that artificially restrict overtime. Unscheduled overtime for emergencies is a problem because it not only shows a lack of planning but also a lack of control over deterioration. Don Nyman (see the Appendix) recommends 6% overtime.

Hourly to support people ratio/Hourly to planner/Hourly to supervisor (span of control): Excess support staff sometimes gets in the way of productivity. One area of savings may come from moving the support staff back to the floor if possible. Other measures such as direct hours to planner and span of control can help you sharpen up your support ratios and optimize the amount of back-up your mechanical staff has.

Jobs waiting (by reason): One of the problems of maintenance work is the number of hand-offs of maintenance jobs in large facilities. The job goes to planning, who determines that it needs engineering, who passes it back to planning, who passes it to purchasing, etc. The waiting time is frequently what kills speed. The customer waits and no one can easily give them an idea of where the job is. Jobs waiting reporting helps highlight the problem areas and can indicate opportunities for reengineering.

Area assignment: When there are problems responding to a customer, one solution is to assign a mechanic to an area. This area assignment is favored by production because it gives them resources and makes them feel that they can take care of problems. The problem is that area assignments are inefficient uses of maintenance people and the goal of maintenance to avoid problems not fix them quickly. In some cases area assignments are essential for safety or for economics (a process that ships $10,000 a minute is probably worth a mechanic). In most cases this number should drop as you get a handle on the unscheduled events.

Open work order hours: An open work order is related to an area assignment. A machine might have an open work order for any work done each month. Usually the goal is to minimize open work orders in most situations because of the loss of control. The exception might be in routine work of a known content and duration (such as line start-up, or policing the parking lot for glass).

Accounted for hours (payroll hours/work order hours)

Use of work orders   *% labor hours*

The first measure after you install a work order system is the ratio of work order hours to payroll hours. It should rapidly increase to the 90–95% range (some say to 100%); that way, you know all of the hours are somewhere in the system (at least). Be sure people are not pencil-whipping the work orders (at the end of the day just whipping through—putting 2 hours on each of the 4 work orders to get 8 hours).

Effectiveness (work order hours/standard hours): Where there are good standards on most jobs, the effectiveness ratio can be useful. It shows how much work is really done. You get 3 hours credit for 3 standard hours even if the job took 16 hours. In this way, slowdowns/problems are subtracted from your output calculation.

Customer Service Measures

Downtime hours (or uptime) (by reason)% Downtime: This is one of the most closely watched numbers in production environments. A utility might have a $3000 to $5000 a minute cost of downtime. The reason for the downtime should also be tracked because set-up, model change, color change, material problems, operator problems, quality discrepancies could also create downtime. The goal of TPM is to reduce all types of downtime, and all types are analyzed together. Don Nyman recommends unscheduled maintenance related downtime to be 2% or less of total operating hours.

David Peterson's capacity metrics: These include measurements of required capacity. Required capacity is the running time needed to meet customer demand (while accounting for set-up and production changeover). Derived from required capacity is percent utilized capacity. This divides the required capacity by total available capacity time. His last measure is actual availability which is actual uptime divided by required capacity.

Breakdown report: The core service that maintenance provides its customers is freedom from breakdowns and quick effective service when a breakdown occurs. Breakdown reports can take many forms from a list of breakdowns with causes and response times to MTBF (mean time between failures) with MTTR (mean time to repair) information added. In all cases, a breakdown should be treated as an educational opportunity to see where (if at all) the system failed.

Number of service calls: This can tell maintenance and user departments how effective maintenance is at foreseeing problems and correcting them before they occur. This benchmark would be factored by significant changes in size, equipment, or mission of the organization being served.

Mean time to respond (MTR, by priority): How long does it take to respond to a service call from the time it is phoned in until the time a service person shows up? In some organizations this is a major way the maintenance department is rated.

Mean time to repair (MTTR): Once a response has been made, how long does it take for the customer to be satisfied? When this is added to MTR you can get an idea how long your customers are unsatisfied.

Callbacks   *% Callback*

The bane of maintenance is rework or callbacks. A callback is defined as a return of a service person to a unit for work on the same system or related system as their original work. Callbacks are a problem in the mechanic, part, procedure, or the design of the asset. In any case, the reason has to be uncovered and fixed. This ratio, trended over time, indicates if the problem is being addressed. This is a David Peterson measure. He states that world class organizations average 3% with the target being 0%.

Maintenance satisfaction survey: This is an ongoing or annual survey of attitudes toward maintenance. This measures the effectiveness of your communication about maintenance.

Look into the work of Eugene C. Wordehoff, an associate at A.T. Kearney, for additional information about maintenance benchmarking. He has published several excellent articles in *Plant Engineering Magazine* on the topic. Also, as mentioned before, David Peterson has done excellent work in this area (one article was "Can Maintenance Measure up to Business" in *IMPO*, March 1996). Building and facility managers have the advantage of widely published benchmarks from the *R. S. Means Maintenance Cost Data* handbook to the *Whitestone Building Maintenance and Repair Cost Reference* (1411 Forth Ave. Suite 1022, Seattle, WA 98101).

9

How to Evaluate Worker Productivity

The problem of measuring and evaluating productivity in maintenance is more complex than in most other fields. One way productivity is measured in maintenance is by the amount of useful work produced by each hour of input.

If a mechanic is working efficiently, accurately, and professionally to remove and replace a garbage disposer in unit 34, we would say she is productive. She might have 3 hours of total input (with travel, parts pick-up, and the work itself) and 1 hour of actual work; and we, as managers, might feel satisfied. But wait—the dispatcher wrote the unit number as 34 when it should have been 43. How productive is she now?

Another mechanic is working on the right unit, with the right materials but is doing a repair which will fail within 2 years (rather than lasting 10 years). This mechanic planned better and works faster and is doing 1 hour of work with only two hours of input. Who is more productive? What about if the repair would last 5 years? These are tough questions that have no easy answers.

Our management systems, human resource departments, and safety departments all impose non-productive time on our crews (some is necessary, which could be another discussion). Within the same organization, two maintenance departments across town from each other could have very different initial productivity levels based on the system imposed activities (meetings, paperwork burden, security requirements, etc.).

How do your maintenance workers work? Does the organizational culture help or hinder the maintenance process? Every culture creates a pattern. The pattern either supports or defies the ability to preserve assets and provide capacity.

77

For example, in some governmental maintenance departments, the pattern created by the culture stands squarely in the way of an efficient process. A maintenance worker there has no authority to spend money to solve a problem (as in make a call to Grainger's and immediately pick up a new exhaust fan). The maintenance person might have to endure a lengthy procurement to get a more expensive, inferior part—and wait weeks to boot. In another case, a large field service company issued $50 bills to the mechanics to help them solve a customer's problem immediately. No call-in for authority, no second guessing, just bring in the receipt and get the money replaced. In both cases, the mechanic's time is impacted by the systems, procedures, and attitudes of the organization. We can see the pattern by studying how mechanics are forced to use their time.

Some of the questions that are posed by management need to be answered by maintenance leadership. They might include the following.

Do we have enough people for the amount of work?

Are these people properly managed and supervised?

Do we support these people adequately?

If we changed the way we supported them, could they do more?

What is the optimum capacity of this work group now as organized?

Would extra training increase productive capacity?

How can we cut 10%, 20%, or more?

We need to know basic information, usually available from the payroll department: How many hours do we pay for daily? among which shift? overtime/regular time?

The first report from the maintenance system is a simple comparison between the payroll hours and the hours that are worked on maintenance jobs. We hope that this number is high (above 90%) without faking it!

The first step in an analysis of productivity determines the number of hours a day or days a year that the specific mechanic is available for work. The form shown in Figure 9-1 takes you through the process of determining available hours.

AVAILABLE HOURS WORKSHEET

DATE DEVELOPED	DEVELOPED FOR:
DEPARTMENT	DEVELOPED BY

Hours per straight time week H/W		X 52	=Hours per year H/Y	
CATEGORY OF DEDUCTION			TOTAL DEDUCTED	NOTES
Weeks of Vacation per year		X HOURS PER WEEK		
Paid Holidays / year		X HOURS PER DAY		
# Days calling in sick whether or not sick or paid last year		X HOURS PER DAY		
# Days calling in personal time whether paid or not		X HOURS PER DAY		
Off site training days per year		X HOURS PER DAY		
Short term disability leave average for crew per year		X TOTAL HOURS PER ALL INCIDENT/ CREW SIZE		
Family leave act average or projected for year		X TOTAL HOURS PER ALL LEAVES/ CREW SIZE		
Standard meetings that cannot be interrupted (HR/W)		X 52		
Average for crew: Jury duty, military service, union time		X HOURS		
Other:				
Other:				
TOTAL DEDUCTS				

Total hours per year	-	total deducts	=AVAILABLE HOURS
	-		=

Figure 9-1. Available hours worksheet.

Work Sampling

Imagine taking random snapshots of your maintenance mechanics. You would find that at the instant of the snapshot a percentage of your crew is involved in marginally or non-productive activities. Work sampling is a formal technique to evaluate the activities of your maintenance workforce. Studies show that 80% of the losses are directly attributable to management attitudes, systems, and procedures. Only 20% can be traced to the worker's motivation, attitude, energy, or desire.

Most supervisors informally do their own version of work sampling as they walk around. This section introduces a formal methodology. Using work sampling management can secure facts and uncover patterns about the operation without watching everyone all the time. It is a systematized spot checking where different observers under the same conditions will get the same results. Under certain circumstances, work sampling can be more accurate and reliable than continuous observation.

Why is Work Sampling Important?

1. Sampling will help you see the real pattern present so you can attack the problems where they actually occur.

2. You must know how deep is your organization's individual pool of lost time.

3. You need to baseline your shop's productivity level before changes are made.

4. After you change the culture or install some labor productivity improvement, you need to be able to measure its effect.

How to Use the Results of Work Sampling

Work sampling itself is not a problem solver. It is more of a problem finder. If properly planned, it will give very definite indications of what should be done. For example, your study may show that excessive time is spent waiting for materials in the morning. Some judicious planning could allow the parts room person to pull standard jobs the night before when the window is quiet. In the morning, 75% of the mechanics can be put immediately to work.

Once problems have been isolated, then means can be used to improve the situation. The formula can be used after the observations have started. You can keep re-interpreting the results for different uses. For example, a 1000-observation study might yield:

Element observed	Number of observations	Occurrence %
All of the work categories	320	32%
Talking	180	18%
Travel	150	15%
Unable to locate	150	15%
Waiting	100	10%
Idle	100	10%
Totals	**1000**	**100%**

In this shop, we can conclude that about 32% of the time (or 153 minutes) is spent working a day. The crew also spends 15% (or 72 minutes) traveling. What useful conclusions can we come up with knowing that 80% of the productivity losses result from management and corporate culture rather than laziness?

This Type of Study was Done in Industry

In the November/December 1986 issue of *IPE*, investigators from Emerson Consultants, Inc. published the results of work sampling 35 typical industrial maintenance departments including all types of maintenance.

Areas and amounts of lost time each day:

- Bargaining agreement time losses (rest breaks, meals, wash-up, including normal plant practices such as get ready, etc.)   78 minutes

- Travel to and from job assignments, transporting materials, tools or the unit itself for service   77 minutes

- Idle time (no job assignment, unsanctioned, avoidable delays)   44 minutes

- Excess personal time (eating, talking, smoking, drinking, resting, in excess of

the provisions of the bargaining
agreement) 35 minutes

- Picking up and putting away tools
between jobs 25 minutes

- Waiting for materials, tools, or for the
unit to be serviced 22 minutes

- Getting job assignment, instructions 21 minutes

- Late starts and early quits at beginning
and ending of shift 21 minutes

The conclusion of the study was that there were 323 minutes per day of non-productive or marginally productive time. During the normal 8-hour (480-minute) day, the average maintenance worker spent 157 minutes (32% to 33%) on the job using job-related tools.

95% of the improvements to labor productivity will come from this 323 minutes per day pool of time. Further study showed that most of the losses were from management getting in the way of maintenance activity.

Would you like to try your hand? Below are the instructions for conducting a work sampling study. Before you start, do the following.

1. Define the scope of the study. You may want a general study to start to see how much time is spent in the general categories. At a later date, you might want to sharpen the focus and study waiting time related to the parts room.

2. Plan a study that will address the problem at hand. Assign people to the study.

3. Review the study with the people in your crews. The people may not like the idea but should be informed of their responsibility to contribute to efficiency. They should be shown how high efficiency will improve their jobs and their quality of life.

4. This is a powerful technique to determine if improvements in procedures, systems, culture, tooling, parts room, supervision have actually increased the time spent working. A survey before and after would demonstrate the impact (if any!).

10 Steps Once the Study is Underway

1. Randomness is the key to the whole study.

2. Vary your routes through the maintenance facility to increase randomness and help "surprise" the people. You can also train other people to conduct the study and randomly vary the observer. You might consider training supervisors to conduct the study in other departments.

3. Select random times from the random time tables. Plan 4–6 tours per day. Allow a reasonable time between tours. If the selected time runs into lunch or other established break, stop the tour. If the random time occurs during lunch, then skip that tour.

4. The second principle for success in work sampling is impartiality. Do not prejudge what you see or mix in outside factors (for example, you know that someone is a "good" or "bad" worker). You are trying to analyze the system, not the people. Do not assume what the people were doing or what they are about to do—only what they are currently doing. *It is important to record a person's activity before you are seen by them.*

5. Use one tally sheet per shift. Use a separate sheet for each observer.

6. Fill out the random times and random tour routes in advance. Enter them in the observation time and observation route rows. Conduct the tours using the routes indicated at the times indicated.

7. Record the number of people at work that day, that is, the number of observations you will make each tour. Enter that number in the available manpower row.

8. During each tour, record the observation elements (what is being done) in the column of the specific observation time. You can use tally marks as you walk through. One element per observation per person per tour.

9. If you observe crew members working in the wrong area or on the wrong job, make your observations where they are. Don't judge where or what they are doing, just if they are working.

10. At the end of each shift, total the observations and transfer to the recap sheet.

The special language of work sampling: accuracy, confidence level, DIN/emergency job, idle, observation (special definition), occurrence, preventive maintenance, law of probability, randomly selected time.

Use of the Random Time Table

The random time table (Figure 9-2) is used to select times to tour your facility and make observations. The table can easily be used a number of ways. The most common would be to start in the upper right and use the times across. You can also use the times vertically or on the diagonal.

Start filling in times from the table onto your observation sheet (Figure 9-3). Reorder the times from random to time order when you enter them on the observation sheet. Each day note where on the random table you end, and start there on the next

9:40	10:40	1:45	7:00	11:00	9:30	7:20	8:50	9:55	2:45	5:00	3:15	10:35
5:15	8:35	3:20	11:50	1:55	12:35	6:40	9:10	9:05	9:30	8:00	12:45	5:15
10:00	7:30	7:30	4:05	9:10	4:35	7:45	5:20	11:50	2:05	6:15	6:45	3:55
11:25	8:30	5:15	8:50	12:40	5:10	8:40	11:20	6:05	11:55	3:55	10:05	10:30
7:15	11:00	12:50	11:20	10:00	5:25	9:00	1:30	5:30	9:30	8:10	12:00	4:05
5:40	10:55	1:05	7:40	6:10	8:10	3:00	12:30	11:10	7:05	6:35	1:25	4:45
9:30	4:00	2:20	6:55	5:00	2:45	2:05	4:50	8:55	7:35	9:30	2:45	9:05
2:10	9:30	8:20	11:05	8:00	5:15	10:45	5:50	1:00	9:35	1:15	11:55	12:30
2:50	9:30	5:20	11:05	8:20	2:45	7:05	2:00	5:00	11:00	11:45	2:50	3:10
2:25	7:40	3:35	11:10	1:50	12:55	1:40	5:10	7:20	10:10	6:55	9:05	3:00
7:40	3:10	7:00	6:10	12:45	8:55	3:50	8:05	9:25	6:05	5:10	6:20	11:00
3:50	12:40	3:20	4:20	5:45	9:55	10:30	7:35	10:50	3:50	6:50	6:00	1:05
11:20	1:35	11:15	5:40	7:00	4:20	7:50	11:15	7:35	10:10	7:50	9:55	2:05

Figure 9-2. Random times.

Work Sampling Tally Sheet	Observer							
	Date							
	Group							
	Department							

								TOTALS
# people this day								
Time of Observation								
Route thru shop								
Emergency Job or DIN job								
Scheduled, corrective or planned work								
PM, short repair, TLC								
Bargaining agt losses, breaks, end of day								
Traveling								
Idle, no assignment								
Excess personal time, after breaks, smoking, eating								
Picking up putting tools away, between jobs								
Waiting, for parts tools, asset, elevator access...								
Getting job assign, instructions, dwgs								
Meetings, safety, production, quality, team...								
Other								
Column totals								

\Handbookworksamp.frp

Figure 9-3

day. Any time that is too close to one previously selected can be discarded for this round and used on a later day's tour.

We acknowledge the pioneering work of L.H.C. Tippett, who introduced the whole field of work sampling in England in 1934. Robert L. Morrow of NYU introduced work sampling here in a paper to the ASME in 1941. More recently, A. Kallmeyer has done extensive teaching on the subject. This section is adapted from the work of these men.

10
Maintenance Budgeting

There is tremendous pressure on maintenance managers to improve their budget performance. One problem is that history can only predict a small part of maintenance expenditures. For example, if a fleet is purchased together then vehicle replacement can be predicted, but it cannot be anticipated based solely on the past. To successfully budget maintenance expenditures, we must divide the whole maintenance demand into its basic parts. Then we have to divide each part into demands to account for different types of maintenance work.

How were maintenance funds spent last year? In a building you might have had $22,000 to replace that section of old sewer, $9,500 for repainting, $14,000 for HVAC, $7,000 for vandalism, $11,500 to respond to minor complaints, and on and on. In a factory the mix would be spread among each machine or other asset throughout the year.

The whole organization is run from a chart of accounts which is created by the general ledger. Every dollar spent or earned should fit into one of the lines. Maintenance itself has several to a hundred or more line items from this budget. The maintenance budget is a complex document. Reproduced below is a typical list of the line items in a manufacturing company's budget reproduced from *How to Manage Maintenance*.

Salaries
salaries
overtime salaries
salaries transferred
extra help

Labor
overtime premium
shift differential
other indirect
labor-projects
labor-breakdown, routine

87

Labor (cont.)
labor-building maintenance
labor-PM
labor-housekeeping
labor-alteration and modification
cation
labor-capital project
labor-transferred

Fringe benefits
SUI & FUI tax hourly
SUI & FUI tax salary
FICA hourly
FICA salary
vacation pay
holiday pay
sick pay
group life insurance hourly
group life insurance salary
long-term disability salary
pension hourly
pension salary
fringe benefits transferred

Travel and Auto/Truck
auto expense—gas and oil
auto expenses—repairs
auto expenses—insurance
auto expenses—permits, licenses
censes
travel expenses and tolls

Materials—Repair
material-projects
material-breakdown, routine
material-building maintenance

material-PM
material-housekeeping
material-alteration and modification
cation
material-capital project
material-transferred
inventory adjustment
redistribution accounts

Other expenses
demurrage on tank for welding
laundry and uniforms
supplies
electricity
insurance-inventory
rentals
taxes on personal property
membership dues
subscriptions and publications
extra meals
licenses (not auto)

Fixed Asset Costs
depreciation-building and machinery
chinery
depreciation-tools and equipment
ment
insurance-building and machinery
chinery
taxes-property

Optional areas
distribution accounts
training
energy accounts
capital replacement accounts

When working with a traditional budget there are several things to keep in mind. Cost reduction is cost avoidance. Cost reduction will show up in a line item above. When you increase production, asset availability, or quality (repeatability) none of the costs in the maintenance budget (listed above) will go down. The impact will be seen on the revenue side of the ledger.

The goal of cost reduction efforts is to impact one or more line items without other line items increasing as much as the reduction in the reduced item. Understanding maintenance means knowing which items you can have an impact on without unanticipated consequences.

Look at the variances for each line item over the last two or three years on a month-by-month basis. Create an average to plot against. Try to detect seasonal trends or trends related to certain projects or events. Factor in the true material and labor inflation rate. The cpi (consumer price index) or wpi (wholesale price index) might rise more slowly than parts prices. OEM's raise prices on parts to offset decreases in equipment selling prices (increase market share) or to improve their whole profit picture rather than just responding to incoming price increases.

Zero-Based Budget

A zero-based budget breaks the overall demand for maintenance services into its constituents, that is, assets or areas. Look at each asset (or group of like assets) to determine the maintenance exposure. In addition to the unit or asset list, a zero-based budget has allocations for certain areas that are hard to define as assets such as misc. electrical system or pavement. Grounds might be covered in areas by activity such as "Turf, mowing" or "Turf, fertilizer" as line items.

All maintenance activity can be traced back to one of the eight demands that follow. A demand is a categorization of where maintenance labor or materials are expended. An example of a demand might be breakdowns or "the boss told me to." Demands are described as level or irregular. PM would be considered a level demand that can be designed to use a relatively constant level of hours throughout the year. Seasonal demands are an example of an irregular but predictable demand. Breakdowns would be an irregular non-predictable demand.

Shops that are craft dominated have a more complicated problem. After the budget is completed they must go back to the individual demands and break-out the labor by craft.

The Eight Reasons for Maintenance Resources

1. *PM*—Preventive Maintenance hours/materials. Based on your facility and equipment size, use, construction, and the standard times of the PM activities, you can predict how much time PM's will take. In a TPM shop some of the PM hours will come from operators. The simplest formula is to multiply the number of services by the time for each service. Also look at the materials used for each service. Include some time for the short repairs that the mechanic will get done during the PM. Since you have some flexibility in scheduling you can consider PM's as a level demand throughout the year.

2. *CM*—Corrective Maintenance hours/materials; also called scheduled repairs or planned maintenance. As your PM inspectors look at each part of the facility and all equipment, they write up repairs (deficiencies). These write-ups become your backlog of scheduled work. The repairs are considered scheduled repairs as long as they don't interrupt jobs in process. You can look at previous years to get an idea of the hours for this activity. Since you have control of the schedule, this demand can be considered level throughout the year. These scheduled repair hours are inserted by equipment, by group of like equipment, or by area.

3. *UM*—User Maintenance (hours/materials), which includes both breakdowns and routine service requests. UM is all requests from users/tenants/customers from the routine broken handle from a broken sink to a $1,000,000 catastrophe. UM also includes small project work below the threshold of capitalization. At the beginning of the year, budget the same amount of hours for UM as the previous year by asset or category. At the end of the year, you can back off on the emergency component of UM as the PM system starts to take effect. For purposes of budgeting UM creates a level demand. In fact, emergencies will tend to bunch. Many operations use outside contractors to level the demand for UM. See seasonal demands for a special case of UM demand.

4. *SM*—Seasonal Maintenance hours/materials. This includes all special seasonal demands. Your entire grounds maintenance effort is certainly driven by season. Review of roofing systems before summer and winter or checking air conditioning before summer are seasonal demands. Some businesses are seasonal. Cleaning the candy cane line before it starts up in July would be a seasonal demand. You can also use this category to pick up some percentage of the seasonally driven emergencies or seasonally driven PM. Budget hours at the beginning of each season by asset or group based on history.

5. *RM*—Replacement/Rehabilitation/Remodel Maintenance hours/materials. In some organizations this category is capital improvement and is handled outside the normal maintenance budget. At some point, units which have not been maintained for a period of time, or have reached the end of their useful life, will have to be rebuilt or replaced. The rebuilding effort should be added to your maintenance budget as a capital replacement line item separate from any current maintenance activity. If your people are doing the modernization to bring units up to PM standards, then the hours will have to be budgeted. Since you have control of the rebuild schedule, you may be able to use rebuilds as a crew-balancing tool.

Maintenance demands for the whole operation (not tracked by individual but by location): After the base demand is cataloged by equipment or area of the plant, look into some of the budget busters below. A well-designed budget can be ruined by excessive social demands generated by visiting dignitaries or a large construction project's effect on the rest of your operation.

6. *Social Demands* (sometimes known as hidden demands because they don't always show up on work orders). Your primary mission is maintenance of the equipment and facility. You may be called upon for other duties in your organization. These duties may include supplying clean-up people, running tours, preparation for visiting dignitaries, providing chauffeur services, organizing picnics, or work on non-organization equipment and facilities (charity work). Estimate your hours for these activities.

7. *Expansion.* Any expansion in the size of your facility, size of your work force, additions to the scope of your control will add hours to your overall requirements. New buildings, assembly lines, major changes to the plant require start-up time. New facilities disrupt current activities as well as taking direct time. Adding satellite facilities will result in additional lost time until systems are well in place. Estimate additional time if an expansion is contemplated.

8. *Catastrophes.* It seems that every part of the country has characteristic catastrophes. Add time for one or two catastrophes. You can review your records for the actual amount of time spent in a typical catastrophe. This can include floods, hurricanes, and even trucks taking out the side of the buildings, fires, etc.

Once complete, this document becomes your zero-based budget. Changes to this starting point need to be justified in terms of higher or lower levels of service. Distribute your zero-based budget to the users, staff, and top management for comments. If your current hours available is a small percentage of your demand, then something will have to be negotiated. You will have to fight for the resources to do your whole job.

How to Start

1. Start the process by building a list of all assets that you maintain. As much as possible, arrange the list by department or cost center. This will facilitate report printing at a later stage. If you have a Computerized Maintenance Management System (CMMS), print an asset or equipment list. This list might have 250 assets or 20,000 assets, depending of the size of your plant.

2. Add to this list areas of the plant and site that require maintenance resources that don't lend themselves to the unit concept. Typical areas include roofs, pavement, electrical distribution system, waste water systems, compressed air piping, exterior facade, doors/windows, etc.

3. Look at the list and see if there are any units that can logically be grouped together. A wire harness assembly plant might have 50 braiding machines of similar usage and vintage. These could logically be aggregated into one line. Putting similar units or areas together simplifies the process and also makes predictions more accurate.

4. Collect any maintenance data available by unit or area for the last year or more. Your CMMS would facilitate this step.

5. We recommend this whole mass of information be designed in a computer spreadsheet like Lotus 123, Excel, or Quarto Pro. Create a template to duplicate the form at the end of this section. The equipment, areas, and groups of units/areas are listed in the template.

6. After the individual units and the general assets are listed, add the global lines (that apply to the whole site) social, expansion, catastrophes. Look into your history or estimate the impact of these areas. The three areas can be added as hours and materials or as percentages depending on the need. If these areas have traditionally been non-work order items, now would be a good time to set up the codes to put them on work orders. Once accounted for, these costs can be studied year to year.

7. Once assets have been inserted into the template, this document becomes the basis for your zero-based budget. Back up the filled in template onto diskette or tape. You have many hours in at this point, so make your backups now and keep them up to date!

8. Review each unit, area, or group and estimate your PM, CM, UM, RM, SM costs and hours. A usable history of costs simplifies this process.

9. Your materials are the sum of all material columns; your hours are the sum of all hour columns.

When management wants reductions to your budget, you have a new level of discussion. All changes need to be justified in terms of higher or lower levels of service on individual assets or areas. Now when cuts are needed you can talk about which assets will be allowed to deteriorate, and which departments will not be served as well.

Asset		PM		CM		UM		SM		RM		Totals	
#	Desc	Hours	Mat'l $	Hours	Mat'l $	Hours	Mat'l $	Hours	Mat'l $	Hours	Mat'l $	Hours	Mat'l $
Totals													

Some organizations use outsourcing strategies where they crew for 75% to 80% of demand and use outside vendors during peak periods. Distribute your zero-based budget to the users, staff, and top management for comments. If your current hours available is a small percentage of your demand, then something will have to be negotiated because deterioration is taking place or your customers are unsatisfied. You will have to fight for the resources to do your whole job.

After all assets and areas are accounted for, add in the inflator for social costs, expansion costs, and catastrophe costs. The final document should reflect your best budget estimate for next year, asset by asset. In the recent releases of the popular spreadsheet products, there is the capability to duplicate the same sheet for several years. This becomes a three-dimensional spreadsheet. Along the x-axis are the cost areas, along the y-axis is the list of assets, and along the z-axis is years into the future. Large assets should be looked at 5–10 years into the future.

11
Life Cycle Costing

The life cycle cost is the total of all of the cost areas for the life of the asset. In overall financial terms, the life cycle cost should be the determining factor in equipment selection and building design. In a production environment, the life cycle cost should be divided by the projected lifetime production for an asset cost per output (such as part, ton of pulp, pound of gold extracted).

Since life cycle costs are cost projections, they are guesses about the future labor costs, parts costs, energy prices, interest rates, and other factors. There are two ways to evaluate life cycle costs which are different only in the way they handle the time value of money. The first method disregards the time value of money and looks only at the total cost:

Life Cycle Cost (LCC) = (Ownership Costs + Operation Costs + Maintenance Costs + Allocation of Overhead Costs + Downtime Costs + Other costs)/ unit production cost

(A factor for usage: in a fleet can the new unit carry more product than the old one, or in a factory is the production rate higher? Not usually an issue in buildings.)

The second method includes the time value of money and weights the investment by when it occurred. For example, if you buy two pumps for $6000 and $10,000 the money has a net present value of $6000 or $10,000 since it was spent in year one. Assume you retire the pumps after 10 years and the salvage values (the amount of cash you receive from the machinery dealer) are $2000 and $7500. O & M costs are $1000 and $800, respectively.

96

	Pump A	Pump B
Purchase price	$6000	$10,000
Salvage value after yr 10	$2000	$7500
O & M costs per year for 10 years	$1000	$800
Using 10% interest rates		
Net present value of salvage	$771	$2891
Net present value of O & M costs	$6144	$4915
Present value in today's dollars	**$11,373**	**$12,024**
Total cost without present value	**$14,000**	**$10,500**

Without the time value of money, it would seem that the more expensive pump is the better deal. With the time value of money factored in, the less expensive pump is the better choice even though the O & M costs are higher and the salvage value is much lower. Complete life cycle costing places each cost (energy, maintenance, downtime) in its projected year and discounts it according to formulas or tables. Details and formulas for this technique are discussed in the chapter on selling to management.

Questions to consider in your evaluation of LCC:

1. *Utilization:* High guesses for utilization will skew the results toward low operating cost units. Low utilization guesses will skew away from high fixed cost units.

2. *Interest rates:* High rate assumption will favor lower initial investment (higher operating and maintenance cost) units.

LCC should drive all acquisition decisions. When the monthly LCC of an old asset exceeds the monthly cost of a new one, then the old one should be replaced.

A Look at Specific Cost Elements

As maintenance managers, we hear "reduce costs or hold the line on the budget" while the asset base is being increased. Good management means providing capacity while reducing and controlling all of the core cost areas.

To get a handle on the costs, *we must identify the costs at an actionable level.* An actionable level is a level where the consumption of the resource takes place. The costs of most maintenance operations can be broken down into five core and several secondary areas. To have an impact on costs operation, you *must* impact one of the five cost areas.

CORE AREA	EASE OF IMPACT
Ownership Costs	difficult to change
Operating Costs	easier to change
Maintenance Costs	moderately difficult to change
Overhead Costs	very difficult to change
Downtime Costs	moderately difficult to change

Consider all programs to reduce costs in terms of the above chart. All long-term opportunities for cost reduction can be expressed in terms of this chart. We stress long-term reductions in the cost structure of your building, factory, or fleet because it is easy to effect short-term reductions in trade for higher long-term costs.

Ownership Costs

As soon as you purchase or lease an asset, you begin to accrue costs. This run-up of costs might predate your use of the asset by months or years. The first line of defense against high maintenance costs is to make sure you need the asset before you buy it, or change the way you own the asset. Any cases where the needed capacity (from beds in a hospital to machine tool open time) can be moved to an existing building, vehicle, or machine will avoid ownership costs.

a. Purchase costs and depreciation
b. Costs of money
c. Lease/rental payments (fixed portion)
d. Insurance costs, deductibles, self-insurance reserves, self-insurance claim costs
e. Permit, license costs, occupancy licenses, statutory costs (costs mandated by laws)
f. Make-ready costs, bringing equipment in line with organization standards
g. Procurement costs: actual cost of searching, bidding, shopping for equipment
h. Re-build/re-manufacture costs/major capital improvements
i. Design fees, engineering fees

j. Permits, licenses to build

The ownership costs vary directly with the size of the asset base. A reduction in asset size through increased availability (less back-up units), increased utilization (add second shift), decreased cycle times (in a factory), or decreased mileage (in a fleet operation) resulting in fewer units required, or reducing the permanent asset base through peak demand leases will all favorably impact ownership costs (as long as assets are actually disposed of).

Ownership costs can be increased by specification. If you specify higher quality finishes or products or heavier duty systems, your ownership costs will increase. Specification of fancy trim and wall finishes, exotic technologies, and options will all increase your ownership cost. Many of these expenditures may be desirable (by lowering other costs or improving comfort /morale), but they still will increase ownership costs. The tradeoffs are essential to calculate. The whole field of guaranteed maintainability is based on these tradeoffs.

Depreciation: Some of the components of the ownership cost (depreciation—depending on the method used, and interest) go down as the asset ages. There is an ongoing discussion about the best method of depreciation to use for maintenance purposes. Your accounting department uses techniques dictated by the IRS to influence (increase/decrease) profits.

Our goal is to simply consume the value of the building, vehicle, or machine over its true productive life. *We recommend that you be aware of the technique used by your accounting department to depreciate your asset categories.* For maintenance management purposes, however, use the straight line method (to simplify your calculations).

Depreciation formulas:

Total depreciation = (Total purchase price - Salvage value)
Yearly depreciation = (Total purchase price - Salvage value)/ Years of life

where

Years of life = the average age in years of a unit when it is retired
Total purchase price = total of all costs including closing costs, prep
Salvage value = how much the unit will be worth when you retire it

Operating Costs

Most assets cost money to operate. In assets such as buildings, trucks, and furnaces the operation cost will eventually exceed the acquisition cost. It is not unusual for a tractor trailer to use its acquisition cost in fuel dollars every 4 years.

Components of operating costs include:

a. energy (oil, electricity, gas)
b. energy taxes (excise, sales, surcharges)
c. other utilities such as water, sewer, steam, chilled water
d. usage charges on rental/leased units
e. wear part consumption (such as tires)
f. consumable such as lube or cutting oil, added hydraulic oil
g. misc. operating costs

Operating costs vary directly with utilization of the asset. Reduction in usage through better scheduling, reduced standby or operation hours, preventive maintenance lubrication which reduces friction in bearings and power transmission, proper adjustment and trim of burners or engines will all favorably impact operating costs. The unique aspect of operating costs is that they respond *immediately* to improvements in your operation.

Maintenance Costs

Mechanical systems wear. All systems need attention from time to time. The cost of this attention is the focus of almost all management cost-cutting efforts. We see that maintenance costs are only one of five core areas

a. inside labor
b. inside parts
c. contractor labor
d. contractor parts
e. service contracts
f. misc.
g. hidden costs of failures

Maintenance costs vary with many factors, including condition of the asset (age, type, and condition), type of service, operator expertise, mechanic expertise, company policy, equipment specification, and weather. It is very difficult to determine all of the contributors to a particular maintenance exposure. Managing maintenance means understanding the major contributors to your maintenance exposure.

Today's maintenance costs reflect wear and tear that took place in the past. Therefore, if we reduce or increase our utilization today, our maintenance costs will not be affected until some future date.

As a mechanical asset ages, the maintenance costs usually increase. Maintenance costs increase significantly when the critical wear point in mechanical systems is reached.

Since many factors influence maintenance costs, investments in various areas will have different effects in different departments. For some organizations operator training might have a major impact, for other organizations specification of heavier duty components will reduce costs most efficiently.

Maintenance costs are avoidable. You invest funds today to avoid costs in the future. A comprehensive preventive maintenance program will increase your maintenance expenditures in the beginning (the investment). The return may begin 12–18 months in the future in reduced interruptions and higher productivity.

Management of the maintenance component of expenditures is by far the most complex of the cost areas. You have to track, analyze, and cross reference all of the repairs The real answers about the exact factors contributing to your maintenance exposure for your operation lies in the mass of data collected over the asset's life.

Two of the hidden costs that should be included when evaluating different maintenance modes are the true costs of downtime (covered in the downtime section) and the hidden costs of emergency repairs. We use the word hidden to mean the disruption to your ongoing operation caused by emergency calls. This disruption adds to the cost of the interrupted activity.

Overhead (of the Maintenance Effort)

The overhead costs tend to be fixed except for major changes in the department's size or role. These cost areas tend to be the last to change and are generally changeable only by management decision. For example, fuel costs will naturally decrease if you change to high-efficiency boilers. Overhead costs change as the result of closing facilities, consolidation, restaffing, etc. Money-saving efforts can cut a percentage of overhead costs.

  a. Cost of maintenance facilities
  b. Heat, light, power, phone for the maintenance shop only
  c. All persons in maintenance department not reported on repair orders (supervisors, clerks, etc.)
  d. Supplies not charged to repair orders (rags, nuts and bolts, etc.)
  e. Tools and tool replacement
  f. Repair and cleaning of maintenance facility, maintenance tools
  g. Computer systems, all expenses

Excellent maintenance controls may be able to reduce facilities, supply usage, non-repair time, and tool loss. Applying sound maintenance procedures to your own buildings and tools will certainly reduce your overhead costs.

Overhead costs begin to accrue before the first asset is purchased and before the first part is made, mile is driven, or student is taught.

Downtime Cost

In a production facility, the downtime cost is usually by far the highest of the core cost areas. A $20 million dollar machine could be producing $100 million worth of shipments per year.

  a. Revenue loss less recoverable costs like materials
  b. Idle operator salary
  c. Replacement unit rental costs
  d. Replacement cost for ruined product
  e. Late penalties, missed JIT window implications
  f. Tangible and intangible costs of customer dissatisfaction, loss of good will, hidden costs, other costs

Downtime calculations are difficult because they appear clearly on someone's budget as a lack of production, not as a definable cost.

Examples of downtime costs:
Automobile assembly—the downtime costs for the main assembly line approaches $5000/ minute.

Hospital—if the ICU goes down and someone dies the lawsuit could cost millions.

Fleet—the downtime cost of a loaded tractor might be $700/ day or more.

Power utility—a peak load shaving turbine might have a downtime cost of $150,000/ hour.

On a construction site—a downtime incident for a bulldozer might be $100/ hour. A crane on the same site might be $2750/ hour or more if it holds up construction.

Downtime costs can be a deciding factor in the decision to invest in a PM or computer system. Related to downtime is the concept of demand hours. Demand hours are the hours that the equipment is in demand. A one-shift operation that runs equipment 52 weeks per year (less 4 holidays) has the following demand hours:

(8 hours/day × 5 days × 52 weeks) – (4 holidays × 8 hours/day) = 2048 demand hours.

To take advantage of this limited demand, many one-shift operation firms reduce downtime by using second or third shift for PM or repairs. Where this is possible, many of the costs associated with scheduled downtime are eliminated. Downtime due to emergencies would still have a impact.

The reasons for downtime should also be tracked. Much downtime is not related to maintenance but rather to set-up, material replenishment, color changes, or operator-controlled situations such as sabotage, or abuse of the unit. This reason for downtime is usually captured as "repair reason" on the repair order.

Additional Cost Areas to Discuss

There are costs beyond the core costs that apply to specific industries and organizations. These costs can dwarf the core areas, where they apply.

Productivity

In production, a responsive maintenance department can improve productivity. This is hard to prove, but anecdotal evidence shows that attention to the worker on the floor (by fixing little things quickly, replacing bulbs, etc.) will improve morale. Improved morale helps workers care for how they use the equipment, which results in fewer breakdowns and increased productivity.

Safety

In the transit business the cost of accidents due to maintenance is important. Airlines carry billion-dollar insurance policies against maintenance catastrophes. The cost of safety must also include cost of damage, medical costs, loss of morale, lost time, lost time for management, and legal fees.

Scrap and Variation

Good maintenance can reduce scrap by making sure that machines won't fail in the middle of runs causing scrap. Also, well-maintained machines produce parts (assuming they can produce the part in the first place) with less variation. One of the several elements of quality is good maintenance practices.

Environmental Protection

In the chemical, oil, and related businesses the cost of environmental clean-up, permits, and fines can be traced back to poor maintenance practices. These costs are added to the costs of having your customers angry about your environmental policies.

Attractiveness to Customers

In the retail field the store has to be designed to help the customer want to shop. Deterioration would scare upscale retail away. Of course, if you operate a surplus store, the deterioration might work in your favor.

Buffer Stock or Work in Process (WIP)

In a factory you build up inventory in front of each machine in case the previous machine breaks down. This work in process has a cost associated with it that JIT (just in time) manufacturing wants to recover. Good maintenance can reduce WIP.

Late Penalties (Early Incentives)

In the trucking world some products must be delivered on time. Good maintenance practices facilitate on-time arrival. In construction they use the carrot-and-the-stick approach with liquidated damages for lateness and a reward for early completion. Good maintenance does not produce early completions, but bad maintenance can block the company's ability to meet its commitments.

There is specialized software that has been available from time to time to help with these calculations. Two older packages come to mind: PERDEC from Oliver Marketing in Montreal, and VIA (Vehicle Investment Analyzer) from Dan Chesire in New Jersey.

12
The Science of Customer Service

Keeping the trucks repaired and safe is not enough. Keeping the building comfortable and free from pathogens is not enough. Keeping the machines running and making quality parts is not enough. Doing maintenance, no matter how well, does not add value to your organization's product or service. The challenge is to focus the knowledge, skills, and can-do attitudes of maintenance people to go beyond just fixing breakdowns. Great customer service goes well beyond fixing things—great customer service solves the customer's problems.

Excellent customer service is the holy grail of all service organizations. Great industrial organizations such as IBM and Xerox built their empires on excellent customer service. In both cases the service representatives were trained to look for and solve problems. Maintenance in all organizations is seeking great customer service.

The important questions of customer service are: who is my customer, what do they do, and how can I contribute to the success of their effort? It may seem trivial, but many maintenance departments are so involved in the personalities of the players that they forget these simple questions. Each type of business has a dominate need and several secondary needs.

The science of customer service has some rules and some laws.

1. The attention of the organization, work group, or individual is on the outcomes for the customer.

2. Thought has gone into the customer's needs, problems, pressures, and history.

3. The mechanic is the key player since they are the ones closest to the customer.

4. The mechanic has the power to solve problems for the customer.

5. Customers are treated as individuals.

6. Non-value added coordination is kept to a minimum.

7. Information systems must provide information that will help service the customer better.

8. "What it costs" is second to "make the customer happy."

9. Benchmarks are shared and understood by mechanics.

10. Benchmarks reflect the service values of the individual customer.

Maintenance departments are called upon to provide different types of service (even in the same organization). There is no one way to provide customers with service. A research scientist wants the equipment to last to the end of his or her experiment, while the facility department might want to save money and have few headaches. The company president is interested in quick solutions to maintenance problems, and the accountants want low cost coupled with good controls and accountability.

Ultimate Customer Service Goal

Customer service is a popular topic for business books. One of the most popular ideas is Ultimate customer service. Ultimate service is an unanticipated level of support from one person (the service provider) to the customer. Maintenance people can provide this level of service.

Sometimes it is as simple as Gary Boucheor's statement that at the VA Hospital in Montrose, NY, at various times maintenance people can be seen pushing wheelchairs and helping people find their way around a 1.3-million-square-foot 50-building complex. In a story related by Ed Litano at Berk Concrete, a

Examples of customer's needs and the actions that will provide a real value.

Business Situation	Service problem	The business	Service values
Oil terminal has a computer system that runs the loading racks, prints bills of lading, does accounting functions, starts and stops the appropriate pumps and opens the correct valves.	Computer goes down in the middle of the night in the winter.	The oil business is such that if your terminal is shut down in the winter the truck driver will go elsewhere to get their oil. That sale is lost permanently. Truckload value= $6-7000. 200 trucks on a busy day.	Reliability, no downtime, safe operation, quick response. High cost of downtime (million dollars a day range). The cost of service is secondary to the response time.
Machine shop making precision turbine blades. Each blade costs $1000, mistakes could scrap a blade with $500 or $750 invested.	3D Milling machine goes out of specification.	Each step adds value to the piece. End customer needs precision and a very accurate delivery window.	No scrap, keep unit in spec, reliable capacity (so schedules will be met) and customers will be satisfied.
Office of the Secretary of State, Wash, D.C.	Air handler bearing squealing whenever it starts up.	This job has decorum and protocol to follow. The physical functioning must be background for the smooth functioning of the office.	Comfortable building, safe indoor air quality, professional and very quick response, discretion, no repeat repairs, no interruptions.
Taxi service	Blown motor	Money is made only while the vehicle is on the street. Spare units are costly, it is a low margin business.	Back on the road fast, cheap repair, no repeat repairs, safety.
Hospital outpatient department	Tiles pulled up and leading edges are up creating a tripping hazard.	The public moves through the building. The patients might not be too steady on their feet. They may be in pain already or might have heard bad news. Hospitals are perceived as having deep pockets in relation to insurance claims.	No calls from the public, public and employee safety, low cost.

truck washer took it upon himself to wash the truck completely (rather than the inside of the drum only) and then gave it a coat of wax to help make it easier to clean in the future!

Other stories also have to do with going above and beyond the call of duty in different ways. Mike Mendralu of U.S. Tsubaki remembered an electrician who was given a job assignment to create a wiring diagram for a special control panel to send to an OEM who was building a 300 ton press for the company. The electrician decided to not only make up the diagram but also pre-fabricate an actual wiring harness and the special panel itself, and shipped it to the manufacturer. The entire assembly could be added to the press in minutes instead of hours. The possibility of hidden problems was largely eliminated, and the quality and materials were up to the company's high standards.

Smith and Wesson makes pistols and accessories, and is basically a precision machine shop. Bob Demcas remembered when an electronics technician was working on a critical measuring machine that was giving them ongoing problems in keeping the calibration. The technician made the adjustment at the end of the day and went home. After dinner that night the technician realized what the underlying problem was, returned to the plant, and made the permanent repair. It turned out that he was right on target. When he showed the staff his idea, he was applauded and got to implement the idea on all of the other measuring machines.

These four stories have something in common. You might think that I traveled far and wide to collect stories such as these. In fact, these stories were picked from 20 given by 20 maintenance professionals in a single class. Every maintenance department has incidences of excellence like these. Find (or remember) these stories and retell them from time to time. When they happen, honor the people involved with a thank you or more (if you can swing it).

Going beyond repair to problem solving is Ultimate level service. Ultimate level service requires that management trust the maintenance workers because many of the examples go beyond the role of the maintenance person. Along with trust is freedom to stray outside the traditional roles for maintenance. In many of the stories, a tightly scheduled person, even if they are trusted, would be unable to help because their day is predesigned. Ultimate service levels also require a friendly atmosphere where these spontaneous events can grow and thrive. The most illusive is for management to create an environment where the maintenance crews are motivated to look up from their repairs to see fellow human beings that have problems, and decide to help.

Communication is a Two-Way Street

How serious is your firm about customer service? DuPont is serious about safety. All accident reports have to be faxed to the president of the company within 24 hours. Whether the president gets involved or not, that act conveys a message about the importance of safety.

In your facility, how are complaints and communications handled? Is the feedback from the customer treated like a valuable commodity, or is it almost ignored?

In the science of customer service, communication is one of the most important tools. For example, in a corporate data processing center and executive office complex, one of the three chillers had to be serviced unexpectedly. With the data center at the top of the priorities, the offices would have to be warmer than usual. Usually this would set off a flurry of complaints. The new building services manager sent building-wide E-mail messages warning of the hot conditions, explained the "why" of the problem (to keep the data center up), and sent out updates every few hours until the repair was complete. After the building was back to normal, he E-mailed a public thank you to his crew, the contractor, and the building occupants!

At Abbotts Laboratories in Abbott Park, every work order that is sent in is responded to with a copy showing the hours and materials to complete the job. The work order is folded into thirds and on the front is a customer satisfaction survey. Follow the example of a major pharmaceutical company and put a quality/satisfaction survey on the back of the user's copy of the completed work order. Questions include: were you satisfied? was the job done in a professional manner? has this problem happened before? was the mechanic on time with any appointment? was the mechanic professional? and any suggestions. While less than 5% are returned (and many of them are compliments), the ones that are are jumped upon and solved. The customer is thanked for contributing to the improvement of the department.

Another firm printed business cards for everyone in the maintenance department. Whenever a job was completed that was initiated by a customer, a business card was left behind identifying the maintenance team member, with a number to call for any problems.

13
Reengineering Maintenance

Reengineering is the reaction to increasing competition and the new opportunities offered by technology. Reengineering is possible because technology has changed the business of business. But no matter how much business and other organizations reengineer, roofs, pavement, and equipment still deteriorate. Good maintenance practices, whether they are lubricating a bearing or doing a complete infrared scan of the buss duct, are still needed. In reengineering we are less focused on what maintenance practices we choose, and more focused on how we initiate the jobs, who does them, what support we provide, and who else is floating around (less non-value added personnel).

The difference between continuous improvement and process reengineering is summed up by Peter Drucker: "There is nothing less useful than to do a little better that which should not be done at all." In reengineering we look at the process of delivering maintenance and decide if a task should be done at all. After the tasks are shaken out, we can apply continuous improvement to improve the tasks that are left.

The first rule of reengineering is to look at the process that maintenance is part of instead of the department or function of maintenance. We want to see the whole process from the customer's initial upset (machine doesn't cycle) to their satisfaction (finished parts cascade out once again). Reengineering starts with the customer's needs and works through the process of delivering satisfaction. It focuses on the handoffs between departments and the idle times versus work time.

Reengineering has been known throughout history. The great changes in the history of organizations like the departmentalization and financial controls of General Motors were reengineering efforts. Look at what Thomas J. Watson Sr. did at IBM. He transformed the field of sales—that was clearly a reengineering of the process of contacting, managing, and satisfying customers.

The people who named reengineering were Michael Hammer and James Champy in their book *Reengineering the Corporation* and previous articles (*Harvard Business Review*). *Reengineering the Corporation* calls itself a manifesto for business revolution. Reengineering radically changes the way you do business. It looks at the process of delivering maintenance to customers. Reengineering ignores departmental barriers to provide an order of magnitude improvement in maintenance delivery.

Many companies have gone through complete reengineering. Reengineering is a billion dollar business. The biggest management consulting firms have major departments in reengineering the whole business one process at a time. In some cases the experience of reengineering in maintenance is not a positive one.

Maintenance has been decimated by ignorant downsizing efforts, sometimes called reengineering programs. Many have lost jobs as a result of sloppy efforts. Dr. Mark Goldstein, in his article "How the Maintenance Profession Can Combat Destructive and Irresponsible Downsizing" says it clearly that outside consultants promise improvements in maintenance, slash the head count irrespective of the long-term impact, and crow about the improvement in the quarterly results. The piper will be paid because entire skill sets and specific building, fleet or plant knowledge have been liquidated. Eventual deterioration of the protected asset, unsafe conditions, increased regulatory scrutiny and fines, increased downtime, and accelerated replacement will be the price.

The other issue is the union. Many unions protect the ability to do certain types of work. They argue persuasively that long training is needed to ensure quality work. Any dilution of a trade's mandate is considered dangerous to its union. Many unions believe it is literally dangerous to have outsiders perform their craft. Maintenance as a field doesn't want amateurs poking around inside our electrical panels or playing with the PLC programs that operate complex machine cells. Reengineering requires specific knowledge of maintenance and a good working relationship with the unions.

The payoff of informed and well-executed reengineering is increased uptime, reduced cost of operation, and possibly fewer maintenance people doing maintenance (they should be reassigned to keep their skills available). One of the factors that can be improved by reengineering is mechanic productivity. The productivity of maintenance mechanics is less than 33% by work sampling and less than 25% by other measures.

One of the leading practitioners of reengineering in the maintenance department is Edwin Jones, a now retired internal manufacturing consultant from E. I. du Pont. He worked extensively throughout their plants documenting their best practices and creating benchmarks. In the article "Why Reengineer Manufacturing Practices" (*Maintenance Technology*, Sept. 1994), he lists the steps he took in the process.

Benchmark
   Identify function
   Develop comparison measures
   Select comparison organizations
   Collect data
   Validate comparison data
   Analysis

Best Practices
   Leadership
   Improvement in reliability
   Preventive maintenance
   Predictive maintenance
   Planning
   Outsourcing and contracting
   People development
   Parts, stores management

Strategic Plan
   Key issues focused on best practices
   Identify stake for each key issue
   Pick a champion for each issue
   Define actions under each key issue

Track performance measures
Adjust based on measures

Execution
Measure Progress
Adjust the Strategic Plan

Ask Questions to Reengineer

In some ways, the process of reengineering is a process of asking questions and debating about, reflecting on, and listening to the answers of a small group or reengineering team. Robert Williamson, a management consultant with Strategic Work Systems, listed 11 questions to consider when approaching maintenance as part of a larger reengineering effort. As with all reengineering efforts, the questions should be answered from the customer's point of view.

1. What maintaining do we perform?

2. Why do we perform them?

3. Why do we perform them in the specific way we do?

4. How are these processes aligned with other related business processes?

5. If barriers were eliminated, who is in the best position to perform these tasks?

6. How could a new configuration of people and technology perform the tasks more efficiently?

7. What type of information support is needed for a small team to make all of the day-to-day decisions about the process?

8. What measures could be used to evaluate the performance of these tasks?

9. What would it take to motivate a work group to take on this process?

10. How could handoffs (allied tasks handled by other groups or departments) be eliminated?

11. How could we redesign the organization to support such a work accomplishment scheme?

Attack Assumptions

Reengineering attacks the assumptions. Any suggestion that contradicts an assumption should look ridiculous on the surface. All of the reasons "why not" flood in. The first step of maintenance process reengineering is thinking the unthinkable. Our assumptions might include the following.

Mechanics are the best people to perform service.

One of the assumptions is that maintenance department people are the best people to provide service. In some cases the best person is an operator, in others it might be a custodian. In other cases the customer should fix it themselves and eliminate the need for a department. We already clear our own paper jams in copiers. Maybe there should be a threshold: on one side the user does it, on the other the skilled maintenance person does it. In all cases the new jobs need to be supported by adequate training and coaching to ensure success. The rewards have to change to reflect the new design.

Maintenance budgets should be a percent of direct labor budgets.

Automation increases the amount of machinery while reducing the direct labor. In new plants the amount of maintenance required has nothing to do with the amount of direct labor.

Mechanics need to travel to get to jobs.

Maintenance could be a network like the current fire brigade or emergency response teams. Maintenance people could be in place and tied together with a network. They would be assembled for ad hoc assignments and then blend back into the original department.

There should be a central maintenance shop.

In some cases the maintenance shop is obsolete. With technology, the mechanics can be dispatched from home and file their reports by fax, E-mail, or direct interface to the CMMS from remote locations.

If it's not broke now, don't PM it.

This is an old thought on the part of maintenance and operations people. Deep in the hearts of maintenance people is the belief that breakdowns are fun and PM is boring. So we ensure breakdowns. Perhaps we should tie maintenance department compensation to uptime! With overtime, we are now paid a bonus for breakdowns!

Keep the driver or operator and the mechanic from becoming too friendly.

Perhaps they should be partners. Their compensation should be tied to output.

Service requests should be filtered before giving them out.
Take 5% of the day just getting instructions, etc.

Service requests flow directly from the user to a multi-skilled mechanic who determines if they can handle the job themselves or if they need help. Perhaps jobs are charged back to originating departments to ensure no department uses too much resource. The mechanic opens the work order, updates CMMS, and orders parts.

You should carry everything possible on the truck, and buy nothing locally.

Perhaps a relationship with local distributors is better than trying to imagine what you will need.

Keep maintenance departments of local companies separate.

Create a maintenance community. Maintenance departments band together in an industry or in a town into an informal group for the purpose of sharing expensive tools, hard-to-get parts, maintenance skills, training capabilities, etc. Are we beyond the barn raising as a community activity?

Mechanics are not smart enough to enter data into computers.
Maintenance people are glorified grease monkeys and are incapable of doing analysis.

Only engineers can design things.
Mechanics can't plan their own jobs.

All these assumptions are interrelated. A mechanic has a longer apprenticeship than a doctor or lawyer. Many mechanics are responsible for multi-million dollar assets. They make daily decisions that the outcome could cost or save the organization hundreds of thousands of dollars. Organizations have a tremendous resource that is only partially tapped. There is almost nothing that the maintenance people can't do!

Equipment should be repaired and not thrown away.

Why fix stuff anyway? Design to discard. Some environments cannot justify too much attention to maintenance. Equipment in some sectors is very reliable, low cost, and obsolete within a year or two anyway.

An in-house department is always the best.

The building maintenance field has demonstrated that an in-house department is not always the best from a cost or service point of view. More flexibility in this area might improve the outcome for all concerned.

Make trips to the storeroom to get parts.
Hunt around for specialized tools.

Why not put cabinets near every machine and put tools and parts right there.

In fact, by thinking creatively, many of the above assumptions will be found to be wrong. Worse than being wrong, these assumptions create barriers to seeing the ways maintenance could work. Champy and Hammer call this kind of thinking "thinking outside the box." Consider what has never been considered before.

Reengineering could change the maintenance department so much that you will not be able to recognize it. The important criterion is the customer getting the service they need and deserve.

Reengineering Versus Continuous Improvement

This book has an entire section on continuous improvement. How does reengineering relate to continuous improvement?

John Campbell, a leading consultant with Coopers and Lybrand, said it succinctly in an interview: if you need a revolutionary change, reengineer; if you need an evolutionary improvement, use continuous improvement.

Continuous improvement and TQI (total quality improvement) help keep a department from falling behind by improving and sharpening the delivery and quality of the service. Never being satisfied with the status quo is essential. Sometimes the problem is too big and too deeply ingrained in the fabric of the organization to be significantly impacted by continuous improvement techniques. Then reengineering is appropriate.

When a problem spans many departments, reengineering is appropriate. When there is a need for tremendous change in a short time, then reengineering is appropriate.

Productivity is the Name of the Game

If productivity is the name of the game, let's look more closely at the levels involved.

Strategy	Level	Tools
micro level—personal improvements	personal	time mgt, training, better layout, tools
continuous improvement	crew	scheduling improvement, better supervision, people skills, planning
continuous improvement	department	same as crew, communication of mission, common vision of maintenance, CMMS
macro level—reengineering	company or unit	procedures, interfaces, downtime, customer relations, quality

Reengineering comes into play where the problem transcends the department or work group. Dramatic changes in overall productivity come from dramatic changes in the way you do business.

Other Areas

When constituted as a department, the maintenance function provides many other services. One of the problems of reengineering efforts is that no one considers that the maintenance department is a catch-all group of multi-skilled individuals. The AMA course book *Maintenance Management* lists some of the non-maintenance functions that must be considered:

1. safety
2. store room control (for non-maintenance items)
3. engineering
4. quality assurance
5. fire safety and fire brigade
6. first aid officer
7. operator training
8. construction or construction management
9. energy conservation
10. recycling
11. regulatory compliance
12. porter activities (moving boxes, etc.)
13. moving furniture
14. pick-up and delivery functions
15. equipment inspection before acquisition
16. personal services
17. errand running
18. community service on behalf of organization
19. event set-up and clean-up
20. relief operators

Return to this list after considering the next chapter, on outsourcing. One of the biggest mistakes in outsourcing is to consider maintenance duties only and not consider this list of 20 items. When all activity that maintenance is involved in is tabulated, then outsourcing sometimes becomes a less attractive proposition.

14
Contracting, Outsourcing

The first goal of using contractors and outsourcing is to get maintenance or construction work done at a higher quality, faster, safer, or lower cost than would be possible with your own crews.

The second goal of outsourcing is to reduce or control the number of FTEs (full time equivalents). This increases flexibility to allow the company to expand and contract with sales. It allows the organization to maintain profit margins during downturns. A contractor's people can be sent home without severance or adverse publicity.

The third goal is to concentrate talent, energy, and resources into areas called core competencies. These are the core reasons for being of the organization, and where the organization feels its survival depends. For example, truck maintenance to a freight company might be a core competency. But truck maintenance to a potato chip maker might not be a core competency even if both had the same number of vehicles.

Once an outsourcing contractor is chosen, the best relationship is to make them into partners or team members. They can help with all kinds of decisions relating to maintenance. A bad contractor will cause damage that will take years to undo and will wipe out savings from taking the lowest bid many times over. As team members, and over time, they will look out for your interests when you are not around.

The degree of outside contracting has no impact on your need for a PM system, for effective root failure analysis, or for a responsive organization. Even with good contractors the maintenance leadership needs to be involved in the root failure analysis and PM.

In a recent survey in *Maintenance Solutions* magazine (Jan. 1996), which addresses the issues of all types of building maintenance, over 66% of the respondents used contractors and their own staff for the maintenance effort. Only 30% used their own staff exclusively, and 2% used contractors exclusively.

Types of work commonly contracted:

1. Seasonal (grounds, snow removal)
2. One-time work (construction)
3. Specialized work such as tank cleaning or water tower painting
4. Low-skill work (floor care, security, grounds)
5. High-skill work (electronics, rebuilding components)
6. Work requiring a license
7. Work where the contractor takes legal responsibility (fire safety systems, alarms)
8. You can contract out any function including the complete operation of your factory, fleet, or rental property.

The best manager balances their needs for work with the true overall economics. They also factor in the organization's policy toward contracting.

Seventeen reasons to hire a contractor:

1. Save money.
2. Improve quality.
3. Lack of skill in-house (high-voltage electrical, generator set repair, elevator, etc.).
4. Lack of appropriate license (even if you have the skills).
5. Lack of specialized equipment (thermography).
6. Reduce liability (elevators, fire systems, asbestos removal).
7. It is sometimes easier for contractors to get environmental and other permits (less visibility).
8. Reduce hazard to own employees (window washing, asbestos removal).
9. You want an outside opinion (a devil's advocate or skilled tradesperson or engineer), or you need an outside "expert" to show you a whole new approach.

10. Training (send your mechanic along on the job to improve their skills).
11. Save time when you are already busy or the job is too large.
12. Don't want to manage job (hiring the contractor to do that).
13. Want the flexibility.
14. Politics (disagree with top management about the number of hours a job should take), other political reasons.
15. You don't want to lose control of existing projects to make room for a large new one.
16. Shaving peak labor requirements, using a contractor instead of overtime, or crewing up.
17. Free up talented people in maintenance to work on core business, get out of unnecessary sidelines.

Eleven reasons not to hire a contractor (common concerns):

1. Loss of control.
2. Too slow or high cost for fast response.
3. Is the job well defined enough to contract out? How will you know if you are getting your money's worth? You don't have time to define the scope of work.
4. Is there a negative image to using contractors when they are representing you to your users?
5. Possible quality problems where the contractor knowingly or unknowingly cuts corners.
6. Problems completing the last 5% of the job (even when you hold back money).
7. Dependency—you can become dependent and not develop critical skills in-house and can't make a move without the contractor.
8. You get ripped off by a con artist.
9. A well-meaning contractor is out of his/her area of expertise in your job.
10. Keeping business secrets is more difficult.
11. Too much time to manage contractors.

The Problems with Low Bid

Low bid is one of the best ways to ensure the honesty of your procurement process, people and the contractors. It also ensures that the organization will not waste money by paying more for similar products. Having said this, low bid is also the bane of maintenance departments. More unmaintainable junk has been bought under the banner of low bid than most people thought possible.

Yes, professionals can work the system and get what they want by manipulating the specifications. It takes time and effort and doesn't always work! In a state transit authority the specifications for a common pick-up truck was 75 pages. They still had to fight for the model that performed best even with the specifications.

In the most limited version of low bid, you are required to purchase anything that meets the advertised specifications regardless of how that particular asset performed in the past. Examples of this can be seen throughout municipal governments' fleet departments. The typical fleet in a local government is made up of a few of everything. Every year they get different truck makes and models, irrespective of which ones performed best in the past. It seems that whoever was hungry for business when the bid came out that year got the bid. Standardization lowers maintenance costs and conversely the variety of municipal fleets forces high maintenance costs.

The more advanced form of low bid, now being practiced by some levels of governments, is the life cycle cost. In this case the cost of owning , operation, and maintenance is taken together to evaluate the bid. This is significantly better (and more complex to write and bid).

Steps for Successful Contracting

1. Decide that you have a problem that a contractor could solve. Gather support, if necessary, for the idea of contracting. If you are new to the field, seek advise from vendors, teachers, consultants, in-house expertise for competent contractors. Create a specification, and solicit initial responses from contractors.

2. Define work to be contracted. The better the definition at this stage, the lower the bids will come in, and the better the job will go. This is called scope of work. This may include drawings, written specifications, interview with users in the area. This step is called the pre-planning step in the chapter on planning.

3. Search for potential contractors. Check your company records for contractors used in the past. Advertising is appropriate for large jobs. The job A & E (if there is one) will also have lists of contractors they have used in the past. Consider this an opportunity to create a partnership with your contract vendors.

4. Customer reference is a good way to initially build your bid list. References should be solicited on all bids. The list should be all or 50% of the customers in an area. Check a few at random. Ask for a dissatisfied customer. On large contracts, visit similar sized projects and speak to the users. Here, it is appropriate ask for methods, procedures, equipment, capacities of the vendor. Ask to see craftsperson training and testing materials.

5. Existing and past contractors who have performed well are on a preferred list. A contractor who has served you well in the past is likely to do so in the future. Make sure there have not been substantial changes such as new ownership, negative changes in financial status, pending litigation (your lawyer can find out about any cases that have been filed).

6. Evaluate initial information (or initial proposals if this is a two-step process) from your vendors. Eliminate the ones that do not fulfill your requirements. Eliminate contractors that are impossible to reach, show up very late, or don't show up at all for a meeting.

7. Communicate your ideas to the contractor (make sure they understand your scope of work). Do you need shop drawings from the contractor? Discuss the quality of materials needed.

8. Get cost estimates, and review the budget available for this job.

9. Based on information, decide to go ahead or not.

10. Selection of contractor and negotiation of contract is the next step. Visit other jobs to see their quality, and call references.

11. How is your chemistry with the key people in the firm? How is your chemistry with the people actually running the job? Both are important. Ask about the contractor's owner's background in relation to the contract. Is the owner active? If you don't know the firm, ask about the background of the manager or supervisor who will be actually managing your job.

12. On bids over $5000, or longer than 1 year: check out the firm financially through your bank, Dun and Bradstreet, their bank (have your banker call), trade references; verify that their bond is still good, and verify their contractor status with the city (or local government).

13. Verify with your insurance agent or lawyer what insurance coverage is required in your area from the contractor. On larger jobs the contractor should provide certificates of insurance for at least liability and worker's compensation.

14. The agreement should include who is responsible for permits and plans.

15. Prepare the area to be worked on. Remove as much as possible to avoid damage. Be as complete in both your discussions and your documentation as possible about responsibilities, who supplies what, where to unload, rules (user contact, clean-up, security, keys, etc.), when and amounts of payments will be made, etc.

16. Manage the contractor by keeping a record and providing feedback through frequent inspections. Identify problems as early as possible. Require paid receipts to prove subcontractors and material vendors have been paid. Get a release of all liens form signed before last payment. On larger jobs with sub-contractors, consult with your legal department about lien laws in your state and be sure you are covered.

17. Evaluate each contractor on a regular basis for quality, service, cost, and fulfillment of contract terms. Write up a short narrative to put into the file about how the job went.

This was partially adapted from an AIPE article (Aug. 1990) by David Vondle, President of Vondle and Associates, Albuquerque, NM. Additional information from *Building Maintenance Management* and *Plant Engineering Magazine*.

Fifteen Tips to Avoid Claims

Disregarding these areas has cost organizations billions! Before the job starts, do the following.

1. Tighten specifications (on both materials and work to be done).

2. In the specification, define performance, and define what a good job would look like. Add a clause like "all work is expected to be done in a professional and workmanship manner. All work will be in compliance with applicable city building codes."

3. Don't always take the lowest bid.

4. Don't necessarily accept the contractor's standard contract form. If you do enough bidding, have one drawn up by your attorney.

5. Include clear statements about how the site is to be left at the end of each work day; who is responsible for locking up, cleaning, and debris removal.

6. Require insurance and have an agreement about what happens when the contractor damages your property (or damages a neighbor's property who then sues you). Require a certificate of insurance covering general liability, property damage, workmen's compensation, auto liability.

7. Include a cancellation clause. You need to spell out how and why you can cancel the contract. Otherwise you may find yourself with a mechanic's lien over an inadequate job that you did not pay the final payment.

8. Include a schedule of extras. A common ploy is to low-ball the bid to get the job and flood the manager with small extras. Look for clauses like "all extras not included in the original price have to be agreed to in writing prior to the commencement of the work."

9. Deduction clauses; spell out what you will charge back and when you will charge it. Examples would be debris removal, clean-up, missing firm completion dates.

10. Pick a term that allows the contractor to fund their mobilization and allows you to shop periodically. Avoid too short or too long of a contract term (for ongoing services).

During the job, do the following.

11. Keep track of documents (keep a fair and complete set of contract documents).

12. Keep good records.

13. Have a functional schedule.

14. Verify that material vendors and subcontractors were paid. You could have paid off the general contractor and still be hit with liens.

15. Aim for early resolution of disputes.

(We thank Ed Feldman, P.E., for his insightful "things to avoid" list and ideas about the bid package; also, some material adapted from the San Francisco State University *Manager's Bulletin* and some from *The Landlord's Handbook*).

15
In-Sourcing

In-sourcing: to transfer maintenance workload to another department in the organization (opposite of outsourcing which transfers workload to outsiders). There are too many tasks that need to be done for the number of hours available in the maintenance department. There are only five choices.

1. Let the asset deteriorate, lower the service level, allow increases to deferred maintenance, and hope no one gets hurt while you are responsible.

2. Improve efficiency so that you have enough people, through continuous improvement/reengineering.

3. Outsource, lease the asset with service included, service contracts, etc.

4. Buy new, better assets that don't need much maintenance, throw away or trade in the assets before the warranty runs out.

5. In-source to other groups within your organization.

In this chapter we will discuss the fifth option. There are other groups within your organization that are in a better position to perform certain maintenance tasks.

Example: At the Opryland Hotel in Nashville, the housekeeping department does much of the basic PM in the guest rooms instead of the maintenance department. They have been trained, have check sheets, and report abnormal conditions to the maintenance department. Since the housekeeper is in every room every day, and is closer to the guests (than the maintenance department), he or she is the logical party to perform these tasks.

Example: A driver of a class 8 heavy truck is responsible for certain safety inspections under the terms of the CDL (commercial driver's license). These safety checks are actually PM tasks geared to detecting safety problems. The driver has a significantly higher stake in the safety of the truck than the mechanic. The driver is the logical person to perform these tasks.

Example: The biggest single application of in-sourcing, and the one that the remainder of the chapter will concern, is operator centered maintenance. This is called TPM, or total productive maintenance.

TPM—Total Productive Maintenance

There is a revolution on the factory floor of selected organizations. The ideas of TPM to make the operator a partner in the maintenance effort fly in the face of commonly held and cherished beliefs. This import from Japan has taken root in factories, refineries, mills, and power plants throughout North America. It forces us to realize that, to remain competitive, we have to use more and more of the capabilities of everyone.

The machine operator is the key player in a TPM environment. There is less reliance on the maintenance department for basic maintenance. Control and responsibility are passed to the operators.

The maintenance department becomes an advisory group to help with training, setting standards, doing major repairs, and consulting on maintenance improvement ideas. Under TPM, maintenance becomes very closely aligned with production. For TPM to work, maintenance knowledge must become disseminated throughout the production hierarchy. The old type, "produce at all costs, damn the torpedoes—full speed ahead," will fall flat on its face with TPM. The operators must have complete, top level support throughout all phases of the transition and thereafter.

TPM uses the operators, in autonomous groups, to perform all of the routine maintenance including cleaning, bolting, routine adjustments, lubrication, taking readings, start-up/shutdown, and other periodic activities. The maintenance department becomes specialists in major maintenance, major problems, problems that span several work areas, and trainers.

Rather than relying entirely on a staff of maintenance specialists to keep equipment in good running order, TPM pushes the responsibility down to the people operating the equipment. "The concept is that the operator must protect his own equipment," states Dr. Suzuki (the Japanese originator of TPM), "Thus the operator must acquire maintenance skills." Maintenance experts may still make periodic inspections and handle major repairs. And design engineers also play a big role. They must take maintenance requirements—and the cost of equipment failure—into consideration when they design the equipment, stresses Dr. Suzuki.

The operator goes through seven steps to reach full autonomous maintenance.[1]

1. *Initial cleaning, review of entire machine, tightening:* Complete cleaning of machine. Repair any deficiencies that become apparent during the complete cleaning. Tighten all fasteners to spec. Review entire machine operation.

2. *Inspection:* Initial inspection follows manufacturer's manuals, engineering recommendations, and equipment history (what has failed). Group is taught how to correct minor defects.

3. *Maintenance prevention:* Reduce time to perform cleaning. Remove source of contamination. Make the machine easier and quicker to service (lubricate, tighten, clean, adjust).

4. *Establish consistent TLC standards:* Specify all tasks and frequencies (daily, weekly, every 1000 pieces, etc.). Set standards for tasks (how clean, what to use to clean, how much and what type lubricant). Autonomous group prepares documentation. Documentation could include sketches, photographs, pictographs, and short concise narratives.

5. *Autonomous inspection by operators:* Inspection is turned over to group. Check sheets are utilized for inspections. Minor repairs

---

[1]We would like to acknowledge the groundbreaking work of Nakajima and Suzuki. Much of the information on TPM in this section is derived from the writings of Seiichi Nakajima—*Introduction to TPM* and *TPM Development Program*, published by Productivity Press.

are completed. Maintenance is only involved in major problems that involve specialized knowledge, skills, or contacts.

6. *Organization to support ongoing TPM:* Systemize the autonomous maintenance activity. Align the organization to support TPM. Use the TPM productivity reports to run the plant. Develop standards for all activity.

7. *Full functioning TPM:* Track the results of the effort and give ongoing recognition to progress. Monitor failure frequency and look for additional improvements. Spend more time on improvements that reduce maintenance effort while increasing equipment availability.

TPM is one of the most effective methods of improving the delivery of maintenance service while increasing the effectiveness of the equipment. TPM could be the production department's answer to the empowerment, job enrichment, total quality programs. The great advantage is that TPM can be incorporated into and can greatly enhance these programs. TPM is based on PM as a motivational technique through autonomous maintenance groups (operators have greater involvement and say about equipment). TPM works only because the operators begin to own the equipment. As ownership spreads, then autonomous maintenance becomes a reality.

The first goal is to maximize overall equipment effectiveness. TPM has a very strict definition of effectiveness. One of the tenets of TPM is that sloppy reading of effectiveness can cover up opportunity for production improvement. When the autonomous aspect of TPM takes hold, the operators help design a shared system of PM for the equipment's complete life (takes into account the age and condition of the equipment). Without this, the PM tasks might not reflect the changes in failure modes over the life of the equipment.

TPM and Explicit Tasks

Task lists will have to be developed to be followed by the operators. These task lists are different from the task lists designed for maintenance professionals. The individual tasks must be explicit, visual and concrete. Drawings, photographs, schematics all help the operator do the tasks as designed. Allow the operators significant input into the task list design.

TPM is attention to and elimination of the six losses of production by primarily autonomous production/operations teams.

Downtime:

1. Equipment failure from breakdowns. It is the biggest element directly the responsibility of maintenance. With TPM, first-line maintenance activity is transferred to operations. Proper design ensures reductions in breakdown-related downtime.

2. Setup and adjustment. The stated goal is single-digit-minute setup times. This allows up to 9 minutes for setup and change-over to new colors, sizes, or packages. Adjustments are simplified or eliminated from the system. Overall reengineering to reduce set-up and adjustment time is expected.

Speed Losses:

3. Idling and minor stoppages due to abnormal operation of sensors, blockage of work on chutes, etc. These slow-downs are tracked and analyzed to see what is really happening. Analysis of root causes and of process are ongoing until the system no longer has losses in these areas.

4. Reduced speed due to discrepancies between design and actual speeds. Design speeds are reviewed and actual speeds are observed. The comparison, if unfavorable, initiates a design and engineering review.

Defects:

5. Process defects due to scraps and quality defects to be repaired. Quality problems are not tolerated. Deep analysis is undertaken until these losses approach zero.

6. Reduced yield from start-up to stable production. The production process is tracked and watched for start-up problems. Stable production should follow start-up very closely.

Measuring Equipment Effectiveness is an Essential Part of TPM

Many organizations do not or cannot capture accurate information about run time, slow-downs, minor stoppages, and defects. TPM relies on good record keeping in the six areas of loss.

Compare these results to TPM standards:

Availability	>90%
Performance Efficiency	>95%
Rate of Quality parts	>99%

Problems

Any implementation of TPM has to overcome real problems. Following the plans already mentioned will minimize the negative effects. At a TPM seminar in a precision manufacturing company, supervisors were asked what real problems they would encounter installing TPM. These problems must be thought about, discussed, and overcome to have an effective TPM effort.

1. Top management sign-off and support throughout a multi-year TPM process. If your management has a short view, they might agree to a multi-year plan and withdraw their support after the first year.

2. Top management might give lip service but does not support it with their deeper commitment and their time. How do we get top management to be boosters of the program?

3. Supervisors might criticize rather than solve problems, they also might complain and doubt the success of the project in front of their subordinates.

4. TPM requires minimal downtime. How do you integrate TPM with customer demands and the sometimes unreal demands of the forecast and production schedule?

5. The workers might object to the perceived "extra" work by a slow down, increases in absenteeism, letting quality suffer, etc.

6. Where does a (small, medium, large) organization get time to do all the training necessary?

7. How do you run TPM in a high turnover situation? Operators don't stick around long enough to get trained.

8. Where do temps fit into TPM? We use temps to operate machines during busy times.

9. How do you solve the problem of inadequate communication between the production group and the maintenance group? Who will really manage the maintenance part of the operator's job? Why will operations take advice about operator maintenance effort when they won't now?

10. Is there willingness and is there interest to accept the new roles of the two groups?

11. Operators don't like to clean equipment and neither do maintenance people.

12. Where do our die setters (set-up people) fit into this scheme?

TPM and JIT

"Total productive maintenance (TPM) is indispensable to sustain just-in-time operations," says Dr. Tokutaro Suzuki, senior executive vice president of Japan Institute of Plant Maintenance. In a JIT system, he emphasizes, "you have to have trouble-free equipment." Prior to the adoption of TPM, Japanese manufacturers found it necessary to carry extra work-in-process inventory "so the entire line didn't have to stop whenever equipment trouble occurred."[2]

In one Japanese plant, adoption of TPM reduced the number of equipment failures by 97%. In other cases, cited by Dr. Suzuki, TPM boosted labor productivity by 42% and reduced losses related to downtime by 69%.

For additional information, books, and videotapes an extensive list of references was printed in the January 1996 issue of *Maintenance Technology* magazine.

[2]Information about the originator of TPM, excerpted from an article titled "Lessons from the Guru's," published in *Industry Week*, August 6, 1990.

16
Guaranteed Maintainability

One way to be effective in maintenance is to choose designs that do not break down frequently—and when they do break down, they are easy and quick to fix. The second part of this strategy is to ensure that the operators can properly operate the asset and the mechanics can repair the asset. This field is called guaranteed maintainability (I first heard the phrase from Ed Feldman).

The first enemy to guaranteed maintainability is buying low bid. Low bid specifications are seldom set up (by design) to exclude major manufacturer's assets just because they didn't perform in the past. The second enemy is mental laziness, where the maintenance people never thought through performance of the assets and cannot identify the best of anything (but they do have opinions about everything).

When designing a new building, for example, there are thousands of decisions that will have an impact on maintainability. Some include surfaces, access, design, components, and installation techniques.

Surfaces: Some surfaces are better than others. In Rutgers University in New Jersey, they installed solid surface material such as Corian© in the dorm bathrooms instead of tile. Solid surface material is more expensive but almost undamageable (and it can be repaired if it is damaged).

Access: In a school, all of the HVAC subsystems including the filter locations were located in 11′ ceilings. The ceilings were drop-in tiles with no catwalks, and were rats' nests with wires, pipes, supports just stuffed up there. Changing filters was a dangerous, time-consuming, and dirty ordeal.

Design: When they redid the Philadelphia Airport restrooms (and now most airports), they put in a serpentine entry instead of doors, electronic flush valves, and the soap dispensers were located over the sinks. Partitions are wall mounted to expedite cleaning. All of this was designed to reduce maintenance, water use, and housekeeping expenses.

Components: In a subsidized housing development, the contractor submitted (and got approved) an imported furnace. No spare furnaces or spare parts were bought. Two years later, the wait for parts was 9 months and the management company was forced to refit the apartments with new units as they broke down.

Installation: When a professional installs a boiler in a home or small rental building, they put in valves on both side of the unit. Without that the system has to be drained to remove the boiler. An extra $25 on the installation can save a mess and extra hours of effort. It also means if a unit goes out, a new one can be slipped in quickly.

One of the aspects of guaranteed maintainability is the ability to get vital information when you need it. Equipment vendors are notoriously variable in the quality, organization, and usability of their support documentation. Lucent Technologies (formerly AT&T and before that Western Electric) now specifies how the manual and support documentation should be constructed in their purchase orders for new equipment. This is also part of their ISO 900X process.

They specify the chapters and contents of the manual:

1. What is the asset?

2. What does it do, including detailed functional specifications?

3. Complete description of how the asset works.

4. What are the components?

5. What should be done if it is not working up to specification?

(This is broken up by component which then follows 1-6 for that component.)

6. What are the safety and environmental considerations (how can it hurt me or the environment)?

Guaranteed maintainability has other aspects that are important:

1. Maintenance professionals need to know how to operate the equipment, as well as operators. At GE Engineered Plastics, the maintenance department personnel were certified operators at the plant.

2. Unfortunately many maintenance professionals do not have good networks of maintenance people in other companies. Product intelligence, repair experience, and reengineering tips can all come from a well-cultivated network.

3. If your business depends on a manufacturer's equipment and there are significant finances riding on the relationship, be sure to visit the factory where the asset was built. Meet the behind-the-scenes engineers and shop people. These people can become a great resource. Be sure to let them know how important their equipment is to your operation. Bring pictures of their "babies" that you put to work.

4. When buying new equipment, use the complete life cycle cost as the cost basis rather than the purchase price. Look at your operation, see how you need the asset to be used, try to imagine future uses and capacity needs, and pick based on the complete picture.

5. Guaranteed maintainability requires experiments in new types of assets, new techniques, and new materials.

17
Maintenance Quality Improvement

The usual definition of quality in production is to consistently produce parts with low variation. Maintenance quality usually deals with the consequences of the repair, not the repair itself. The emotional context of the response is also tied up in maintenance quality (a surly, dirty maintenance technician is low quality even if their work is superb). Quality is another name for the concept service value discussed in the chapter on customer service.

In some circumstances	maintenance quality might	= no downtime
In others	maintenance quality	= no scrap
	maintenance quality	= fast start-up
	maintenance quality	= safe operation
	maintenance quality	= on-time delivery
	maintenance quality	= lowest cost
	maintenance quality	= quick response
	maintenance quality	= no calls from the public
	maintenance quality	= comfortable building
	maintenance quality	= safe indoor air quality
	maintenance quality	= no repeat repairs
	maintenance quality	= keep unit in spec
	maintenance quality	= no interruptions
	maintenance quality	= satisfied user

Every maintenance operation should define quality in a way that is useful to its operating environment. The late W.E. Deming was considered the quality guru for the last generation of Japanese quality experts. In fact, the quality award in Japan today is called the Deming award. He had much to say about quality in manufacturing. It should be no surprise that Deming's points apply to maintenance also.

W.E. Deming's Points

The following points were first discussed in 1950! They have been adapted to discuss maintenance issues.

1. Create constancy of purpose toward improvement of the delivery of maintenance services with the aim to stay competitive, stay in business, and provide stable employment. Maintenance deterioration usually takes a long time. Any effective maintenance strategy must also have a long horizon. Constancy of purpose means resources are allocated for good maintenance practice and are not taken away with every bump in the quarterly results.

2. Our maintenance departments often are the last areas of the organization to realize the need for change. The department is dragged kicking and screaming into the new corporate culture. Deming says "awaken to the challenge and adopt a quality philosophy." Nowhere else is high quality so closely related to safety, high self-esteem. Quality is intertwined with the very history and culture of the crafts. Take responsibility for and leadership in change. Looking toward the future I see maintenance departments providing leadership for the rest of the organization.

3. Build quality into process. Quality comes from skilled and knowledgeable mechanics given good tools, adequate materials, and enough time to do the job. Quality comes from choosing well-designed equipment that doesn't need much maintenance. Quality comes from proper installation so that the work that is needed is easy to perform and easy to get to (enough room, light, air to work—where possible). Quality comes from pride in a job well done. Cease dependence on inspection to achieve quality. Lead by example with ceaseless training, coaching, and systems analysis. When defects occur, concentrate on the system that delivered the defect rather than having a preoccupation with finger-pointing.

4. Move toward a single source for each item, and toward a long-term relationship of loyalty and trust. End the practice of awarding business on the basis of price alone. Instead, minimize total cost. A revolution in purchasing is at hand. More and more organizations are looking at the total costs of a part or the life cycle cost of a machine. Some savings are illusionary and hurt the overall goals of the organization. A low cost bearing might be the most expensive bearing you ever buy.

5. Keep up the improvement to the system to deliver maintenance, to improve quality and productivity, and thus constantly reduce costs. In today's market the way it used to be done is not good enough anymore. All improvements and growth flow from dissatisfaction with the status quo. Build measurement into the maintenance information system. Continually strive to improve both the visible and the invisible performance.

6. Training should be mandatory for mechanics the way it is for doctors or teachers. Our factories and facilities have today's levels of technology, but maintenance people have yesterday's set of skills. We must train to bridge to gap. Special effort should be given to the people on your staff who deliver the on-the-job training. These informal trainers need instruction and back-up materials in how to teach adults. Institute a vigorous program of education and self-improvement: world class maintenance departments make a commitment to invest 1–3% of their hours in training for all maintenance workers. Technologies are changing—skills must change too. A world class auto manufacturer mandates 96 hours of training per year for everyone. A high-tech manufacturer requires 110 hours.

7. Institute leadership. The aim of supervision should be to help people and machines do a better job. Supervisors should serve their subordinates by removing the impediments from productivity. The supervisor should ensure that the five elements of maintenance are present: the mechanic, the tools, the parts, the unit to be serviced, and the permissions—and all should converge at the same time. The supervisor should also be the lightning rod for disruptions from management and production (unless there is an emergency, the mechanic will not be disturbed because interruptions reduce quality and worker satisfaction).

8. Break down the barriers between departments. Everyone's expertise is needed for constant improvement. With scarce resources, we must include knowledge from other departments and groups to come up with the best overall solution for the organization. Maintenance problems are complex and have financial, marketing, purchasing, quality, and engineering ramifications. The best solution to a problem might not be the best maintenance solution (like run until destruction to fill an important order). Information for the best solution might come from another department and another expertise.

9. Eliminate slogans, exhortations, and targets for the workforce asking for zero defects, new levels of production. Such exhortations create adversarial relationships. A bulk of the quality problems belong to the system, not the people. Stable processes create quality. Create stable processes producing quality outputs, and the people will feel the way the slogan says without coercion and alienation.

10. Eliminate work standards, quotas, and management by objectives (MBO). Work standards and quotas are associated with management styles that treat the maintenance worker as someone needing to be told exactly what to do and how long to take. Standards are useful for scheduling and to communicate management's expectations. It is difficult to not use them as a production whip. That is a disaster in maintenance situations because we want the mechanic to take the time needed to fix everything they see (within reason!), not just the original job. We must trust the mechanic to look out for our interests particularly when we are not there. The problem with MBO is that it focuses on visible, measurable aspects of maintenance. Many of the real issues of maintenance concern aspects of the environment that are hard to measure.

11. Remove the barriers that rob the worker/engineer of his/her right of pride of workmanship. Tradespeople must be allowed to feel pride in their jobs that are well done. Maintenance managers and supervisors must not allow anything to stand in the way of that pride.

12. Put everyone in the organization to work to accomplish the transformation. This transformation is everyone's job. This transformation requires the talents of all the employees. It requires all of the talents of each person. When a hotel chain had the house-keepers meet with the architects (for a new hotel) the result was concrete suggestions to improve the designs that reduced maintenance costs and resulted in a better designed room for the customers.

Deadly Diseases and Obstacles to Success

1. Hope for instant pudding. Change of fundamental processes take time. In the current U.S. culture, it is hard to imagine instituting a change in processes that could take 5 or 6 years. In actuality, if you start with a typical reactive maintenance department, it could take you 5 years or more to create a proactive partnership between maintenance and production.

2. The supposition that solving problems, automation, gadgets, and new machinery will transform industry. Maintenance problems are people problems. The systems, attitudes, and approaches are at issue. The paradigm of maintenance as a necessary evil, or of maintenance workers as grease monkey slobs, must be transformed. The transformation starts in the minds and hearts of the maintenance department and then flows to the rest of the organization.

3. Emphasis on short-term profit; short-term thinking fed by compensation systems that focus exclusively on short-term performance. Top management squeezes maintenance to reduce costs below the level that is necessary to avoid deterioration. The asset will then deteriorate, and long-term integrity of the process will be compromised. Maintenance requires long-term planning and commitment.

4. Mobility of managers and job hopping. At one beverage bottler, the average tenure of the maintenance manager was 18 months. Some lasted as few as 9 months. Everyone came with bright ideas and wanted to prove themselves. The result was a complete lack of focus on long-term goals and plans. As each manager tried to cut costs, the negative results impact fell to the next player. This job hopping in management without a master plan dramatically exacerbates the short-term view.

5. Management by use of only visible figures, with little or no consideration of figures that are unknown or unknowable. For example, when you invest in training for your maintenance crew where does the increased asset show up? When, after spending hundreds of thousands of dollars on a long, expensive trial-and-error development process, a firm finally develops expertise in a new process. This expertise (a new asset) is nowhere on the balance sheet. It is important to measure and also to realize that much of what goes on in maintenance is unknowable.

6. "Our problems are different." Actually many people's problems are the same. In the PM area, while no two plants will have the exact same schedule, the problems will be the same. In our public sessions, maintenance managers in widely different industries, of varying sizes and sophistication, marvel at the similarity of the problems.

7. Search for examples. We think that if something worked in another machine shop or foundry it will work in ours. Since maintenance has no strict rules, examples from our industry may not be useful or even relevant. The flip side is that we can learn critical knowledge from maintenance professionals in all lines of maintenance.

8. "Anyone that comes to try to help us must understand all about our business." The sad truth is that if the solution to your problem was commonly known in your industry, you would probably know what to do.

9. Every system produces defects. There is a fallacy of zero defects. The goal is a stable system that produces an appropriate level of quality.

10. "Our trouble lies entirely within the workforce." Your maintenance system is a stable system to produce a certain number of defects. Changes in the workforce are irrelevant to the output. Only changes to the system can have an impact.

11. False starts with inadequate planning, wavering top level support, and lack of follow-through kill quality improvement programs in most places. These false starts kill CMMS installations, TPM, and RCM too! Serious thought and planning are needed before starting. Commitment must start in the highest levels of the organization. Buy-in at each level must be earned, worked at, and appreciated before proceeding to the next level.

12. "We installed quality control." Quality is a way of life. It is a daily diet. You don't install it, you become it.

13. The unmanned computer is a danger of computerization of maintenance. The computer is a great tool that, like any great tool, is frequently overapplied. Allow the people to have their say and make sure the computer answers to someone (a real person), and that they can overrule the machine. Make it easy to remove equipment as it retires, and modify task lists as conditions change.

14. The supposition that it is only necessary to meet specifications. Many of the important aspects of a component are not included in the specifications. You never know which attributes are important until you try changing vendors and find out that your production depends on qualities of a particular vendor's products that are not covered by the specifications.

15. Inadequate testing of prototype machines, buildings, vehicles, and whole processes. By starting manufacturing on inadequately tested prototypes, we strain the system of improvements. There will be so much ground to cover before everything stabilizes that the product will be half-baked for a long time. To leapfrog this phase, exhaustive testing should be built in.

18
ISO 9000 and Its Relation to Maintenance

ISO 9000 is an international standard for assuring that an organization has an approved quality system. Many organizations now require that their vendors be ISO 900X certified. They want assurance that their vendors' quality system is sound and will be *likely* to produce good parts. ISO 900X does not dictate how to do things but how to have a sound methodology.

For the U.S. market, the American National Standards Institute has published its own version (the ANSI ASQCQ90 series). The ANSI standard is the same as ISO 9000 except it adopts standard U.S. usage, spelling, and conventions. ISO 9000 stipulates the areas to be controlled without telling the organization how to control them. ISO 9000 certification is by site, not by organization (each plant must get its own certification).

There are three types of ISO 9000 organizations (9001, 9002, 9003). ISO 9000 is a general description of the whole standard and ISO 9004 is a manual for developing a quality assurance system.

ISO 9001: Facilities which design/develop, produce, install, and service products or services to customers who specify how the product or service is to perform.

ISO 9002: Facilities that provide goods or services to designs or specifications by the customer.

ISO 9003: Final inspection and testing.

The important question is: how is maintenance involved in the ISO 9000 process? Maintenance is not covered directly by

146

the ISO 9000 specification. It is indirectly covered as the organization's function that assures capacity, and also assures the machine is capable of consistent output (assuming it ever was). Many organizations implement ISO 900X with the least possible extra work and without involving maintenance. This is what is referred to as a bare-bones implementation.

Maintenance is a resource for the implementation team and an advisor in key areas. The following tips are partially adapted from comments by Michael Fink, superintendent at Arkansas Eastman and an ISO 9000 certification veteran (*Maintenance Technology*, Jan. 1993).

1. The maintenance department can be dragged kicking and screaming into the ISO 9000 certification process or it can lead the way. Decisions have to be made if the plant is to pursue a complete certification or a bare-bones certification. The maintenance leadership realizes that the ISO process is an excellent lever to move the plant toward better maintenance practices. Once something is documented in the ISO 9000 manual, it must be supported and done.

2. ISO 9000 can happen to maintenance, or maintenance can learn what it is about and direct the process. The manuals should be read and understood, the classes taken seriously (as if your job depended on it), and the team and consultants should be worked with closely.

3. The ISO 9000 language calls for the identification of a definition of critical equipment to the quality of the output. This definition will be applied to each piece of equipment to determine if it is critical and therefore included under the scope of the ISO 9000 specification.

4. Check each piece of equipment against the definition. Working in the team, maintenance lends its expertise and knowledge about the assets to the process. Maintenance can ensure that support equipment is included in the definition if it is critical to quality (such as process steam boilers, etc.).

5. Help the team document the tasks necessary to ensure that the critical equipment stays within production specifications.

Documentation is the core of the ISO 9000 process. The team will document the equipment list and maintenance tasks needed to support the list.

6. Help the execution process. Be one of the leaders and boosters of the whole process. Help ensure that the processes documented to support the quality output of the equipment are carried out as specified.

Initially, facilities decided to go for certification because they did business or wanted to do business with companies in Western Europe. Many of the global manufacturers started to require all of their vendors be certified. Now even local organizations are looking at certification. If your facility decides to go for ISO 9000 certification, the process is quite rigorous.

The first step is to design and document a quality system that will ensure that your product or service will be consistent and will satisfy the needs of the customer. The second step is to implement the design and perform self-audits to ensure compliance with the design. Documentation of each step in your process must comply with what is actually being done. The third step is to hire a third-party registrar. The registrar will audit your processes and procedures to ensure that they comply with the ISO guidelines and that you conform to your own documentation. The last step is ongoing review of your processes and audits of compliance. These are initially carried out by your own staff and also periodically followed up by the registrar.

There are 20 areas that must be documented by the ISO 9001 standard. ISO 9002 requires 18 areas (#4, #19 are not required). ISO 9003 only requires 12 of the areas (#3, #4, #6, #7, #9, #14, #17, #19 are not required).

1. There must be management responsibility. One person is responsible for quality, and is high enough in rank in the facility to develop, monitor, change, and enforce the quality system.

2. There is a quality system, including a manual that is followed, documented means to identify customer requirements, measurement and testing gear to ensure quality is being maintained, and a system to evaluate the process capacity. The system should be well understood. Adequate records should be kept.

3. Contract review to understand customer requirements, to set up a process for dealing with variance in advance. Reviews of contracts are documented.

4. Implement controls to ensure that the design process will produce a product or service that satisfies the customers needs. The system for control will define interfaces, ensure training, assure competence in the areas being serviced, provide a provable system for gathering input and translating it into design product, ensure that the product meets the legal and service requirements.

5. Document control to administer the creation, publication, distribution, use, and revisions to documents used in connection with the quality effort, process of production or design, and other critical areas.

6. Purchasing control that assures that purchased products meet the specifications. Some of the elements of purchasing control are documents that include data which thoroughly describes the products or services, reviews of the descriptions to assure accuracy, choosing vendors on the basis of documented capabilities, and keeping records of inspections, and performance of the vendors.

7. Purchaser-supplied product must be tracked, protected, and non-suitable product is segregated and reported back to the supplier.

8. Product identification and traceability is essential in certain industries (such as drugs, beverages, aircraft parts, etc.). The product should be traceable through all of its steps of production.

9. Process control means each step in the process is documented with references to equipment, measurements, work steps, code compliance, monitoring of critical parameters. All changes to the process are documented and published so that other shifts know the current process that is expected from them.

10. Inspection and testing is one of the most important elements of the whole picture. Inspection includes specifications, techniques, and steps for incoming materials, in process work, and final products.

11. Inspection, measuring, and test equipment needs to be calibrated to be in compliance with appropriate standards. Scheduled calibrations, cleaning, handling would be documented. Storage and safety of test gear is also ensured and documented.

12. Inspection and test status of every product through every step is to be assured by the quality system. Untested, tested, test-underway product has to be tracked and kept separate.

13. Control of non-conforming product has to be thought about, documented, and segregated from other product. A means of identification has to be established, and procedures for correction or disposal published.

14. Corrective action should be documented and the procedures put in place to detect the causes of non-conformance and to formulate corrective action.

15. Handling, storage, packing, and delivery is designed to protect the product from damage, theft, destruction. Secured areas for storage and packaging designed to protect the product until the customer gets it. Modes of transport are chosen that can transport the product without damage, in a timely manner, to support the needs of the customer.

16. Quality records are required. These records are controlled by the quality manager, are defined to serve the product and customer need, have controlled access, and are kept for an agreed upon and documented amount of time.

17. Internal quality audits are required to ensure that the system is operating as designed. These audits can be surprise or scheduled and will cover all of the other 19 points. The auditor cannot have responsibilities in the area being audited. The procedure for the audit will be predefined, and the results will be kept on file for future review.

18. Training is to be documented. The ISO specification calls for a documented process of evaluating if there is a deficiency or void in the skills or knowledge of the facility personnel. The system will document that the training was procured, given, and that the results show that the void is filled. Post-training monitoring is conducted, and records kept on each individual employee.

19. Servicing is tracked by the quality system. The service should be appropriate to the client and the product or service provided.

20. Statistical techniques are powerful tools in the review of quality. The facility is reviewed for the possible uses of statistics. These reviews are documented and tracked.

In an interview with maintenance leaders who have completed the ISO certification process (Donald Overfelt from Lucent Technologies, Bob Posey from Stanley Tools, and Harold Greene from NTN–Bower), several points were made.

The first step for each maintenance organization is going to be different. Even in these companies the level of ISO implementation varied greatly. Lucent started the process (the maintenance part) by writing specifications for the whole process of maintenance. They explained the use of the work orders, PM tickets, history files, and every other part of maintenance. From then on, they had a description that had to be followed.

All had comments on new emphasis on calibration. At Lucent, all measuring devices had to be calibrated in a specified way, at a specified frequency, following a specified procedure. This now included maintenance department meters, tapes, rules, mikes, calibers. In the past, only a few items were looked at. They stressed that the written procedures for each item had to be current.

While the PM program is not specified in the ISO specification, if they wanted to have a PM system it had to be specified, documented, and actually done. The critical piece from all three manufacturers was the drive to document what was done and do what they documented. They felt this was one of the great benefits of the certification because they had more clout to get the PM done as scheduled.

The issue of training and certification was an important part of their programs. Each person authorized to run a machine had to be checked out on the machine following an established procedure. This created certified operators. Also, each mechanic had to be checked out on each machine that they worked on. The test, training materials, and retest interval was established and documented. This ensures that only trainer mechanics open the machines. The airlines and military have used this concept for mechanics and pilots for years.

The process at Stanley was to hire a consultant, complete several levels of training, perform a preliminary audit, make adjustments, hire a certification agency (two used Lloyds of London), have a certification audit, and have periodic re-audits to stay certified. The audit used a point system, where there were major items that would stop the audit and minor things to be corrected. A major item (that would stop the audit) would be an out-of-calibration measuring device used on customer's product. A minor fault would be out-of-date documentation where the operator could correct it immediately.

While on the topic of the International Standards Organization, there is another standard on the horizon. The new standard is ISO 14000, which is a proposed series of environmental standards. This is expected to be finalized in 1996.

As in ISO 9000, the European countries may require certification to supply them. Our EPA is considering a different level of oversight for ISO 14000 certified companies. It may also reduce legal liability because it would show any court that the company was acting responsibly in environmental issues.

Certification issues would include (there will be others): establish a policy which includes a commitment to prevent pollution, clear accountability and responsibility for environmental issues, set target levels, commit resources to achieve target levels, set up and communicate an emergency preparedness process for environmental problems, and encourage contractors and vendors to seek certification with ISO 14000.

Information can be obtained from CEEM Information Services in the report "What is ISO 14000?"

19

Techniques for Continuous Improvement in Maintenance

Continuous improvement is defined as ongoing incremental improvement in maintenance performance. If a revolutionary improvement is required to meet competitive pressure or accommodate severe budget cuts, then reengineering is usually indicated. Where maintenance is already effective, then continuous improvement is the prescription.

Continuous improvement helps avoid stagnation and complacency. There are opportunities in every maintenance operation. An internal study done by a major maintenance provider in Canada estimates the opportunity:

Percentage of possible savings of maintenance budget dollars

39%	Reengineering of equipment and maintenance improvements to equipment
26%	PM improvement and correct application of PM
27%	More extensive application of predictive maintenance
7%	Improvements in the storeroom

Continuous improvement means either ongoing reductions in:

1. labor (production operator, maintenance mechanic and contractor)
2. maintenance parts, materials
3. raw materials
4. energy, fuel, other utilities, including water
5. machine time
6. capital

153

7. management effort (reduce headaches, non-standard conditions requiring management inputs)
8. overhead

or

1. improve reliability (uptime) and repeatability of process (quality)
2. improve safety for the employees, the public, and the environment.

1. *Measurement:* A necessary prerequisite to continuous improvement is establishing ways of measuring the maintenance effort. Without measurement, it is difficult to determine if an operation is truly improving. The process of setting up measures is called *benchmarking.* Consult the section on benchmarking for suitable metrics.

2. *Information and investigation:* To improve maintenance, examine all sources of data for opportunities to reduce the inputs or help the improvements. In particular, review the maintenance incident history. An incident could be a breakdown, a series of breakdowns, PM's for a machine, a series of minor adjustments, or other maintenance activity. Review the asset, area, or system from the six different points of view covered below.

3. *Action:* Based on all of the factors, what action provides the best long-term outcome? Action can be in the domain of changing any of the facts, including the process, maintenance procedure, product, or even business situation. Put into action improvements in one or more of five areas.

Below, we will review the investigation and action possibilities for continuous improvement.

Areas to Investigate
1. Economic
   A. What is the cost of the incidents.
   B. What is the cost of the downtime (total down hrs × rate).
   C. What is the cost per year (cost/incident × # of incidents).

    D. What is the return on investment of a projected improvement.

    E. How much should we spend to fix this.

    F. What is our investment in this asset or process.

2. Maintenance
   A. How disruptive is this breakdown.
   B. Is this breakdown caused by an action of a maintenance person (iatrogenic).
   C. Does this breakdown cause mechanical or electrical problems elsewhere.
   D. What is the honest opinion of the maintenance "old timer" experts.
   E. Is the root cause a faulty or inadequate maintenance procedure.
   F. Is the root cause related to inadequate training in any maintenance skill.
   G. Compare down hours to Mean Time to Repair (MTTR). Are we responding fast enough.

3. Statistical
   A. How often does the incident occur.
   B. Is there a pattern or trend.
   C. What is the mean-time-between-failures (called MTBF).
   D. What is the MTTR.

4. Engineering
   A. What was the mode of the failure.
   B. Is a structural analysis of the broken parts indicated.
   C. What happened just before the failure.
   D. Why did the breakdown take place.
   E. Was there a failure of the PM system.
   F. Do the PM task lists look at this failure mode.
   G. Are we looking at the root cause or a symptom.
   H. Is the root cause related to an inadequate part specification.

5. Operations
   A. How does this event impact operations.
   B. Does the event force the failure of other parts of the process.

C. Can we bypass the problem with a backup or standby unit.
D. Is there a scrap or start-up exposure.
E. Is this indicative of a failure of the operations system.
F. Is inadequate operator training a cause or a contributor.
G. If there is a TPM program, was this a failure of the TLC (tighten, lube, clean program).

6. Marketing/Business
   A. What is the impact of this failure on the customer (internal/external).
   B. Can we afford to have this type of event happen.
   C. Can this event impact quality.
   D. Does this impact morale.
   E. Is there an impact outside our sphere of influence (environmental, competition).
   F. How high should the priority be to deal with this problem.
   G. Is there a regulatory or legal impact to this event.
   H. Is there a logical business decision indicated (outsourcing the part, sell off product).

You can see from the scope of the list above that no matter how knowledgeable individual maintenance professionals are, they cannot know all of the ramifications of a major maintenance event. If continuous improvement is seriously pursued, multi-departmental teams will be necessary on an ad hoc basis to attack problems.

Possible Actions

1. Modify maintenance or PM procedure. Add tasks to catch the particular failure mode earlier on the critical wear curve. Increase the technology of the tasks such as adding a thermographic survey or vibration analysis. Increase frequency of the tasks. If economic and business analysis shows that we are spending too much on PM in relation to other costs, do the opposite of the above (reduce frequency, depth, etc.). If this is a wear problem, then investigate better cleaning, improved lubes, better alignment processes. Design easier-to-do maintenance procedures and

reengineer the equipment to suit. Can the MTBF be managed by replacing the component more frequently than the failure frequency where the component clearly wears out (PCR).

Questions to consider when reviewing maintenance or PM procedures:

- Are there unnecessary steps in the whole maintenance process (step through it)?
- Are any maintenance problems related to certain procedures?
- Would different materials or parts reduce cost or improve reliability?
- Would a change in scheduling improve efficiency (2nd shift maintenance, for example)?
- Could PM be streamlined by job engineering to improve efficiency or increase reliability?
- Have mistakes occurred from defects in communications (fallen through the cracks)?

2. Modify machine, building, or vehicle (maintenance improvement). Improve the machine so that it doesn't break or need adjustment. Improve the tooling. Make it easier and faster to do the maintenance tasks. Automate some of the maintenance tasks. Remove the source of the problem (redirect the dirt so it doesn't fall on the cylinder). Add automated lubrication systems. Add instrumentation to read out vibration, temperature, or other information.

3. Modify the business or marketing plan. Discard the product; go to full-service leases; increase the lead time; buy your competitor; sell out to your competitor; buy a maintenance contractor; drop a pick-up location; get out of the city; sell the building and lease space; raise prices; sell custom units; have your competitor make your parts/you make theirs; give headache processes to vendors who are expert in that process; hire a top engineer/production person or consultant to help smooth the process.

(For factories, there is the following additional option to consider.)

4. Modify part/product. Make the part easier to produce. Improve the tooling. Change the shape, size, material, finish. Reduce (or increase) the number of steps. Reduce (or increase) the part's specifications. Simplify the steps. Job out part.

5. Modify the process that produces the product. Improve the whole process. Improve incoming materials. Filter the oil to the boiler. Improve the process to allow greater variation in incoming materials. Make the transfers more bulletproof. Look for improved technology. Look for a whole new and more reliable process.

Continuous improvement is an antidote to the constant pressure of competition. It is essential that management realize that continuous improvement in the maintenance department is everyone's business and can only be achieved with everyone's input. The goal of continuous improvement is the gradual elimination of the need for maintenance. The inputs drop and drop. New processes, new assets, new products will keep this an ongoing business.

Setting the Stage

In years past, the engineer, manager, or superintendent was responsible for the ideas for improvements. Maintenance people were "hands" hired to do what they were told. Today when organizations are lean and mean, we need all of the person. Leadership facilitates contribution.

The following are ideas to facilitate the flow of cost reduction/reliability improvement ideas.

- Common complaints are good places to look for possible improvements. Ask what things people are complaining about. Design projects around those complaints.
- Invite informal ideas. Spend time with the employee refining the idea and packaging it for presentation to the maintenance group and then to management.
- When an idea is tendered, follow up and keep the initiator of the suggestion informed about what happens. As much as possible, let the suggester stay involved throughout the whole process.

- When an idea is rejected outright, handle the negativity carefully. If you think it is still a good idea, see where you can do some of the suggestion as a pilot project.
- In looking for ideas, follow the money: big gains come from areas where big money is spent.
- Remember the president of Sony who said that his company's attitude was to bend over backwards to implement a suggestion of an employee even if it seemed to have marginal usefulness in order to bolster the confidence and morale of all the employees.
- Make ideas the currency of your whole maintenance effort. The verbal and monetary rewards should follow the ideas.

20
Improving Maintenance Reliability—RCM

One of the best models for continuous improvement is the application of the techniques of reliability centered maintenance (RCM). RCM is one of the powerful ways to improve maintenance because it addresses the core of the customer need, that is, reliability. The techniques are an outgrowth of the investigations into reliability by the aircraft manufacturers, airlines, and the government.

Using these techniques the aircraft industry was able to reduce maintenance requirements for the 747 to 66,000 hours from 4,000,000 hours for an older generation DC8 (from *RCMII* by J. Moubray). The reliability and safety also increased by a factor of 5 times.

RCM is a five-step process. The process is usually a team effort, with members from both operations and maintenance. It is facilitated by an RCM specialist with good knowledge of the process.

1. Identify all of the functions of the asset. At first this might seem trivial. On a second examination there are many secondary functions that are important. Functions are divided into primary, secondary, and protective. Each function is defined by a specification or performance standards.

For example, the primary function of a conveyor is to move stone from the primary crusher to the secondary crusher. The specification calls for 750 tons per hour capacity. Secondary functions include containment of the crushed stone (you don't want pieces falling through the conveyor).

An important set of functions are hidden safety functions. In this case, an example is the thermal cut-out on the motor. The function of that device is to interrupt the current to the motor when the internal temperature reaches 150°C.

2. The second step is to look at all ways the asset can lose functionality. These are called functional failures. One function can have several functional failure modes. A complete functional failure would be that the unit cannot move any stone to the secondary crusher. A second failure would be it can move some amount less than the specification of 750 tons per hour. A third functional failure is if the conveyor starts moving more than 750 tons per hour and starts to overfill the secondary crusher.

Each secondary function also has losses of functionality. In our example, the conveyor could allow stones to fall to the ground, creating a safety hazard.

The hidden safety function can lose functionality by not working at all, by cutting out at too high a temperature, or by cutting out at too low a temperature.

3. Review each loss of function and determine all of the failure modes that could cause the loss. In our example, the list might be 20 or more failure modes to describe the first functional failure alone. Failure modes of our rock conveyor include motor failure, belt failure, pulley failure, etc. Each functional failure is looked at, and the failure modes are defined.

Use some judgement to include all failure modes regarded as probable by the team. All failures that have happened in the past in this or similar installations would be included as well as other probable occurrences. Take particular care to include failure modes where there would be loss of life or limb, or environmental damage.

It is essential that the team identify the root cause of the failure and not the resultant cause. A motor might fail from a progressive loosening of the base bolts which strains the bearing causing failure. This would be listed as motor failure due to base bolts loosening up. Motor failure, for example, would have many entries from all of the different root causes. Others might include shorting of windings, bearing failure, shaft failure, etc.

It is important to include failure modes beyond the normal wear and tear. Operator abuse, sabotage, inadequate lubrication, improper maintenance procedure (reassembly after service), would be included. The failure modes of the hidden safety functions that would cause a complete loss of function could be dirt inside the cutoff preventing activation, contacts sealed shut, etc.

4. What are the consequences of each failure mode. Consequences fall into four categories. These include safety, environmental damage, operational, and non-operational. A single failure mode might have consequences in several areas at the same time. John Moubray, in his significant book *Reliability-Centered Maintenance,* says "Failure prevention has more to do with avoiding the consequences of failure than it has to do with preventing the failures themselves."

The consequences of each failure determine the intensity with which we pursue the next step. If the consequences include loss of life, it is imperative that the failure mode be eliminated or reduced to improbability.

A belt failure on an aggregate conveyor would have multiple consequences which would include safety and operational. A failed belt could dump stone through the conveyor superstructure, hurting everyone underneath. The failed belt would also shut down the secondary crusher unless there is a back-up feed route.

The failure of the drive motor on the conveyor will cause operational consequences. Operational consequences have costs to repair the failure itself as well as the cost of downtime and eventual shutdown of the downstream crushers. Other failures might only have non-operational consequences. Non-operational consequences include only the costs to repair the breakdown.

Failure of the hidden safety functions usually causes nothing unless there is an associated failure. A boiler safety relief valve is only an issue when the primary shut-off systems have failed. Our thermal overload switch will only activate if the motor is overworked for some length of time or there is another failure causing the motor to overheat.

5. The final step is to find a task which is technically possible and makes sense to detect the condition before failure or otherwise avoid the consequences. Where no task can be found and there are safety or environmental consequences, then a redesign is demanded.

For example, if it is found that the belts start to fail after they are worn to 50% of their original thickness, then the addition of a task called belt inspection might be indicated. But if the belts fail rapidly after cuts or other damage then a full time electronic sensor would be necessary. The task frequency (say, monthly inspection) would not be often enough to detect the impending failure. This is the practical consequences of the PF curve. The time between possible detection and failure should be at least twice the inspection frequency. In all cases where safety or environmental damage is the main concern, the task must lower the probability of failure to a very low level.

We also have to be conscious of the fact that some failures of sensors or protective devices are hidden. A failure is said to be hidden if it occurs and the operators would, under normal conditions, *not* notice the problem. For example, if a belt thickness gauge fails, then (unless the design is fail safe) the operators would have no way of knowing that the sensor is out of service. Now without protection a second failure (of the belt) could more easily occur and cause an accident.

In operational failure modes (such as the motor failing), the cost of the task over the long haul has to be lower than the cost of the repair and the downtime. If the PM task costs $1000 a year and a breakdown cost $2500 and downtime cost $4000 to repair, then the breakdown has to be avoided more than every 6 years (15.38% probability in any random year).

$1000	<	$2500 + $4000	×	15-16%
Total PM cost. Should also include corrective activity	must be less than to be justified	Breakdown costs + total downtime		probability of a breakdown in any year

21
Accounting Issues of Maintenance

In the eyes of the accountants, maintenance is a pure expense—a necessary evil, if you will. In the general ledger (the accounting bible for your organization), maintenance expenses reduce profit. Anything that reduces profit is bad. Unless we create a different view, we can never be anything other than an expense to be reduced and controlled.

In addition to their gatekeeper role, the accountants categorize every financial transaction that takes place in the organization. This categorization is essential to determine profit or loss. Maintenance activity is frequently in the gray area where these decisions are made. For example, many accounting discussions deal with the decision to capitalize or expense a major repair. In one case, the repair is an investment in the organization and becomes an increase in the value of an asset. In another case, the same repair is an expense, a reduction to profit, and a reduction to taxes.

The accounting department uses double entry bookkeeping techniques. That means that every transaction has two sides called debits and credits. A parts purchase will increase (credit) A/P (accounts payable—what is owed) when the invoice is posted and also increase (debit) the Expense account by the same amount (debits always equal credits). When a machine is purchased the invoice is booked to A/P (credit—liability account) and also to equipment, an Asset account (debit asset account).

The accounting balancing act for a maintenance part purchase:
Debit—maintenance and repair (expense account) = Credit—Accounts Payable (liability account)

All maintenance work must be accounted for within the general ledger (GL). The accounts mentioned above (expense, asset,

164

A/P) are accounts in the general ledger. The GL is the score card of the organization that reports the condition of the finances, profitability, assets, and liabilities. A repair to a machine is fundamentally different than a rebuild/improvement to the same machine. These different types of transactions must be differentiated and reported.

Some categories that accounting might be tracking:

- maintenance materials such as repair parts
- maintenance supplies such as rags, absorbent pigs
- contract labor
- maintenance labor, fringe benefits, overtime
- labor and materials for grounds
- housekeeping services
- maintenance support like supervisors, staff, manager
- costs to fabricate tools, jigs fixtures, whole machines
- non-recurrent labor on large repair jobs
- structural repairs to building
- contractor charges on large repair jobs over $50,000

The accounting department also verifies that the transaction doesn't exceed any of the predetermined limits (if limits are exceeded then special authorization might be needed). The following shows example amounts that a maintenance department can spend (from a large manufacturer) without special authorization.

Type of charge	Limit before higher authorization is needed
Office furniture	$500
Computers, instruments	$1000
Small tool purchase	$500
Capital spares program purchase	$5000
Any MWO is eligible for capitalization	$5000
$ limit on any individual MWO	$40,000

Costs need to be charged to the department that incurred them. Tracking and charging all costs is the function of the cost accounting group within accounting. Cost accounting is one of the most complex functions of the accounting profession. In a production facility, cost accounting determines the accurate cost of each product line. This is true (and handled differently) in service businesses (what is the true cost of a service call), fleets (what is the cost of the Memphis–Chicago run), property management (how much does this building cost to run per square foot).

In cost accounting, overheads (such as energy, phone, and support departments, such as maintenance, purchasing, receiving, etc.) must be allocated to the activity which consumes it. For example, a plant that manufacturers wire harnesses uses injection molding equipment which uses large amounts of energy. The molding department might use 1/2 or 2/3 of all of the electricity of the whole plant. Cost accounting might calculate this and add the additional overhead to the cost of each plug molded. Cost accounting's role is to fairly assign overhead costs to products.

Many fights are the result of illogical or incorrect assignments of overhead. An equipment leasing division of an aerospace company was stuck with the overhead (they called it the burden rate) rate of the entire company (which made it competitive in aerospace but non-competitive in the leasing area.) They could not convince the accounting management that the sophisticated engineering, computerization, and other elements of aerospace made no sense in the leasing arena.

Activity-Based Costing (ABC)

Many organizations have grappled with the costing of overheads (like maintenance) to make up a product cost. Some of these questions directly affect the maintenance effort. Traditionally overhead was costed on a direct labor-hours or by material purchases basis. Imagine the impact of using direct hours in an environment where automation is reducing hours each year. Some maintenance budgets are based on a percentage of direct labor. As the product line becomes more automated, the maintenance budget goes down (instead of up)!

In both cases, misallocation of overhead causes the calculation of the cost of the product to be off. The skewing of costs tends to increase market share of the products not carrying their weight because they are underpriced. In traditional product-driven organizations, mature products tend to support the new product introductions. Simple products tend to support custom engineered products, products with high labor support products with high materials. ABC techniques focus the attention of financial management toward the proper driver for the overhead. For more details on ABC consult the Appendix (Ron Giuntini).

Accounting and the Maintenance Parts Inventory

Maintenance stores traditionally do not exist as an asset of the organization by the rules of accounting. To prove this assertion, let's trace two different material purchase transactions.

Bookkeeping for Raw Material Purchases Where an Asset is Created

You place an order for bolts with AAA Bolt Company for your product.

1. Bolts come in to receiving; invoice comes in the mail from vendor.

2. Invoice is booked to accounts payable (credit which increases a liability account); raw materials are booked to raw materials (a debit also increases an asset account on opposite sides of the ledger to balance).

3. Invoice is paid which is booked to the cash account (credit which reduces the cash asset) while the accounts payable (debit reduces a liability) is reduced by the same amount.

4. The net result is that the asset cash is traded for the asset raw materials.

5. No effect on profit until the raw material is used and the product is completed and shipped to the customer or the warehouse.

Bookkeeping for a Maintenance Parts Purchase Where No Asset is Created

You place an order for bolts with AAA Bolt Company that might be needed to fix your product assembly line during shutdown 5 months from now.

1. Maintenance parts come in to receiving; invoice comes in the mail from the vendor.

2. Invoice is booked to accounts payable (again a credit to a liability account), maintenance parts are booked to expenses, maintenance parts (a debit to an expense account increases it) profit is reduced immediately.

3. Invoice is paid which is booked to the cash account (credit to cash-reducing an asset) while accounts payable is reduced by the same amount (debit-reduction to liability account).

4. The result is that the asset cash is consumed by an expense.

5. Profit is reduced when the item comes in the door in step 2 and the first booking occurs.

6. The item might be *sitting on the shelf* in the store room but there is *no asset on the books.*

There is another type of inventory in the stock room that creates a problem for accounting; that is, parts removed from machines that are rebuilt by both outside contractors or by your own people in their spare time. This inventory is on the shelf either for $0 or a fraction of its value (usually they are on the shelf for the rebuild cost only since the original cost was charged previously). Ron Guintini, a leading expert in maintenance stock, calls this inventory cloaked. It is cloaked from analysis by management, purchasing, or accounting. In some cases, hundreds of thousands of dollars could be involved.

When the organization needs more cash or additional profit, one place to look is the maintenance parts in the storeroom. As mentioned above, these parts are a physical asset, but not usually a financial asset (on the books). Because they are not on the books, any revenue stream from the inventory (say, by selling some of it) shows up as extra profit. Metaphorically, if you squeeze the storeroom and dispose of valuable parts even at a loss, profit and cash drip out! This explains the zeal with which the maintenance stock is attacked.

Actual Cost of an Hour of Maintenance Labor

The actual cost of labor includes direct wages, overtime, benefits, and indirect costs. It must be spread over the time actually spent on chargeable repairs. Benefits include the costs of health insurance, FICA (employer's contribution), pensions, insurance, and any paid perks.

Indirect costs include indirect salaries (all support people who don't show up on work orders, materials (ones not individually charged such as bolts, grease, etc.), supplies (uniforms, bulk materials, soap, etc.), costs of the shop/yard (utilities, depreciation of facility and tools, insurance, taxes), allocation of costs of corporate support, costs of money, and hidden or other indirect costs.

Charge-Out Rate

Accounting also maintains charge-out rates for all job classifications. The charge-out rate includes hourly wages, fringes (retirement costs, insurance, FICA, vacations, etc.), and overhead (supervision, clerical costs). This number is important because it is the comparison number for outside contractors bidding jobs against inside people. To calculate the charge-out rate:

Step 1: Calculate wages with benefits.

Direct hourly wages:	$15.00
Benefits @ 25% of wage	$ 3.75 (vacations, sick and other leave added later)
Total cost per hour, wages and benefits	$18.75

Step 2: The total cost per hour must be increased to cover time paid for that doesn't appear on work orders.

Consider your operation of the *2080 hours* straight time (52 weeks $\times$ 40 hr./w) available per year.

Annual Hours purchased = 2080 less:		
vacation	160 (4 weeks)	
holiday	64 (8 paid holidays)	
paid sick leave	40 (5 sick days)	
Other: jury, guard, training, union	32 (4 days)	
Total deductions	296	(296)
Annual time available		**1784**

Actual cost of work time $18.75 $\times$ 2080/1784 = $21.86

Step 3: Adjust rate for unreported hours. If all of the available hours were charged on work orders, then the true cost of labor would be $21.86 plus the overhead. However, even in a closely controlled shop, some time is left unchargeable (not on work orders). We will add a 10% factor for time working on non-chargeable jobs.

Cost per direct labor W/O hour: $24.29 = $21.86 / .90

Step 4: Calculate the cost of maintenance overhead or maintenance indirect costs. In the following example the actual costs are divided by the W/O hours to get an overhead cost per hour

Category	Amount
1. Indirect salaries	$131,500
2. Bulk materials	$ 27,200
3. Supplies	$ 21,500
4. Cost of shop and yard	$ 42,900
5. Other (specify) depreciation on tools	$ 29,700
6. Other (specify) misc costs	$ 21,500
Total indirect costs for last year	$274,300
Total work order hours for last year	29,050
Indirect cost per hr = Total indirect costs/Total WO hours	$9.44

Step 5: Add the labor cost to the overhead cost to get the true cost of maintenance activity.

Overhead	+	Labor	=	Charge-out rate
from step 4		from step 3		
$9.44	+	$24.29	=	33.73

An interesting sidebar is the impact of overtime to the equation. In many situations, the overtime has no impact on the charge rate because overhead does not increase as much (you don't need more utilities, space, computers, or supervisors) for limited amounts of overtime. For many companies, using a modest amount of overtime is an attractive way to get jobs completed without hiring contractors or additional full-time people.

True Costs of a Maintenance Inventory

Consider the distinction between the price of an inventory item and the cost of that item of inventory. This is the same argument we pursued with the true cost of labor. Some firms put all of the burden on the labor hours (including the overhead costs for the parts). There is a problem with this. When the burden is carried by the hours, jobs with a large proportion of materials will be too cheap, and jobs with high labor content will be too expensive. When you compare against outside shops, the difference will be striking. Creating alignment of where to charge overhead is the field of activity-based costing.

There are certain costs associated with having, buying, and using inventory. A complete analysis of these costs prompted one military base to replace their storeroom with a private partnership led by W.W. Grainger.

There are two costs that are part of the costs of inventory. The first cost is called the COA (cost of acquisition). The second cost is the COP, the cost of possession. The COP is discussed later in this section.

The COA is an invisible cost that is only discussed in passing in the industry as an element of the EOQ formula (discussed later in the chapter on the maintenance stock room and inventory control). The paradox of this invisibility is discussed in Dr. Mark Goldstein's text *Getting a Quick Return on Your CMMS Investment.*

"That means that the average costs of purchasing (COA) an item ($42.50) are running 40–205 times the costs of maintaining that item on the shelf (COP). . . . While it is mandatory to control the amount of inventory that you have on hand, through optimization tools such as JIT, management has even more profoundly rising concerns. The costs of acquisition, and of most important postponement that they are forced to pay for MRO items are so avoidably punitive, as to make the optimization of the MRO supply chain management a (more) highly competitive plant goal, than the optimization of physical inventory maagement!"

Goldstein research shows that the COP multiple is 2.3 times the prime interest rate (plus inventory tax). We will see below the accuracy of that estimate when we look at the specific costs for an east coast manufacturer.

Parts room overhead has to be spread over the parts used, added to the part price to yield the part charge-out rate. These costs, which are unique for each operation and can be unique for each location within an organization, include the following.

1. *Cost of Money:* Your organization could invest the money now tied up in inventory at market rates and get a secure yield from 5% to 10% of the average value of the inventory.

2. *Expenses of Warehousing:* This includes depreciation on building space and shelves, an allocation of utilities, building maintenance and security, life cycle costs on material handling equipment, forms and paper, office supplies, and machinery. At a large city airport, the parts room space could be rented to a concessionaire for $30/ square foot per month. It was a constant battle to keep the space. Cost is usually figured at 2.5% to 6.5% of the value on the shelf.

3. *Taxes and Insurance:* Some localities tax the assets of the organization, most have real estate taxes, this category also includes casualty insurance. Costs vary from 1% to 3%.

4. *People Costs:* Full-time stock clerks, allocation of other clerks, pick-up/delivery people, supervisors, and purchasing agent. Overall parts volume has a great influence, and the cost is from 14% to 40%.

5. *Deterioration, Shrinkage, Obsolescence, Cost of Returns:* Parts sometimes become unusable, disappear, are obsolete, or must be returned for warranty or incur a re-stocking fee. Costs vary from 4% to 15% (higher if shrinkage is a significant problem).

In *Fleet Management* (McGraw-Hill, 1984), the author, John Dolce, states "it takes 12%–15% of a company's annual parts expenditures to support people costs...Support dollars...should be approximately 25% of the on the shelf inventory."

Inventory Carrying Charge Calculation

Organization: Springfield Manufacturing

Total Inventory at the End of Last Year	$391,000
Total purchases last year	$782,000
Cost of money avg. 12% for yr, 6% per turn	$23,460
Cost of warehousing 2000 sq feet × $6 sq foot per year	$12,000
Cost of taxes and insurance	$7820
Salaries stock clerks (1 @ $25,000)	$25,000
Allocation of other people (25% of purchasing clerk)	$7500
Deterioration, shrinkage, obsolescence, costs of returns	$22,624
Other costs (rebuilds not charged, depreciation of fork truck)	$7500
Total cost of carrying inventory	**$105,904**

Multiplier for charge-out cost = 1 + (carrying costs / total purchases)

1.135 = 1 + ($105,904/$782,000)

Example of the True Cost of Inventory

The charge-out ratio is a multiplier to change from parts price to parts cost. For example, the price for a hydraulic cylinder might be $495.00 net. The charge-out cost (entered on the WO part cost column) would be $561.83 (at the Springfield charge-out ratio of 1.135).

Charge-Back

One way to get and keep the attention of your user groups is to use charge-backs. Each month, all maintenance activity is charged to the department or cost center where the work was done. The issue that charge-back addresses is use of maintenance resources instead of departmental resources for projects, process improvements, office improvement, car washing, and other non-maintenance activity. In a major oil company that is known for its frugal nature, charge-back demonstrated an executive office building that costs twice as much per square foot for maintenance as do other buildings of the same type because of excessive maintenance use for non-maintenance activity.

It could be heard in all types of enterprises that maintenance costs too much (which it might). A light goes off in someone's mind when they realize that maintenance is mostly driven by requests from users. In some organizations, user call-ins of small projects are by far the largest component of the maintenance demand. The bright idea is to charge-back maintenance costs to the originator. The next step is that the user departments, after seeing how much they spend, spend all of their time arguing about where the numbers came from (no way did your people make five adjustments to the temperature last month). The moral of the story is if you choose to use charge-backs, be sure your record keeping is easy, accurate, and can be audited.

To accurately calculate the charge-back rate, the labor and materials have to be burdened.

Labor charge-back rate: $34.00 (according to the preceding calculations)
Part mark-up: 113% (again based on last year's performance)

Example of Charge-Back

XYZ INTERNATIONAL CORP
Falls River Facility
Charge-back Sheet   1st Q-1997

Dept.	Labor	Mat'l	Number of requests	$/Request	Total
Admin.	6575	4050	252	$42	$10,625
Distribution	72000	65500	1009	$136	$137,500
Warehouse	24100	20500	219	$204	$44,600
Sales dept	4400	5500	211	$47	$9900
Charge-back Totals	**107075**	**95550**	1691	**$120**	**202625**

Statistics:

% Materials	47%
Average # of requests per day	26
Cost per request	$12

22
Maintenance Information Flow

How do you find out about work to be done? Where does your work come from? Do any requests for work fall through the cracks? Is the part of your process that accepts maintenance requests, digests them, and hands them off efficient? How much thought has gone into the flow of maintenance requests recently?

A study of your information flow might show where requests for work could fall through the cracks. It would also show where you are not capturing the minimum amount of information to efficiently dispatch maintenance workers.

Under traditional models, all requests for service should converge in the maintenance control center and be reviewed by a single person or group. If you use a team structure, then the team point person on planning and scheduling should review the requests. Emergencies should be handled as directly as possible. The goal of the system is to serve the customer efficiently with minimum overhead.

The control center provides several functions that have to be handled for the smooth delivery of maintenance to take place. Under new models of management, a center might not exist at all, but the functions mentioned below still must be done.

1. Receive and triage emergency calls. Advise callers how to make areas secure and safe (if appropriate).
2. Dispatch people who are likely to have the skills, tools, and materials to repair the problem (or can obtain what they need) to urgent calls.
3. Receive, prioritize, evaluate, and screen incoming work.
4. Plan new jobs.
5. Visit sites, determine the scope of work, identify problem areas.

6. Review jobs in process when necessary.
7. Administrate outside contractors.
8. Ensure record keeping is up to date for all work performed.
9. Ensure that CMMS input is up to specifications.
10. Order materials and specialized tools for upcoming jobs.
11. Ensure regulatory compliance where needed.
12. Ensure safety equipment and techniques are available where necessary.
13. Account for all resources (labor, materials, outside vendor work, engineering, etc.) spent on jobs and be sure information makes it into the permanent files.
14. Keep scheduling boards or work lists.
15. Keep data files safe (both paper and computerized).
16. Possibly house and maintain the MTL (maintenance technical library).
17. Serve as a communication center, mailboxes, radios, etc.
18. Prepare reports on performance for the maintenance people and for management.
19. Develop procedure charts for operator-led maintenance.
20. Schedule project work, coordinate construction.
21. Update database for new equipment and asset acquisitions.
22. Keep budget data and prepare budgets.
23. Develop PM routines, maintain RCM files.
24. Guide cost reduction processes, reliability improvement programs, and serve as a repository of research developed from those efforts.

This list has to be constantly reexamined in light of new technology and staffing realities. The company's E-mail capability, paging, radios, laser scanning capability, voice recognition, intranet, MAP backbone, LAN's, WAN's, and CMMS should be used to their fullest capabilities.

The gurus of business reengineering—Hammer and Champy—talk about inductive thinking. In their model, when we find a new solution, we should then go looking for problems to apply the solution to. In short, the maintenance control center processes may stay the same, but due to new enabling technologies, the players and processes might radically change.

There are several ways that the maintenance function is notified about work needed. Some of these sources are internal (PM tasks, corrective maintenance) and the rest are external to the maintenance department. Even in a traditional department, the mode of communication could be written work request, scrap of paper, phone call, page, E-mail message, fax, collaring a mechanic, note slid under door, output of a monitoring system, or other mode.

Sometimes the jobs require major dislocations and month-long commitments, and other jobs are 5 minutes without disruption to ongoing activity. Some are planned events (known about days or weeks ahead of time) and some are unplanned. Finally some jobs are emergency, critical, important, routine, fill-in, or some other priority. The entire maintenance process starts with a request to fix or work on something. This section is concerned with the work request from the users.

One can imagine how work must have been requested when maintenance as a separate discipline started to develop. Originally craftspeople repaired their own machines—no request, as such, was needed. The mechanic was right there. Also, these machines were fairly simple, cycled slowly, and were fairly robust for their required use. As the division of labor continued and the manufacturing work was divided into smaller and smaller units, the skills to fix the machines were concentrated into fewer and fewer people.

A foreman, floor boss, or superintendent initiated a request for a repair by yelling for a repair person. As time went on and plants got bigger and bigger, yelling was not as effective. Maintenance people were summoned by page, by being collared at lunch, or by going back to a known place after jobs were completed to be redeployed. Many old timers, even today, remember when, if the maintenance crew members were in the coffee room, the plant manager was satisfied. He or she knew that nothing must be broken down (or the mechanics would be out there fixing it).

To-do lists and work orders were soon developed to manage the mess (the driving force was breakdown, and it was messy). When more sophisticated systems were introduced into large departments (in 1965, with MIDEC, a product of Mobil Oil's predecessor in the fleet world; and in the late 1950's, in the U.S. Navy Bureau of Docks and their Sears-designed PM system), this informal way of capturing information was not acceptable.

Maintenance information flow is independent of the size of the function. Large organizations might have whole departments providing each function. Small ones can literally use four clipboards for all maintenance activity.

1. *Customer has a need.* They communicate the need to maintenance via: written work request, scrap of paper, phone call, page, E-mail message, fax, collaring a mechanic, note slid under door, output of a monitoring system, or other mode. The request is converted to a standard format and put into the in-bin.

2. *The in-bin.* Incoming work requests are taken from the in-bin and are reviewed for authorization, parts ordering, planning. Time/date stamping should take place when a job is placed into the in-bin. The in-bin should be scanned frequently for high priority work that no one is yelling about. One of the Pentagon building managers reported that he received a routine request to mop up some water. When they got there the flood was so bad it had filled an elevator shaft! The in-bin is the funnel into the rest of the system. (See Figure 22-1.)

One function of this step is planning. Planning might be as easy as making sure that the materials are in stock on a machine repair, or as complex as coordination of contractors and employees on a model change in an auto plant. Once the work order is past this function, it will be scheduled for action.

Work orders that you have not decided to do should be sent back to the requester with an explanation. Some organiza-

In-Bin Funnel

To the rest of the maintenance system--next-step backlog (if job is ready or waiting if not)

Figure 22-1

tions keep files of wish list jobs in case some resources are freed up.

3. *Waiting.* After jobs are reviewed from the in-bin, some are ready to go directly to backlog. Other jobs are stuck "waiting." There are many reasons for waiting defined by status codes. These status codes follow the job through all of the steps before the job is released to the backlog. Status codes could include: waiting to be planned, waiting for engineering, waiting for P.O. to be issued, waiting for material to be received, waiting for weekend (or night or summer), waiting until shutdown, waiting for fill-in assignment.

4. *The "backlog" clipboard.* This is all work available to start. All backlog jobs have been authorized, parts are available, priority has been set, and planning (if required) has been done. Once the job is issued to the contractor or staff mechanic, it then moves out of backlog to the pending clipboard.

Managing the backlog is an important way to manage an entire maintenance department. It is essential that the available backlog not fall below a few days nor increase greater than 2 weeks per tradesperson. Too low a backlog will encourage the tradespeople to stretch out the work that is left to avoid layoff. Too high a backlog will cause the customers to suffer unreasonable delays for routine work requests.

5. *The "pending" clipboard.* Jobs in the pending clipboard have been issued to maintenance employees or to contractors and should be completed in a short time. This clipboard should be reviewed regularly for jobs that are stuck. A stuck job that stays stuck is a problem for the manager, and the user. Extra effort (sometimes an outside contractor is needed) is needed to unstick these jobs.

6. *The file.* All completed jobs are costed and posted to the various re-cap sheets for the asset filing system. All sketches, photographs, and little booklets (that come with thermocouple, PLC modules, pumps, etc.) go in the file. Costing and root failure analysis notes also go into the file in the appropriate section.

Typical benchmarks of the effectiveness of a maintenance information flow system:

1. Respond to 99% of emergency service calls within 3 hours. Carry it to pending state. Secure building, machine, and user.

2. Respond to 99% of routine service calls within 2 days. Carry it to pending state. Secure machine, building, and user.

3. Clear pending clipboard weekly.

4. 99.8% of work completed is the work requested.

What the Customer Asked for is Not Always What the Customer Wants

Very quickly, system users realized that the work requested was not the work done. Think about a doctor. If you go in with a pain in your back, the problem could be anything from an ergonomic problem with your water skiing technique to a heart attack, from a cancer in your kidney to a deer tick bite.

The doctor uses the "work request" (my back hurts—fix it!) as one data element of many. Listening to the patient is essential in medicine and just as essential in maintenance. We listen because the core of what the user is complaining about will lead us to the real or root cause.

Our users don't always know enough to request the right service. How many times is a no-heat call related to another problem (a window open, thermostat off, etc.). To this day, the problem remains: how do you efficiently run a maintenance effort when the user's request does not reflect the real problem?

Keeping the Database Clean

The second issue is in the integrity of the database that is being collected by the work order system. If bad data from work requests are allowed into the system uncorrected by the mechanic, then we will not be able to rely on the conclusions drawn from all the data. The problem has always been that the mechanic will not always correct their work request-work order with what actually happened.

When visiting maintenance facilities, I like to review an unedited batch of work orders being entered. My favorite is one where the work requested is "pump XYZ broke!" The mechanic says in the comments field "OKAY" and reports 2 hours spent and no parts used. An investigation shows that the valve was closed and it took the mechanic 2 hours to find this operational problem.

The data would have gone in as a pump failure. Too many of these types of problems render the database invalid!

The work request is the beginning of the process

Need → Work Request → Work Order → Satisfaction

The maintenance work request is the most common document to communicate the need for maintenance work to the maintenance department. A work order is prepared from the work request, referring to the request. In many cases, the work requested is a field of the work order. Through the work request, the maintenance planner plans the job, determines who to send, what skills are needed, what tools and materials to bring, and other important elements of the job.

The data in the work request are valuable. In aggregate, the information gathered from years of work requests can be used for diagnostic purposes, but not directly. Your maintenance request database can be used the same way. Out of 100 calls of pump broken, the following was found:

34%	Pump was actually broken
32%	Pump was off
12%	Pump was on and okay, no product to pump
10%	Inbound or outbound valve closed
06%	Other problem
04%	No power due to trip elsewhere

The mechanic should check the switch, valves, gauges (if there is product in tank), and power before touching the pump. The work requested leads us directly to the cause by an analysis of probability. Experienced maintenance people have enough personal experience to know to check the easy stuff first!

The work request should answer the following questions posed by Jay Butler in *Maintenance Management*:

Is there a possible safety or environmental implication to this failure?

Where should the maintenance person go?

What is wrong? What was observed? What unit is to be worked upon?

Who should they see when they get to the area?

What was happening just before failure?

What does the caller think is the problem?

How critical is the unit or process?

Note what time and date the call came in.

Front End Filter

The second possibility is called a front end filter. This technique was pioneered by Xerox when they started to place complex copy machines in office environments. The users were very unsophisticated about the inside mechanics and electronics of copiers. This technique can cut your work load and increase customer satisfaction.

1. They studied all of their service calls. In the process, they isolated the calls that did not require a service person to correct. These calls included: out of paper, out of toner, unplugged, simple jam, dirt on the glass, etc.

2. The list was refined and studied by experienced service technicians, sales representatives, and psychologists. For each problem, they designed a question so that the answer would lead the callers to solve the problem themselves.

3. The result was about 15 questions that were designed to lead the callers to a problem they could solve without demeaning or angering them. The questions were respectful of their knowledge and their position. For example, instead of asking if there was any toner (you can almost hear the "stupid" being appended by the dispatcher) they would ask what level the toner was in sight glass. Of course, if the toner couldn't be seen in the sight glass, then the logical action would be to fill it.

4. Early machines were installed with a key. The person who was given the key was identified as the key operator. The key operator was shown details of clearing jams, refilling toner, adding paper, and other inside information. They were shown the 15 questions and how to seek out the answers themselves.

5. Questions concerned what kind of paper is being used, where the unit was plugged in, how recently was the imaging glass was cleaned, etc. In each case, the question was phrased in a way to gently lead the person to look deeper into the situation themselves.

6. The 15-question system worked because they took the time to show the key operator how to solve their own problems. It empowered the customer. Customer satisfaction went up with the 15-question system. If someone other than the key operator called, the dispatcher would ask for the key operator if they were available.

Idea for action: If you have a dispatcher (or someone who answers the phone and fills out service requests, in any form) consider setting up questions as a front end filter. In some ways, your maintenance problem is more complex than Xerox because the scope of equipment covered is greater (even though the individual equipment is simpler).

Set up a Rolodex with the common problems alphabetized: such as heat under *H*, leak under *L*, etc. The dispatcher can go through the Rolodex and respond with the critical questions that will filter out non-maintenance causes and collect important data that will help whoever responds to the call. (See Figure 22-2.)

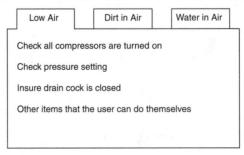

Figure 22-2

Request for Maintenance Services

DATE	TIME		WHO TO SEE		
REQUESTOR	PHONE		PHONE		
LOCATION			SAFETY ISSUE		
ASSET #	QUESTIONS ASKED	☐	ENVIRONMENT ISSUE		
ASSET DESCRIPTION			PRIORITY/ CRITICALITY		

DESCRIBE REQUEST (USE FORMAT: WHAT WAS OBSERVED, WHAT WAS HAPPENING BEFORE FAILURE, WHAT DO YOU THINK THE PROBLEM IS, WHAT SHOULD BE DONE)

MAINTENANCE ACTIONS:	MWO ISSUED #	PLANNING:
PROBLEM FOUND:		SAFETY/ENV ISSUES
REQUESTOR NOTIFIED:	WHEN:	HOW:

Figure 22-3

Your users who routinely call in for service will get used to collecting the information and begin to collect it in advance. Note that the questions are published to the appropriate users.

To reduce your calls, transfer some responsibility to the operators. Questions could be mounted on the machine to lead the operator to the answers themselves.

Won't Start	Funny Noise	Oil on Floor
Electricity on?	Foreign object in hopper?	Sump drain free?
Guards in place?	Oil in sump to line?	Reservoir overfull?
Lock-out?	Bearing hot? Call maintenance.	Seal leakage? Call maintenance.
Air on?	Motor hot? Call maintenance.	Spill while filling? Clean up.
Material in hopper?	Chain jammed? Un-jam chain.	

23
Capturing Maintenance Information

"The palest ink is better than the best memory." This Chinese proverb should be the subtitle of the work order system. The best maintenance professional's memory is subject to the halo effect (weighted more heavily by the near past) and by not being there 24 hours a day.

In this book, when we discuss a work order system, we mean a universal work order system or a comprehensive work order system. All hours are tracked by the universal or comprehensive system. This specifically includes PM activity, short repairs and adjustments, all repairs, emergency work, rebuilds, capital projects, small projects, shop work and clean-up, bench work, rebuilding spares, and non-maintenance work assignments including errands, personal services, community service—I hope you get the picture. The system is only as effective as its completeness or universality.

The information captured in the work order is the basis of all future analysis. Every hour worked by an employee or a contractor should be accounted for. Taken together, the work orders are your unique history. By paying your mechanics or contractors, you buy a history. Without adequate capture of information you will have a history that you cannot study and learn from. The famous quotation applies: "those who do not study history are doomed to repeat it." Having work orders easily available for research or review pays dividends in root cause analysis and continuous improvement programs.

The work request discussed in the previous chapter can be the top or front page to a work order. The maintenance action section would be replaced by the work order itself. Many

CMMS's have the work request as the first page to the work order.

Maintenance dollars spent in the past are maintenance experience. The experience is only important if it is remembered, applied to problems, and published so that others can learn without making the same mistakes. It is essential to review the unit history to avoid useless work.

One of the recurrent issues today is the time and expertise needed to fill out the work order document and enter it into the information system. Organizations will go to great lengths to insulate their maintenance mechanics from this chore. Keeping maintenance people from the computer is another manifestation of the old grease monkey paradigm. Keep the maintenance grease monkeys away from the computer or they will mess something up. You hear: "They aren't smart enough to know what to do." You cannot have a profession of maintenance until these anachronistic ideas are dead. The generator of the data should be the enterer of the data. Advances in technology such as scanning, voice recognition, and radio connected remote terminals will render this whole discussion moot.

If you don't have a work order system (fewer and fewer organizations) you will have to design one to get the benefits discussed in this text. Since CMMS requires a work order, most firms will take both plunges (CMMS and work order) at the same time. Some computer systems allow some modification to their canned forms. Below are some guidelines for designing your own work orders.

Design of Work Orders

If you design your own document, be sure to include the following.

1. Wide lines so that the mechanic can write in easily.

2. Size it to fit on a clipboard if you expect the mechanic to write on it (or buy special clipboards).

3. Put your organization's name on top to be more professional.

4. Use check-off boxes where possible to reduce writing and improve completeness.

5. Consider using bar codes to speed data entry. Bar codes also increase accuracy.

6. Consider a paperless work order system where everything is typed or scanned into terminals around the organization (or use handheld terminals that are radio linked to the network). Software is available that can read check marks, and has limited ability to read and translate handwriting to machine format.

7. For large facilities without networked computerized work order systems, designate and use fax machines to transmit work orders around the facility. Never forget that the prime reason for work orders is to provide written communication with the tradesperson. The message would be "Jack, please go to fax #391 for your next job." In a computerized network, the mechanic would walk to a terminal and punch up the work order themselves.

Consider the *incident report* as an adjunct to the work order system. The report is an aid to directed thinking. We need these aids because in the hustle and bustle of everyday life there is no time for thought. In the incident report, four issues are outlined:

1. description of what happened
2. what was happening just before the incident
3. what was done
4. how can this incident be avoided.

Going through these steps for major incidents helps use the incident as a teaching opportunity. Of course, the work order can also be set up to serve this purpose.

There is a famous story of using your mistakes for learning with Henry Ford and an engineer on the Model T. The engineer made some kind of mistake that cost Ford $10,000,000 (which was a large sum of money at that time). Henry Ford passed by the man's desk and found him packing up his belongings. Mr. Ford asked him what he was doing and the man replied that he was going to save Ford the trouble of firing him, he was quitting. Ford said, "you're staying because I just paid $10,000,000 to educate you!" The man stayed on for years and made many contributions to Ford based on the lessons he learned that day.

Our failures are our opportunities for learning. The work order system or incident report is our way of preserving the past so it can teach us to improve in the future. There are many good reasons to collect the data. Without it we have no way of knowing if equipment is under- or overmaintained. We can't tell if the equipment needs a simple modification, or a complete overhaul. We can't even determine if there is a pattern of operator abuse.

The Computer-Generated Work Order

The trend in all computer systems is to limit the amount of external paper needed to drive the system. Computerized work order generation is an opportunity to cut the outside papers and use the computer to track repairs in process. There are distinct advantages and disadvantages to this strategy.

Advantages:

1. Data entry is more accurate since the computer will not generate bad codes.

2. Task lists or PM codes can be attached to the work orders based on files set up for each asset or each class of assets.

3. Differential diagnosis items can be added. This could transfer useful information to less experienced repair mechanics.

4. WO's that are lost in the shop are still on the system and can easily be reprinted.

5. You can get a computer's view of the jobs open. This is especially true if you post hours and parts as they occur.

Disadvantages:

1. A large effort is spent in managing the paper flow from the computer. If a PM is required and the equipment is not available, the work order has to be stored and "managed."

2. The ease of generating work orders might mean unnecessary paper being generated and excessive open orders on the system. In an "after the fact" system, only complete work orders or large jobs are entered.

3. There can be a laziness if what is reported turns out not to be the actual problem. The system will charge the breakdown to the component mentioned in the work requested rather than make the change to what the mechanic actually found.

4. Some supervisors might defer to the system generated work loads and not solve problems the way they did before (let crises occur and blame the system).

Data Entry Strategies

Every vendor of software has unique tactics and strategies to get the data from the document into the computer. They range from auto-field duplication (so you don't have to re-key repetitive information) to mouse-driven point and shoot (where you highlight the selection and push one of the mouse buttons).

Data entry into the CMMS consists of hundreds, or perhaps thousands, of decisions about how to handle, what to call, or how to express the work that was performed. Total commitment to continuous training is necessary for the right data to be entered. If people using the system and entering data know how the system works, they are more likely to be consistent. It goes without saying that choices made while setting up the codes during CMMS installation are critical since the system's effectiveness is going to depend on the quality of those choices of codes.

The second issue is data entry audits. In accounting systems it is common for the machine to generate batch totals to ensure that all the data were entered and that no transpositions occurred. This batch audit technique is an excellent idea and has been carried through to certain maintenance systems. All maintenance systems should have some facility to audit the completeness of the input.

By far the most perplexing and continuous problem with maintenance is the accurate and timely gathering/entry of data. This problem breaks into the following subproblems:

1. incomplete work orders (problem reported: pump broke—response: OKAY)
2. inaccurate information, inconsistent coding of the same activity
3. missing transactions
4. inaccurate reporting of part numbers/quantities
5. obtaining and cheaply entering accurate utilization data
6. unreadable/incomplete log sheets
7. problems caused by broken meters (cycle counters, odometer and hour meters).

There are now technological solutions to some of these problems. Automated and semi-automated data acquisition options include the following.

Time-keeper card-type systems: These are handheld computers that can keep time against work order numbers, and can include component codes, work accomplished, and repair reason codes. The day's work can be downloaded to a PC which will post the hours to the open work order file.

Laser scanning data entry: All of the information on the repair order is entered into the computer via scanning bar coded pads preprinted on the work order. This eliminates the need for typing and can be used very effectively in a multi-lingual environment.

Software used with page scanners: Software exists that can read handwriting, check-offs in boxes, typing and bar code patches all from the same sheet. This can expedite the reentry of work order data.

Handheld computers can be used to great effect in the storeroom and throughout maintenance. In the storeroom, counting and logging entries during your annual physical inventory can be made easier and more accurate by taking inventory with a handheld computer. You enter the part number and quantity for each part. Many systems also incorporate scanning for increased speed and accuracy. Accurate reporting of part numbers, quantity, physical inventory is a barrier to overcome for effective use of the computer in maintenance.

Since parts make up a great percentage of the maintenance dollar, accurate data entry is extremely important to control this expense. Unfortunately, part numbers tend to be long and complicated. There are also inaccuracies from entering the wrong quantity, and from a mistake in physical inventory. Mistakes in any of these areas could cause stockouts and unnecessary downtime.

Laser scanning of part labels: Using bar code labels and a laser wand attached to your data entry terminal, you can simplify and improve your parts reporting. Whenever a part is used, its label is transferred to the work order. The label is scanned at data entry along with a quantity (numbers can be preprinted on pads or the work order. The handheld scanners mentioned above also come with scanning wands to aid physical inventory.

Have the CMMS Trap Garbage by Conducting Automatic Data Audits

One of the most important functions of the data entry section of the CMMS is audit of the data entered. One major system goes through 31 tests of the data on the work order before acceptance. One class of tests concerns the masterfiles.

1. Valid asset number?
2. Valid item to work on against that asset (does not allow changing the fuse of a toilet!)?
3. Valid repair reason?
4. Valid location?
5. Valid mechanic?
6. Valid work accomplished?
7. Valid part number?
8. Valid work order number (has the number been used before)?
9. Valid meter reading?

A second class of tests concerns calculations or tests on existing masterfiles and detail files.

1. Is meter (if used) in range?
2. After deleting quantity of part number, is quantity on hand positive?

3. Correct craft assignment?
4. Cannot work more than 24 hours in a day per person (with 16-hour alarms).

After the tests are complete, certain calculations are performed using the work order data and the masterfiles.

1. Multiply part cost from masterfile by quantity on work order.
2. Multiply chargeout rate in masterfile for mechanic times work order hours.
3. Total parts costs.
4. Total labor hours and labor dollars.
5. Total all dollars include outside and miscellaneous costs.

The tests and automatic calculations ensure that accurate information is used to update the permanent detail files.

One of the selection criteria for a perspective CMMS is an efficient data entry section or "front end." This "front end" is the section of the system that you will spend the most hours with. Slow, cumbersome, and unnecessary data entry strategies will cost you hours per week, every week you use the system.

Many software designers have incorporated innovative strategies to facilitate data entry. Some of the ideas are very simple, such as the ability to repeat some of the previous line's data. Other ideas are more subtle, such as the ability to save the details of a repair order, assign them a name, and use them over and over in the future as a unit (in some systems referred to as a kit, in others it is called a planning package).

Primary use of the work order is communication with mechanic. Work orders give the mechanic written instructions, which include the location, a description of the problem, who to see, authorization to proceed, and contact phone number.

In simplified format, the whole information collection and storage system should report problems, simplify solutions, help control costs, aid in organizing maintenance, reduce labor and parts, help decide specifications, and finally save dollars and other resources. The work order can also be a tremendous benefit in the following areas.

1. *Cost collection:* All of the labor and parts are recorded on the work order. Charging all time and materials is a more accurate view of a process's real profitability. Cost is allocated by asset number, system (such as compressed air distribution system), department, or by plant.

2. *Invoice or chargeback document:* The work order can be used as the basis of a chargeback system. Chargeback is an excellent way to highlight user departments that use an inordinate amount of maintenance resource. Some maintenance departments do some work for affiliated outsiders (such as a fleet garage working on owner operator trucks) and the work order is an invoice. In field service the work order is the primary charge document.

3. *Proof of safety and proof of compliance with codes and statutes:* When there is an accident or some other catastrophe, officials will want to see your maintenance records. While having records will not remove the pain of a catastrophe, being able to prove that you did what you could will show the authorities and your management that you were doing your job. Properly executed PM records can significantly reduce your share of the liability if you are involved in litigation. A plant, building, or fleet with a well-documented maintenance effort will be safer and is more likely to be in compliance with codes and statutes.

4. *Job, project, and mechanic management, scheduling and control document:* The work orders show the progress of a larger job and facilitate control. On a larger job you can write all the subtasks on write-up forms and use the forms to manage the job. Companies will visually create a PERT or CPM chart with the work orders to help manage the shutdown or major repair.

5. *Lets the user know what is happening on their job:* Work orders aid communication with the user. Sending a copy of the document after it enters backlog shows the user you are working on their problem and improves user relations. We recommend that you send another copy to the user when the job is complete. In this case, E-mail saves trees.

6. *Provides history for various assets and their components:* The history derived from the work orders tells you which type of compressors last longer, which valves are easier to rebuild, and many other useful bits of information. It only takes a few years before you have enough history to make decisions. History can be used to compare different conditions (such as different PM intervals or different lubrications).

7. *The work orders might jog your institutional memory:* The work order history also is useful when you have not done a repair for an extended period and the work order from an earlier repair is available. It could be helpful for trouble shooting, locating vendors, planning, and research into techniques.

8. *Authorization for parts and special tools:* If you use contractors, this allows them to buy materials against your account and secure special tools. In large organizations, the work order is authorization to the storeroom to release the part. It could also be the initiator of a material requisition if the item is not in stock.

9. *Can become a training document for less skilled workers:* By reviewing what was done for each type of complaint, the new mechanic can see what kind of problems they will have to fix and what parts were used. Summaries of actual work orders can be used to design a training course by pinpointing the most likely repairs for training. If the work is formally planned, then the work order will have job steps that might also be useful for training.

10. *Provides data for root failure analysis:* When a recurrent problem plagues you, the work orders will show you how often, how expensive, what parts were used, and how long the repair took. This information helps in several ways to track down the root cause and identify the financial exposure. Without a work order you have to rely on the memory and conversations of the mechanics (who might be on several shifts).

11. *Insurance recovery:* If you have a fire or other claim, the insurance company can use the writeup to determine the amount to pay you. If you claim that an insurable event caused disruption to maintenance activity, then an analysis of the work orders might provide the proof needed by the insurance company.

12. *Warranty recovery:* Many items that you buy have warranties. The work order can sometimes be used to document the problem for warranty recovery. In some cases, a large group of premature failures could initiate warranty recovery for all items in that batch.

13. *Feedback to planners:* The work order tells the planner what actually happened and how long it took. This feedback can improve the effectiveness of the planning function. For this reason many facilities that have planners close out work orders and enter the data into the computer.

14. *Source document to reconstruct what really happened:* We want to get to the bottom of every expensive or disruptive failure. A properly filled out work order will facilitate this analysis.

15. *Source of data about hidden demands:* The maintenance department is frequently recruited into playing roles such as driver, pick-up person, furniture mover, personal servant, boat polisher, picnic set-up/clean-up crew, security crew, and so on. These are special jobs that need to be tracked. A southern manufacturer saved significant money by hiring a limousine company to pick up visitors at the airport rather than using the mechanic. The savings was in time lost, parking, disruption to ongoing jobs, auto costs, and insurance. Most of the passengers preferred being picked up by professionals.

16. *Evidence in court:* The maintenance information system provides data for court hearings on liability claims, injury claims, breach of contract claims, etc.

Reviewing the unit history will uncover answers to the following questions: Is the machine or a component of the machine inadequate for the job (perhaps the job changed since the last time)? Has the operator been abusing the unit (either in ignorance or on purpose)? Is there a problem in the way that the shop has been performing maintenance?

Fat envelopes spell trouble. One very simple method of using a work order system is to weigh the envelope! In this method the work orders are filed in envelopes or folders by asset, unit, area or system. Pick out the fattest files for analysis.

Four Types of Work Orders

Each of the four types of work orders serves a different purpose.

Maintenance Write-up Form (MWU)

This form would be filled out by the PM inspector or a sophisticated user. It could also be filled out by a maintenance office worker from a work request for a called-in complaint. This is the most common type of work order. During an emergency, the documentation would be completed when the job is completed. Maintenance departments need to conduct periodic training sessions in correctly filling out this document. The better this is filled out, the easier and more accurate will be any subsequent analysis.

An example of a discussion work order is shown in Figure 23-1 (this is designed to bring up the elements of a work order for discussion).

The Maintenance Log Sheet (MLS)

This form (Figure 23-2) could be carried by the craftsperson for all the short repairs that they do in the course of their day. Intermittent work stoppages are usually corrected by short, unrecorded adjustments and repairs. This is the form to document (could be attached to the machines) these small problems. It could also be carried by a regular vendor to record all of the little things that they complete. Also, major problems are sometimes covered up by numerous short repairs or adjustments. These little things add up to significant dollars on an annual basis. Not completing short repairs is also the main reason for user dissatisfaction and subsequent maintenance manager hassle.

The Standing Work Order (SWO)

This type of work order can be used for jobs that are done routinely with known labor and materials. Examples would be a machine startup every Monday morning or a walkthrough done

Maintenance Work Order number		SPECIAL			Reason		Priority
Date:	Time:	REQUIREMENTS	SCHEDULED		PREVENTIVE		100
					CORRECTIVE		
Maintenance request number:		LOCK-OUT			USER PROJ		90
Downtime: Y N					REHAB-REBUILD		
		PERMIT			MODERNIZE		80
Start down: Back in service:					INSTALL NEW EQ		70
Customer:		CONFINED			EFFICIENCY IMPR		
		SPACE			USER MOD		60
Phone:		JOB PLAN			CLEAN		
Charge back to account:		ATTACHED			GROUNDS		50
					OTHER		40
Description of work completed:			UNSCHEDULED		BREAKDOWN		
					PER SERV		30
					DAMAGE		
					VANDALISM		20
					COMPLAINT		10
					OTHER		

skill	Unskilled	Helper1	mechanic2	Electrician	Computer	License	Engineer	Contractor	Other

Date	Person	Time	Action Taken	Downtime	Material used	Quan

HOURS * RATE	OTHER:
HOURS * RATE	OTHER:
MATERIALS	TOTAL CHARGES
CONTRACTOR CHARGES	ALLOCATION:

Notes/ Failure analysis/ Ideas:

Figure 23-1

Date	Asset #	System worked on	User Name	Work request	Priority	Reason	Parts	Time

Figure 23-2

every week to look at the grounds and pick up litter. A single standing work order might be good for a week, month, or longer. These are usually routine jobs that aid the user.

The String PM Work Order (SPMO)

String PM's are the only work orders that are not unit based. The string is a group of like assets that are strung together for PM purposes. A vibration route or filter change route are examples. The string PM is one of the hardest to account for in computerized maintenance systems. Usually the CMMS allows the input of only one asset number.

Note: The PM task list is a form of work order. The time and materials necessary should be captured on a PM work order form.

Field Descriptions from Work Order Forms

All work orders are divided into the following three sections.

Header: Information known before the repair starts. Includes all known data at the time, and information from the files on the asset (if the asset is known).

Body: Feedback from mechanic. When there is a planning effort, then the middle of the work order is broken into two sections: 1) the work plan, description of materials, tolls, permits, and access information; and 2) feedback from the mechanic about what happened in relation to the work plan.

Summary: Information added up in dollars, hours, or both. Also include comments, failure analysis information, and other information to review.

Header of Work Order (information is usually known before the repair begins)

Work order number: Non-repeating number, either preprinted on the form or generated by system.

Date, time opened: The moment the work order is received into maintenance, it is time-stamped. For user writeups the date should be when the request was received by the maintenance department. Many systems automatically time/date stamp all incoming work requests. This is an excellent idea since one useful benchmark is the average response time for each priority work. By reliably entering this time, you can fairly calculate the response time and design a system to service the user better and more quickly. On standing orders the date opened is when the work order was initiated, and the date closed is when the work order is full. Usually a new one is opened at that time. A standing order might stay open a whole week, month, or longer, depending on how the information is used.

Maintenance request number: Some work is initiated from a request that is phoned, faxed, or E-mailed into the maintenance dispatch function. If that request is numbered or known, enter it here.

Asset number: Every asset, machine, area, and system should have a unique number that is easy to see and is used by all functions in your organization.

Location: The location of the work should be included on all work requests. In some cases this is the location of the asset. In other cases it is the location on the asset (for a large asset). This eliminates the problem of the mechanic going to the wrong area to do a repair. Location is where the unit is that requires work. Address or building number should be included as required. If you have a CMMS the asset number should bring up the location. If not, a look-up table or the asset number has the location coded in.

Customer and phone number: Unnecessary hours are spent each year trying to get access to equipment, units, and locked rooms. In other cases, the mechanic needs to talk to the operator or production supervisor to clarify the requested service. All work requested needs a contact person to be included in the package. The work requester needs to be authorized and preferably trained in the maintenance request system. Depending on your organization, the requester will be you, your user, or staff. Remember—the better the request, the less likely a wild goose chase will follow.

Date wanted: Realistic date wanted helps regularize the maintenance effort. It is when the job is requested. There should be restrictions against unreasonable "date wanted," such as ASAP (perhaps ASAP authorizes overtime or, if necessary, a contractor, and charges the work order to the requestor's budget).

Charge-back to account: The accounting system might require different types of repairs be coded or charged to different accounts. Also, many service requests are paid by the user's budget. These charge-backs need to be well documented since some of the charge-backs will be questioned.

Special requirements

Lock-out: Where a special lock-out procedure is attached to the work order, this block is checked. Otherwise, follow the lock-out instructions on the machine (or wherever they are kept).

Permit, confined space: If any permit is required (either in company, or from a local authority), this block is checked and the permit is attached.

Job plan: Check if one is attached.

Downtime? Y/N: Do we want to track downtime on this asset? If checked Y (yes) then keep "time down" and "time back in service." If checked N (no) then ignore those blocks. Where downtime is important, the time down is recorded in the date opened/time block. As usual, the request should be time stamped when it is received by maintenance. Production might spend an hour trying to get the equipment back in service before calling maintenance. The difference might be a significant source of conflict. Time back in service should be when the unit is completed and available for production. The actual definition of this time is an item of discussion between production control, production and maintenance.

Reason for Write-Up (or Repair Reason)

All work orders are initiated for a reason. These reasons should be noted on each work order. The reason is captured for future analysis. Repair reason allows you to analyze your demand for work and, hopefully, make those adjustments that will save you money! Possible reasons to initiate a work order (you can specify your own codes for unique repair reasons) are as follows.

Code	Description
Scheduled (activity that was known about at least 1 day in advance and can be planned includes safety inspections, PM, corrective work, rebuilds not part of a breakdown, project work, line changeovers, equipment make-ready and start-up)	
PM or Preventive	1. PM (preventive maintenance) task list activity such as inspection, lube, adjustment, and survey (an initial PM inspection)
CM or Corrective	2. Corrective maintenance (includes scheduled maintenance known 1–2 days in advance, when PM worker finds a potential or impending problem)
UM-R	3. User maintenance—routine work or standing work order (Known work done every week)
UM-P or User Proj	4. Project work requested by production (usually small jobs, can be planned) Larger projects are considered RM type maintenance.

RM or Rehab-Rebuild	5. Rehabilitation maintenance , rebuild, capital project from management decision
RM-M or Modernize	6. Modernize equipment to shop spec
RM-I or Install new Eq.	7. Installation of new equipment
RM-E or Efficiency Impr	8. Efficiency improvement, energy conservation
RM-U or User Mod	9. User-initiated modification
CL or Cleaning	10. Cleaning machines and shop, sweeping up, etc.
GN or Grounds	11. Grounds, including cleaning, mowing, exterior, snow removal

Unscheduled Activity

UM-B or Breakdown	12. User maintenance—breakdown (requiring immediate action)
	UM-B could be jam-up, slow down, leak, quality problem, immediate safety danger, etc.
PS or Per serv	13. Personal service, errands, minor jobs around office
D-R or Damage	14. Reported damage (someone made a mistake and broke something and reported it)
D-U or Vandalism	15. Unreported damage, no report, includes vandalism, sabotage
Complaint	16. User complaint but not a breakdown
Other	17. Other, or other breakdown, including code violation, safety audit, OSHA inspection, PM inspector finds imminent danger or breakdown (cannot be scheduled)

Examples:

1. In a fleet operation the repair reason percentages came out like this:

Wear and tear	32%	Equipment abuse	6%
Accident	18%	Improperly repaired	4%
Lack of PM	15%	Lubrication failure	2%
Operator error	12%	Other	2%
Operations damage	8%	Material failure	1%

If this is *your* fleet, run out and get some training or some kind of facilitation for the drivers and operators. Any positive changes to the maintenance picture would be lost in the operations and driver problems.

2. In a recent review of repair reason in a fabrication shop, we found the following hours:

Reason for repair	1995	1996	1997
1. PM activity	0	560	940
2. PM—Survey*	0	40	40
3. CM—Corrective maintenance	0	2978	2695
4. UM—R** Routine work or standing work order	4706	4245	1675
5. UM—P small user projects	1200	1225	1675
6. RM—Management decision	1323	4580	1521
7. Vandalism/damage	690	345	267
A. D-R Reported damage	120	240	290
B. D-U Unreported damage	810	585	527
8. UM-B Customer/user complaint	5970	2250	1556
9. OB- OSHA Inspection	611	240	58
	14,620 hours	16,703 hours	11,687 hours

Explore the several trends in these labor statistics: which trends are you convinced are real and which could turn around in a "bad" year.

Priority: Priority helps assign work where there is more work than people. It ensures that vital work is not overlooked in the rush of urgent (but unimportant) jobs. Priority systems have a habit of being abused so that users can get their work done faster (if they write up their own work orders). Typical priority codes for a one factor system include the following.

100. Fire, safety, health (*clear and present danger* with automatic overtime authorized until the hazard is removed)

80. Breakdowns that stop all production, shut down whole building, overtime authorized

---

*Survey is a complete walkthrough of a facility to see the "big" maintenance picture.
**Routine work has a known duration and well-understood content such as 2 hour start-up, shift assigned as an area mechanic. It can also have a known work requirement like mopping a hallway, changing a die. There is usually only a limited amount of maintenance work per se involved.

70. Fire/safety/health (potential danger to user, public, employees or environment), statute or code violation, OSHA violation, EPA, DOT, NRC

60. PM activity, potential breakdown of a single process or part of building including core damage, or loss (all types of minor leaks, decay that will get worse)

50. Efficiency improvement, machinery improvements, project work, energy conservation, reengineering

40. Comfort; change use

30. Cosmetics

In some systems, the criticality of the equipment affects the priority. The RIME (relative importance of maintenance expenditures, Albert Raymond Assoc.) technique includes criticality as a factor and where two factors are multiplied. In RIME the two factors are asset importance and the maintenance work (such as high priority breakdown repair and low priority is housekeeping).

Another system by Patton Associates is the NUCREC which adds customer ranking to the mix. Important customers are serviced first. A third system adds aging to the equation. As a job gets closer to its due date the priority increases. All of these systems can be made to work if your organization has the discipline to apply the rules consistently and limit the number of work processed through the back door.

Description of Work Requested: (Normally this information would be obtained from the work request attached to this work order.) In a one-step system, the work completed would become work requested. In that case, work actually completed would be written under the action taken blocks.

It is important to train your requesters to report the observed condition with as much supporting detail as possible. People tend to blame machines, systems, and components if they have failed in the past. They also tend to blame the part of the system that they are least comfortable with. This can mislead the mechanic. Instruct the requestors to note what was happening just before the breakdown on their request.

Example: A process control company whose systems controlled oil terminals received a call (in the middle of the night, of course) that the main computer shut off and the blending subsystem was off line. The mechanic arrived at the terminal and spent several hours doing a complete system analysis. Everything seemed fine except that the PLC output was not turning on the MC for a pump on one of the products being blended. The mechanic finally noticed that the gauging system seemed out of range. The second product's pump was not activating because the source tank for the first product was empty! The system was acting exactly as designed. The description of broken main computer misled the mechanic for several hours. Good descriptions of work report to the maintenance department only the observable rather than the inferred information.

If you design your own work order be sure there are separate fields for work requested and work performed. This will ensure the integrity of the database by keeping the customer's perception level out of the history file as a cause or effect. We want to collect the work requested for another reason. At some point we will be able to build a table between the job requested and the job performed. We can then give the mechanic the highest probability problem based on what was requested (independent of whether or not the requested work is "right").

Skill Level: Certain jobs require higher skills or special licenses. The manager/planner should evaluate the skill needed. Some localities require licensed trades for certain jobs. If a contractor is used the work order would be attached to the Purchase Order. The same information is required for in-house or contracted jobs.

Labor Estimate, Material/ Tool Requirements (on standing work order): This block should include a description of the work to be done, broken into logical steps. Each step is estimated. The labor estimate is determined by experience, observation, or study. The tool requirements alerts the technician what tools are likely to be needed. This block should include a description of the estimated materials used and their costs. Small supplies would be included. All major or recurring parts should be included. Materials for standing jobs should be known in detail with accuracy from observation or engineering study.

Body of Work Order (information known only after the repair is complete)

Time: The mechanic should write down the time they arrive, leave, and the hours of the job. If a mechanic leaves a job to get parts, they should clock out and start a new line—getting parts. They clock in when they return to the job site to complete the work.

Date, Person, Time, Downtime, Action Taken, Material Used: On the standing order, the job will be done several times on different days and logged to the same document. Each time the job is done, the date, initials of the mechanic, the elapsed time of the job itself, total elapsed time the unit is down (when "Downtime Required" block is checked), and the material used is logged to the open order.

Description of Work Completed: Quick description of work to be done. A typical job might be described as: 1) look into problem, 2) get parts, 3) install parts. Use the same standard for expressing time (use hours/minutes or hours/tenth). In the string PM work order, include a quick description of the string like "Vibration route in the press department."

Task, Materials: On a string PM work order, each task is written out with the materials needed. Attach extra sheets for longer strings. Keep in mind that string PM is usually a few simple tasks on several machines.

Materials Used, Quantity: Whenever parts or materials are used, they are recorded on the work order in these columns. Include the total price for all parts used. When an item is drawn from stock, then you have to look up the price and insert it. Put in the part number where replacement parts are used.

Notes, Failure Analysis, Ideas: Frequently the mechanic fixes something or finds something not anticipated by the work requestor (a broken pump call results in replacement of the power supply module on the PLC rack). This allows the mechanic to feed back what was really found to be part of the permanent record for the asset. These notes are essential for root failure analysis.

Summary of Work Order (calculations after job is turned in)

Totals (at the bottom of the time and downtime columns): Add up all the hours and write it in the totals row. If you use a contractor, put in their chargeout rate (if known, otherwise put in the total dollars). Extend the material total and add that to the labor total for a work order total. Total up the downtime.

Rate: The charge rate is the burdened labor rate for your facility. In some cases, each level mechanic or trade has its own charge rate. The charge rate should include labor, fringes, overhead, and a factor for the ratio of payroll hours to work order hours. Charge rate is also the amount that the contractor charges you per hour. For contractor jobs where the hours are unknown, fill in total $ this work order only.

Total Charges: This row allows you to calculate the total cost of the work order. There is room to total both material columns for a total material cost. The grand sum is the cost of the work order. Add cost of labor, materials, and any contractor charges to get the total cost of the work order.

Idea for Action: Set up a file for copies of work orders for major repairs. Review all major repairs asking the question "Is there anything we could have done to avoid this repair?" Write your ideas on the copy and file it. Once or twice a year, review the file and see if there are any hidden trends or ideas that you can use to improve your operation.

24
What is PM?

PM is a series of tasks performed at a frequency dictated by the passage of time, the amount of production (cases of beer made), machine hours, mileage, or condition (differential pressure across a filter) that either:

1. extend the life of an asset (example: greasing a gearbox will extend its life), or

2. detect that an asset has had critical wear and is going to fail or break down (example: a quarterly inspection shows a small leak from a pump seal; this allows you to repair it before a catastrophic breakdown).

Additional details about PM follow in the next several chapters.

How to install and run a PM system: If you want to set up a new plant, fleet, or building on PM follow the ideas in this chapter. If there is some question in your mind about what is missing from your present PM effort, check the check sheets.

PM task list development: A good task list is half the battle. Being sure that you are doing the right tasks, at the right frequency, is essential. Also in the task list chapter is PCR (Planned Component Replacement) also called scheduled replacement. One of the tools in your pouch is PCR. A technique made popular by the airlines, PCR can improve reliability in some circumstances.

TLC (tighten, lube, clean): Also in the PM task list development chapter. Start with the basics. Caring for your equipment is the core of the PM approach. This does not require any fancy equipment or techniques, just basic care. Much of the benefit from PM flows from TLC.

Predictive maintenance: This is the application of advanced technology to detect when failures will occur. It can increase your returns and give you more time to intervene before failure.

PM activity has been proven in study after study to lower the cost of operations and improve reliability. In a 1985 article published by ASME called "Progress and Payout of a Machinery Surveillance and Diagnostic Program," the authors Hudachek and Dodd report that rotating equipment maintained under a PM model costs 30% less to maintain versus a reactive model. It further states that adding predictive technologies adds significant additional return on investment.

Common PM Tasks

Type of Task	Example
1. Inspection	Look for leaks in hydraulic system
2. Predictive maintenance	Scan all electrical connections with infrared
3. Cleaning	Remove debris from machine
4. Tightening	Tighten anchor bolts
5. Operate	Advance heat control on injection molding machine until heater activates
6. Adjustment	Adjust tension on drive belt
7. Take readings	Record readings of amperage
8. Lubrication	Add 2 drops of oil to stitcher
9. Scheduled replacement	Remove and replace pump every 5 years
10. Interview operator	Ask operator how machine is operating
11. Analysis	Perform history analysis of a type of machine

These tasks are assembled into lists and sorted by frequency of execution. Each task is marked off when it is complete. There should always be room on the bottom or side of the task list to note comments. Actionable items should be highlighted.

These tasks should be directed at how the asset will fail. The rule is: the tasks should repair the unit's *most dangerous, most expensive,* or *most likely* failure modes. *Caveat:* There will still be failures and breakdown even with the best PM systems. Your goal is to reduce the breakdowns to minuscule levels and convert the breakdowns that are left into learning experiences to improve your delivery of maintenance service.

PM systems also include the following.

1. Maintaining a record keeping system to track PM, failures, and equipment utilization. Creating an equipment baseline for other analysis activity.

2. All types of predictive activities. These include inspection, taking measurements, inspecting parts for quality, and analysis of the oil, temperature, and vibration. Recording all data from predictive activity for trend analysis.

3. Short or minor repairs are completed during the PM. This is a great boost to productivity since there is no additional travel time, set-up time, job assignment time, waiting time, or idle time. Short repairs are pure productivity.

4. Writing up any conditions that require attention (conditions which will lead or potentially lead to a failure). Write-ups of machine condition.

5. Scheduling and actually doing repairs written up by PM inspectors.

6. Using the frequency and severity of failures to refine PM task list.

7. Continual training and upgrading of inspector's skills, improvements to PM technology.

8. PM systems should contain ongoing analysis of their effectiveness, and the avoided cost of the PM services versus the actual cost of the breakdown should be periodically looked at.

9. Optionally, a PM system can be an automated tickler file for time- or event-based activity such as changing the bags in a bag house (for environmental compliance), inspecting asbestos encapsulation, etc.

One point that is commonly missed is that PM is a way station to the ultimate goal of maintainability improvement. PM can be an expensive option because it requires constant inputs of labor, materials, and downtime. The ultimate goal of maintenance is high reliability without the inputs.

Which situation below describes your organization best?

1. Your organization has a successful PM system and wants you to learn some additional ideas to make it more efficient or more effective. In this type of organization a well-written report to your boss with some concrete examples from your current operation might be all that is needed to start improvements.

2. Your organization hopes you will learn enough to put in a PM program or upgrade an ineffective one. They say they will support you. You think you can count on them to stay out of your way (at the very least). This type of organization will require more work.

3. Your organization either doesn't want to change (and says so) or they say they want to change and you're sure that they have no intention of backing you up. You might like a PM program (or to improve the existing one) and feel like you might be trying to swim upstream. You will have a difficult job that no one may ever thank you for completing. You have to make a difficult choice.

4. None of the above.

Six Misconceptions About PM

Misconception 1: PM is a way of trying to determine when and what will break or wear out so you can replace it before it does.

Reality 1: PM is much bigger than that. It is an integrated approach to budgeting, failure analysis, eliminating excessive resource use, and permanent correction of problem areas.

Misconception 2: PM systems are all the same. You can just copy the system from the manual or from your old job and it will work.

Reality 2: PM systems must be designed for the actual equipment as set-up, age of the equipment, product, type of service, hours of operation, skill of operators, and many other factors.

Misconception 3: PM is extra work on top of existing workloads and it costs more money.

Reality 3: PM increases uptime, reduces energy usage, reduces unplanned events, reduces air freight bills, etc. There are hundreds of ways the PM saves the organization resources. The only time it is in addition to the existing workload is when you put it in.

Misconception 4: With good forms and descriptions, unskilled people can do PM tasks.

Reality 4: With good forms and training, unskilled people can do some of the PM tasks successfully. For greatest return on investment, skilled people must be in the loop.

Misconception 5: PM is a series of task lists and inspection forms to be applied at specific intervals.

Reality 5: Newer PM strategies require control of the equipment for enforced downtime because they initiate activity on condition (initiate task list when temperature exceeds 20° above ambient).

Misconception 6: PM will eliminate breakdown.

Reality 6: In the words of a PM class, "PM can't put iron into a machine." In other words, the equipment must be able to do the job. PM cannot make a 5 hp motor do the work of a 10 hp motor.

One problem in factories, fleets, and buildings is that PM systems fail because *past sins* wreak havoc on anyone trying to change from a fire fighting operation to a PM operation. Even after running for a few months, there are still so many emergencies that it seems you can't make headway.

You face *unfunded maintenance liabilities.* The only way through this jungle is to pay the piper, modernize, and rebuild yourself out of the woods. This is where the investment must be made. Any sale of a PM system to top management must include a non-maintenance budget line item for past sins.

Remember: the wealth was removed from the equipment without maintenance funds being invested to keep it in top operating condition.

Costs of a PM System

One time:

Modernization of equipment to PM standard
Pay for system to store information
Data entry labor for data collection
Labor to train inspectors
Labor to set up task lists, frequencies, standards
Purchase any predictive maintenance devices with training

Ongoing:

Labor for PM task lists, short repairs
Parts costs for task lists, PCR's
Additional investments in predictive maintenance technology
Funds to carry out write-ups (maintain the higher standard of maintenance)

25
How to Install and Run a PM System

There are many details that make the PM system work. In this chapter we will look at some of the details: where to get PM task lists, how often to perform tasks, who should be involved in the PM effort, how to get the cooperation of the users, and many other essential details.

One of the great contradictions of maintenance (pointed out by John Moubray in a series of articles in *Maintenance Technology* magazine, March–June 1996) is that the more effective the PM effort, the fewer are the breakdowns. Fewer breakdowns means less data for in-depth analysis. Statistical analysis, root failure analysis, and mechanic experience in dealing with failure all suffer from a dearth of data. The conclusion: the more effective the PM system is, the greater lengths we will have to go for data for analysis. Nowhere is this more clear than in the aircraft industry where they use everything from computer models to wind tunnels to guns that shoot chickens at the engines for failure data!

Where to Get the Original PM Task List

The task list consists of the items to be done; the inspections, the adjustments, the lube route, and the readings and measurements. Sources of task lists are:

Manufacturers

Third-party published shop manuals

State law

Regulatory agencies, such as EPA and DOT

Equipment dealers

Your experience

Trade association recommendations

Skilled craftspeople experience

History, review of your records

Consultants

Engineering department

Laws

PM Frequency: How Often Do You Perform the PM Tasks?

The first source for inspection frequency is the manufacturer's manual. Ignoring it might jeopardize your warranty. The manufacturer's assumptions for how the machine is being used might be different than your usage. For example, one manual recommended a monthly inspection for a machine. When the manufacturer was questioned, it came out that the assumption was made of single-shift use. The factory used the machine around the clock and was getting excessive failures even with recommended PM frequency.

Some manufacturer's maintenance manuals are concerned with protection of the manufacturer and limiting warranty losses. Following that manufacturer's guidelines may mean you will be overinspecting and overdoing the PM needed to preserve the equipment.

Certain inspections are driven by law (EPA, OSHA, State, DOT). You have a certain amount of flexibility in the timing of these inspections. Consider scheduling them when a PM is also due. *While you have the unit under your control, you also perform the in-depth PM and any open corrective items to improve efficiency.*

Your own history and experience are excellent guides because they include factors for the service that your equipment sees, the experience of your operators, and the level and quality of your maintenance effort.

For almost any measure to be effective the PM parameter (such as cycles, days, etc.) must be driven from the unit level (unique parameter table for each unit) or from the class level (like units in like service). For example, a pick-up truck would have a very different frequency than a dump truck even though they are both trucks.

Failure Experience (Both Frequency and Severity) Feeds Back into Task List

The task list should be designed to capture information about or direct the attention of the inspector toward critical wear areas and locations. If you are inspecting an expensive component system, many inspections might go by without any reportable changes. Depending on the economics, you may want to continue to inspect in order to capture the change when it happens. Always continue to inspect life safety systems.

You have to design standards to increase and decrease the number of tasks based on the failure history. Some organizations use the standard that if they don't get a reportable item every third PM (3:1), then they are inspecting too frequently. Don Nyman uses the standard of 6:1 in *Maintenance Management*.

Task list items that are directly concerned with life safety are not included in this analysis. An OSHA mandated inspection of an overhead crane hook might have a ratio of inspections to observed deficiencies of 10,000:1 or greater. Do not include statute-driven inspections either (boilers, sprinklers, etc.).

Types of PM Clocks

The PM inspection routines are designed to detect the critical wear point and push it into the future as much as possible. Since we cannot yet see the wear directly, the goal is to find a measure that is easy to use and **is more directly proportional to wear.** Traditionally, two measures were used: utilization (cycles, tons, miles, hours), and calendar days. Other measures mentioned below are not only possible but, in some cases, more accurate.

Days: This is the most common method. The PM system is driven from a calendar. (Example: every day, grease the main bearing, every 30 days replace the filter, etc.).

Advantages: Easiest to schedule, easiest to understand, best for equipment in regular use.

Disadvantages: PM might not reflect how the unit wears out, units might run different hours and require different PM cycles (example: one compressor might run 10 hours a week and the other might run 100 hours).

Meter Readings: This is one of the most effective methods for equipment used irregularly. (Example: change the belts after the compressor runs 5000 hours.)

Advantages: Relates well to wear, is usually easy to understand.

Disadvantages: Extra step of collecting readings, hard to schedule in advance unless you can predict meter readings.

Production or Use: This is the second most common method. The PM system is initiated from usage such as PM after every 50,000 cases of beverage, or overhaul the engine every 10,000 hours or 500,000 miles. Some theme parks even use guests through turnstile.

Advantages: Utilization numbers are commonly known (how many cases we shipped today). The parameter will be well understood, should be very proportional to wear, not hard to schedule after production schedule is known, harder to predict labor requirements future month or year.

Disadvantages: Information system might not accept this type of input, extra labor to take readings or collect data.

Energy: The PM is initiated when the machine or system consumes a predetermined amount of electricity or fuel. The asset would have a meter or some other method of directly reading energy usage. This is an excellent indirect measure of the wear situation inside the device and the overall utilization of the unit. You probably are already collecting some energy data for other reasons. Energy consumption includes the variability of rough service, operator abuse, and component wear (increased friction). Used extensively on boilers, construction equipment, and marine engines.

Advantages: Very accurate measure of use in some equipment, raises consciousness about energy usage.

Disadvantages: Need to wire watt-meters or oil meters into all equipment to be monitored, hard to schedule ahead of time without a good history, extra labor to take readings or collect data.

Consumables: An example would be "add-oil." The additions to hydraulic, lubricating, or motor oil are tracked. When the added consumable exceeds a predetermined parameter, then the unit is put on the inspection list. This is a direct measure of the situation inside the engine, hydraulic system, gear train, etc. Wear and condition of seals are directly related to lube consumption.

Advantages: Will alert you if there is a leak.

Disadvantages: Very specialized, very hard to schedule in advance, hard to collect accurate data.

On-condition Measures (such as Quality): The PM in this case is generated from the inability of the asset to hold a tolerance or have consistent output. It could also be generated from an abnormal reading or measurement. For example, a low oil light on a generator might initiate a special PM.

Advantages: Responds well to customer needs.

Disadvantages: Almost impossible to schedule, cause is frequently not in the maintenance domain, response might be too late.

Four Types of Task Lists

Unit Based: This is the standard type of task list where you go down a list and complete it on one asset or unit before going on to the next unit. The mechanic would also correct the minor items with the tools and materials they carry (called short repairs).

Another variation of unit PM is Gang PM, where several people converge on the same unit at the same time. This method is widely used in utilities, refineries, and other industries with large complex equipment and with histories of single craft skilling. In a TPM-run factory, the operator is responsible for the unit PM. A mechanic might be responsible for an annual, in-depth PM.

Advantages: The mechanic gets to see the big picture, parts can be put in kits and are available from the storeroom as a unit, person learns the machine well, mechanic has ownership, travel time advantage (only requires one trip), gets into the mindset for the machine, easier to supervise than other methods. Mechanic can discuss the machine with operator as an equal partner.

Disadvantages: High training requirement, higher level mechanic needed even for the mundane part of the PM, short repairs can force you behind schedule, and if PM is not done, no one else looks at machine.

String Based: Your list is designed to PM one or a few items on many units in a string. Each machine is strung together like beads on a necklace. Lube routes and vibration routes are examples of string PM. If the units are located together it might be easier to look at one item on each unit. The inspector's efficiency would be higher since they would be focused on one activity.

Most inspection only PM's are designed this way. Almost all predictive maintenance is handled by various types of strings. Only a few computer systems support string PM and allow a charge to be spread to several assets.

Advantages: Low training requirements, lower level mechanic required, job can be engineered with specific tools and exact parts, route can be optimized, stockroom can pull parts for entire string at once, lends itself to just-in-time delivery of parts, easier to set time standards for a string, good training ground for new people to teach them the plant, allows new people to get productive quickly.

Disadvantages: Some loss of productivity with extra travel time for several visits to the same machine, don't see the big picture (the string person might ignore something wrong outside their string), boring to do the same thing over and over, no ownership, hard to supervise, if a mistake is made (such as wrong lube) it is spread to all assets on the route quickly.

Future Benefit: This type of task list takes advantage of closely coupled processes. It is commonly considered in the chemical, petroleum, and other process oriented industries. Since manufacturing is looking more and more like continuous processes then it will become more popular there also. In future benefit PM, you PM the whole train of components whenever a breakdown or changeover idles one essential unit. It is usually easier to extend downtime for a hour than it is to get a fresh hour for PM purposes.

Advantages: Little or no additional downtime, take advantage of existing downtime to PM for a future benefit, can become a contest against time, easier to manage, can be exciting.

Disadvantages: Might not have enough people, disruptive to other jobs interrupted when the call came in, cannot predict when your next PM will be done so you can plan but not schedule.

Condition-Based PM: The PM service is based on some reading, measurement going beyond a predetermined limit. If a machine cannot hold a tolerance, a boiler pressure gets too high or a low oil light goes on a PM routine is initiated. Used with statistical process control to monitor and insure quality.

Advantages: High probability that some intervention is needed, involves the operator, brings maintenance closer to production, supports quality program.

Disadvantages: Might be too late to avoid breakdown, usually high skill needed, can be planned but cannot be scheduled, many variations are not maintenance problems.

Example of a condition-based PM: Condition is low oil light on in a large truck.

Task List:

Top off oil if low

Check history file for excessive oil use or recent related work (oil filter change)

Send in sample of oil for analysis

Check for oil leaks

    Check oil temperature sender

    Examine oil cap

    Examine cylinders, cylinder seals

    Examine oil filter, oil filter seal

Check oil pressure sender

Other specific oil related checks

Cost:

1 hour: cost of materials: $4.50

In addition, since the unit is under our control, we might also perform any other levels of PM that are almost due.

Access to Equipment

One of the most difficult issues of maintenance is access to equipment (because the customer wants it or needs it). Access problems fall into two categories: political and engineering.

Political access problems are problems that stem from political reality. The equipment is not in use 24 hours, 7 days. But it is in use whenever you want it for PM. The reason you are not given access might be because production control has assigned no time for the PM, the maintenance department might be distrusted by production, etc. Here are some ideas for political access problems.

1. Go back to the planning department to discuss requirements. Do not wait until you need the unit the next day. In some cases, the production schedule might be set weeks ahead of time. Lay out your PM requirements for each asset for a year in advance including the hours of downtime.

2. Circulate PM success stories from your plant or from the trade press to everyone in production management. Keep doing it until they believe you.

3. Conduct a class in PM and breakdown with examples of broken parts and show how PM could have avoided the problem.

4. Use production and downtime reports now in circulation and highlight downtime incidents that could have been avoided by PM effort.

5. Most importantly, conduct yourself with integrity. When you *do* get a window for PM or corrective work, give equipment back when promised; show up when promised; if there is a complication, communicate with everyone before, during, and after.

Engineering access problems are easy to spot. These access issues stem from equipment that cannot be taken out of service because it is always in use. Consider transformers, environmental exhaust fans, single compressors, etc., in this category.

A partial antidote for both access problems might be in noninterruptive maintenance.

Interruptive/Non-interruptive: This is a variation on the unit based theme for machines that run 24 hours a day (or are running whenever you need to PM them). The unit based list is divided into tasks that can be done safely without interrupting the equipment (readings, vibration analysis, adding oil, etc.) and tasks that require interruption. The tasks can be done at different times. The interruptive list may require half as much downtime as the original task list.

The next step is to reengineer the machine so almost all of the tasks can be done safely without interruption.

Advantages: Same as above, with reduced machine downtime

Disadvantages: Same as above, except slightly less productive since the machine may require two trips.

Steps to Install a PM System

1. Create PM task force. This is a group that includes craftspeople (include the shop steward in union shops), a staff representative, data processing representative, and engineer.

2. Decide on the goals of the task force. Set objectives.

3. Pick a catchy name for the effort, like PIE (profit improvement effort), DEEP (downtime elimination and education program, QIP (quality improvement program). Stay away from "PM" since it has negative connotations for many people.

4. Get training in computers for members of task force if they are not computer literate. Include typing training. Get them access to computers and able to use word processors, spreadsheets, E-mail, and any relevant organizational level networks or systems.

5. Get generalized maintenance management training for the entire task force. This will save time and effort by laying groundwork so that they can share a language and create a new vision of maintenance.

6. Identify the maintenance stakeholders (anyone impacted by how maintenance is conducted). Analyze the needs and concerns of the maintenance stakeholders. Look at each group and see how they contribute to the success of the organization. Include production, administration, accounting, office workers, tenants, housekeeping, legal, risk management, warehousing, distribution, clients, etc. At least look at how your proposed changes will benefit each group.

7. Inventory and tag all equipment to be considered for PM. Compile and review your list of equipment. Compile a list of all of the assets (or units) that you are responsible for. This list is a starting point for the PM program. Inquire if lists exist in plant engineering or accounting.

8. Select a system to store information about equipment, select forms for PM generated MWO and check-off sheets.

9. Design first drafts of the measures or benchmarks to be used to evaluate the PM system's performance. These measures will be revised as the process goes on.

10. Draft SOP (standard operating procedures) for the PM system. This document will also be revised many times over the first year.

11. Have task force members or shop personnel complete data entry or preparation of equipment record cards. Rotate job so that many (everyone?) in the department has experience with the system before you go on line.

12. Consider using contractors to replace the hours lost on the floor by the people doing the data entry. It is essential to build a critical mass of expertise in the system.

13. Another essential is daily audits of all data typed into the system. Have someone highly skilled review all data going into the system. The best way is to take the list and crawl around and physically verify all of the information and name plate data.

14. Select people to be inspectors. Allow their input into the next steps. Consider using inspectors to help set up system. Consider choosing them for steps 1, 2, 7, 11, 13, and others.

15. Get key personnel training in RCM (reliability centered maintenance) and failure analysis. This will help them and the program immeasurably. The training will show them how to root out useless tasks and include important tasks on hidden functions.

16. Determine which units will be under PM and which units will be left to breakdown (BNF, or bust and fix). Remember that there is a real cost associated with including any item in the PM program. If, for example, you spend time on PM's for inappropriate equipment, you will not have time for the essential equipment.

Cost to Include in PM Program:

Cost of Inclusion = Cost per PM × Number of PM services per Year.

To decide which units to include in the PM system, apply the following rules to each item.

A. Would failure endanger the health or safety of employees, the public or the environment?

B. Is the inspection required by law, insurance companies, or your own risk managers?

C. Is the equipment critical?

D. Would failure stop production, distribution of products, or complete use of the facility?

E. Is it the link between two critical processes?

F. Is it a necessary sensor, measuring device, or safety protection component?

G. Is the equipment one of a kind?

H. Is the capital investment high?

I. Is there spare equipment available?

J. Can the load be easily shifted to other units, or work groups?

K. Does the normal life expectancy of the equipment without PM exceed the operating needs? If this is true, PM may be a waste of money.

L. Is the cost of PM greater than the costs of breakdown and downtime? Is the cost to get to (to view or to measure) the critical parts prohibitively expensive?

M. Is the equipment in such bad shape that PM wouldn't

help? Would it pay to retire or rebuild the equipment instead of PM?

17. Schedule modernization on units requiring it. Investigate retiring bad units if possible. A bad unit or asset left on the system will demoralize the most dedicated inspectors.

18. Select which PM clocks you will use (days, utilization, energy, add-oil). A clock is designed to reflect wear on an asset. Clocks on items in regular use or subject to weather are usually expressed in days. An irregularly used asset might be better tracked by usage hours or output tons of steel, cases of cola, etc. Some equipment such as construction equipment is best tracked by gallons of diesel fuel consumed because the hour meters are frequently broken.

19. Decide what Predictive Maintenance technology you will incorporate. Train inspectors in techniques. Even better, provide the information and a budget to the task force and let them pick the technology. Most equipment should be rented before buying. Inexpensive training is available from most vendors and distributors.

20. Set up task lists for different levels of PM. Factor in your specific operating conditions, skill levels, operators experience, etc. Consider all of the strategies including unit based, string, route maintenance, future benefit, as well as non-interruptive/interruptive. Publicize your successes.

21. Categorize the PM tasks by source (recommended by Ron Moore, of RM Group). Categories might include regulatory, calibration, manufacturer's warranty, experience, insurance company, quality, etc. This will be a great aid when you look back to see which ones to eliminate or change.

22. Provide the PM inspector with the following to perform the tasks:

  A. Actual task list (usually a work order) with space for readings, reports, observations

B. Drawings, performance specifications, pictures where appropriate
C. Access to unit history files, trouble reports
D. Equipment manual
E. Standard tools and materials for short repairs
F. Consider a cart designed for the PM's and common short repairs
G. Any specialized tools or gauges to perform inspection
H. Standardized PM parts kits
I. Forms to write up longer jobs
J. Log-type sheets to log short repairs

23. Assign work standards to the task lists for scheduling purposes. Observe some jobs to get an idea of timing. Let some mechanics time themselves and challenge them to reengineer the asset to cut PM time.

24. Engineer all the tasks. Challenge yourself to simplify, speed up, eliminate, combine tasks. Improve tooling and ergonomics of each task. Always look toward enhancing the worker's ability to do short repairs after the PM is complete.

25. Determine frequencies for the task lists based on clocks chosen. Select parameters for the different task lists.

26. Implement system, load schedule, and balance hours. Extend schedule for 52 weeks. Balance to actual crew availability. Schedule December and August lightly or not at all. Allow catch-up times.

Staffing the PM Effort

"A successful PM program is staffed with sufficient numbers of people whose analytical abilities far exceed those of the typical maintenance mechanic" (from August Kallmeyer, *Maintenance Management*). We want high skill and knowledge people with positive attitudes because they will be able to detect potentially damaging conditions before they actually damage the unit. Your best mechanic is not necessarily your best PM inspector.

Six Attributes of a Great PM Inspector

1. Can work alone without close supervision. The inspector has to be personally reliable since it is hard to verify that work was done.

2. Interested in the new predictive maintenance technology; should be trained in techniques of analysis and in the use of these modern inspection tools.

3. Will know how to (and want to) review the unit history and the class history to see specific problems for that unit and class. Also, the type of person who will complete the paperwork.

4. A mechanic is *re*-active in style; PM inspector is *pro*-active. In other words, the inspector must be able to act on a prediction rather than *re*act to a situation. He/she is primarily a diagnostician, not necessarily a "fixer."

5. Because of the nature of the critical wear point, the more competent the inspector, the earlier deficiency will be detected—allowing more time to plan, order materials, and helping to prevent core damage.

6. PM inspectors should not be interrupted, and should ideally be segregated (while in the PM role) from the rest of the maintenance crew. PM inspection hours should represent 10–20% of the whole crew.

How to Ensure the PM's are Done as Designed

One of the toughest problems to solve is how to ensure that the inspector is actually doing the inspection on the task list. Horror stories about maintenance catastrophies frequently feature task lists that were signed as completed but obviously not performed. For most people, PM tasks are boring and mind numbing. The challenge of leadership is to inspire the people in PM roles to want to do the tasks well. The inspector mentioned below (or for that matter mentioned anywhere in this section) can be a regular mechanic, operator, or helper (if appropriate) on a part-time basis, or a full-time PM technician.

1. Does the inspector know that PM impacts reliability, safety, costs, and output? Inspectors in nuclear plants or in airlines know full well the impact of missing a PM (and even then it happens).

2. Have your top management address maintenance crews about the criticality of PM. You might have to write the speech. People attend to what they think management thinks is important. Let them hear it from the boss's mouth.

3. Present the job as important. If people feel that PM is stupid, boring, and low-priority fill-in work, they are less likely to put themselves out.

4. Let your PM mechanics design the system and tasks themselves. Have them trained in reliability, TPM, and general maintenance management, and then let go of the reins.

5. One problem is lack of specific skills. An individual might be lacking a specific item of skill or knowledge to effectively perform the task. Be explicitly sure the PM people are fully trained. A test for PM certification might be appropriate.

6. Improve the relationship between the mechanic and the maintenance user. Where there is an operator such as a driver, machine operator, or a building contact person, instruct the mechanic to make personal contact. Some PM task lists include a task "talk to operator and determine if equipment, building, truck, etc., has operated normally since the last visit."

7. Make it easy to do tasks. Reengineer equipment to simplify the tasks and route the people to minimize travel.

8. Simplify paperwork.

9. Improve accountability by mounting a sign-in sheet inside the door to the equipment. Be sure the people who do the tasks sign a form and are included in discussions about the equipment. When people know they might be quizzed about an asset, they are more likely to complete their PM tasks. When people know that after a breakdown an inquiry is conducted and the PM sheets are reviewed, it motivates them to complete their tasks.

10. Make PM a game. One supervisor got a small amount of money and went to the local fast food restaurant and bought 50¢ gift certificates. Each week he hid eight 3×5 cards (that said "see me") inside equipment on the PM list. He traded the cards for the certificates. He knew when a card wasn't found (PM wasn't done). His comment was "What people would do for 50¢ they wouldn't do for $17.50 per hour!"

11. PM professionals like new, better toys (sorry—better *tools*, not toys). Technology has opened up the field for sophisticated, relatively low-cost PM tools. They might include $700 for a pen-sized vibration monitor, $500 for a cigarette pack-sized infrared scanner, or $1500 for an ultrasonic detection headset and transducer. If appropriate to the size and type of equipment, these tools will motivate the troops and increase the probability that they will detect deterioration before failure.

12. In any repetitive job, boredom sets in. Consider job rotation, reassignment, project work, and office work (like planning, design, analysis) to improve morale.

Implicit verses Explicit tasks

Task lists designed for use by long term and fully trained maintenance professionals can be skeletal, without filling in every detail. An HVAC mechanic could reasonably expect to understand a task like 'fire unit and check for proper controls operation.' These tasks are implicit. They depend on, perhaps, years of experience and relevant training.

Explicit tasks are completely spelled out. They are designed for less skilled maintenance workers or non-maintenance workers doing maintenance work. Explicit tasks should include pictures, descriptions, where to look for advice, whom to call, when to bring in maintenance, etc. As you can imagine, explicit task lists are much more expensive to develop. An explicit list given to a skilled maintenance professional might cause some resentment, while an implicit list given to a novice will be ineffective for the overall PM effort.

It is importany that your task design account for who will do the tasks and whether you have control of testing and training the inspectors.

26
PM Task List Development

The task list is the heart of the PM system. It reminds the inspector what to do, what to use, what to look for, how to do it, and when to do it. In its highest form it represents the accumulated knowledge of the manufacturer, skilled mechanics, and engineers—for the avoidance of failure. Task lists are designed by manufacturers, skilled mechanics, engineers, contractors, insurance companies, governmental agencies, trade associations, equipment distributors, consultants, and sometimes by large customers.

All task list items are designed to perform one of two functions. The two functions are the core of all PM thought, and they are: either extend the life of an asset or detect when the asset has begun its descent into breakdown (before it actually breaks). It is also the assumption of the design of PM tasks that when a problem is detected during inspect, scan, take readings, etc., the maintenance system will respond with a corrective action.

Activities you might find on a task list include the following.

Life Extension	Detection
Clean	Inspect
Empty	Scan
Tighten	Smell for...
Secure	Take readings
Component replacement	Measure
Lubricate	Take sample for analysis
Refill	Look at parts
Top-off	Operate
Perform short repair	Jog
	Review history
	Write-up deficiency
	Interview operator

231

Trash Compactor		Date:		Surveyer:
Site Name:		Address:		
Make and Model:		Location:		
Skill	Fre-quency	Task		
2	M	1. Lock out trash compactor while working		
2	M	2. Check motor bearings for noise, vibration or heating		
2	M	3. Clean motor and pump		
1	W	4. Clean compactor		
1	M	5. Check level of hyd oil in reservoir		
1	M	6. Clean vent breather on reservoir		
2	M	7. Check pump seals for leakage		
2	M	8. Check hydraulic hoses condition and leakage		
1	W	9. Check inside bin and chute for obstructions		
1	W	10. Secure any loose guards		
1	M	11. Check condition of compactor, floor mounting, bolting		
2	M	12. Check condition of electrical connections		
2	W	13. Check operation of electric eye		
2	M	14. Check operation of safety control and equipment		
1	W	15. Check for fire danger		
1	A	16. Check condition of paint		
2	M	17. Check operation of pump and compactor		
2	W	18. Check rams for free operation		
		19.		
		20.		
Comments:				

Tasks are designed to prevent three types of failures: dangerous failures injurious to the public, employees, or to the environment; expensive failures, including downtime and large breakdowns; and frequent failures that happen continually and are disruptive to the work environment.

In RCM (Reliability Centered Maintenance) breakdowns are divided into three levels or grades by the consequence of the breakdown. In fact, to paraphrase John Moubray in his *RCM II* book, the problem is not failure at all, it is the consequence of failure.

1. Breakdowns where the consequences are loss of life or environmental contamination, such as a boiler safety valve or the rupture of a tank of volatile chemicals.

2. Failures where the consequences are operational downtime, such as loss of cooling water to a data center or the breakage of the chain in an auto assembly line.

3. Failures where the consequences are repair costs, such as the breakdown of one of several milling machines in a machine shop.

Each task (line item) should be considered carefully before inclusion because inclusion creates a cost for the long term. The quick way to evaluate task economics is to relate the task cost per year to the avoided cost of the breakdown. In category 2 or 3, economic analysis is the way to determine if a task should be included. In category 1, the task must be included or the asset re-engineered to remove the threat of failure.

For categories 2 and 3:

Breakdown costs:
(probability of failure in 1 year) × (cost of downtime + cost of repair) <

PM Costs:
(cost per task × # services per year) + (new probability of failure in 1 year × (cost of downtime + cost of repair))

Example:

Breakdown Costs: Compressor breaks down once every 2 years (50% probability each year) under the existing bust'n'fix plan. It costs $15,000 each time to repair and causes $60,000 of downtime.

$$0.5 \times (\$60,000 + \$15,000) = \$37,500 \text{ per year}$$

PM Costs: Service compressor 12 times per year at a cost of $1000 per service (parts, labor, downtime) plus a bi-annual scheduled overhaul costing $16,000. New probability of failure with PM: 1 failure in 20 years = .05

$$((\$1000 \times 12) + (\tfrac{1}{2} \times \$16,000)) + (.05(\$60,000 + \$15,000)) = \$23,750$$
which is clearly less than the cost of the breakdown mode.

Developing Task Lists

Most people in maintenance enjoy imagining what tasks would be appropriate for an asset. They enjoy using their experience and knowledge for this kind of problem. They might or might not use the manufacturer's manual. The problem is that there is no linkage between individual tasks and the failure modes that they are to address.

Other people take the manufacturer's task list as an absolute given (which it is while you are under their warranty). There are several observations about manufacturers' task lists.

1. Some manufacturers have tremendous knowledge about the failure modes of their equipment based on deep analysis of thousands of units under all types of conditions. You can certainly start with these lists since they have significant brainpower invested. Certainly the lists from the large automotive or HVAC companies would fall into this category. Even these lists can be fine tuned.

2. Profit drives the task lists of some manufacturers. They want you to over-PM their asset so that they avoid warranty claims. These are the same manufacturers whose recommended spare parts list includes parts you are likely never to use.

3. In most cases ignorance is the biggest issue. Many small machine manufacturers and some large ones do not use their own equipment. Their engineers might know about the design issues of a pump but they never fixed them or worked with them in service. Actually, a big user of the equipment gets to know far more about the equipment than does the manufacturer. Since you see the equipment every day, you get to fix it, you get to be stuck with the results of your actions, you learn what it takes to keep the equipment running. In this case, your knowledge is far more valid.

4. The last issue is that you might use equipment in an unusual service. The manufacturer might be very conscientious, like members of group 1 above. You are using their equipment "outside the envelope." You might be using it more hours per day, higher capacities, for different materials, connected to another asset, or under unique controls. I'm reminded of a pick-up truck being used to run a sawmill. It was chopped up and welded into the machine. The truck maker could not predict this type of service, and consequently you could not rely on their list.

Some issues to consider in picking tasks and assembling task lists:

1. Complete description of task.
2. Drawings to show how task is done.
3. Are there lock-out tag out or confined space entry or possible spillage or release of gases?
4. Specifications and recommendations about task.
5. Type of task list (unit, string, future, etc.).
6. Skill level required.
7. Is a special license needed?
8. Is there a legal liability issue?
9. Can this task be done by the operator or can the task be in-sourced elsewhere?
10. Is a contractor a better choice for this task?
11. What component is being worked on.
12. Will doing this task impact any other task (such as changing oil impacts topping off)?
13. What failure mode is being addressed and how?
14. What is the value of the failure?
15. Idea of the time between detection and failure if this is an inspection.
16. Number of components that this task is addressed to.
17. Planned frequency.

18. Why this frequency?
19. What clock is best for this type of task (days, utilization, energy, condition, other)?
20. How long will the task take if you are already at the unit with the tools on a cart?
21. Special tools required.
22. Is this task seasonal?
23. Parts needed.
24. Value of parts needed.
25. Is this an interruptive or non-interruptive task?
26. Will others outside of maintenance have to be notified?
27. What is the total cost of this task, what is the yearly cost?
28. How does the yearly cost of the task compare to the cost of the failure mode avoided?

TLC (Tighten, Lubricate, Clean)

TLC—tender loving care—means tighten, lubricate, clean. Keeping equipment trim and clean will extend the life and reduce the level of unscheduled interruptions. This approach or strategy is appropriate for all maintenance departments, even those with no support from top management or maintenance customers. *TLC is the simplest way to reduce breakdowns.*

The climate seems to be against TLC. As firms experience downsizing and destaffing, one of the first services to go is TLC. When we read the latest trade journals and listen to the latest papers at conferences, we hear and read that time-based (or interval-based) PM is obsolete. At a recent American Institute of Plant Engineers annual meeting, there were 35 papers or sessions presented, and none of them spoke about improved TLC. Yet studies find again and again that dirt, looseness, and lack of proper lubrication cause the bulk of the equipment failures. TLC is the core of TPM's increased reliability. The following are examples from *TPM Development Program*, by Nakajima.

- One company found that 60% of its breakdowns were traceable back to faulty bolting (missing fasteners, loose, or misapplied bolts).
- Another examined all of its bolts and nuts and found 1091 out of 2273 were loose, missing, or otherwise defective.

- The JIPE (Japanese Institute of Plant Engineering) commissioned a study that showed 53% of failures in equipment could be traced back to dirt, contamination, or bolting problems.
- Other costs can be impacted by effective TLC. One firm reduced electric usage by 5% through effective lubrication control.

Cleaning

Dirty equipment creates a negative attitude that adversely impacts overall care. Inspectors cannot see problems developing, and mechanics don't want to work with the equipment. Dirt can increase friction and heat, contaminate product, cause looseness from excessive wear, degrade the physical environment, cause potentially lethal electrical faults, contaminate whole processes (as in clean rooms), and demoralize the operator.

Cleaning is a hands-on activity. Someone who cleans a machine with their eyes open will see all sorts of minor problems and ask themselves questions about how the equipment works and why it is designed the way it is. They will also increase their respect for the machine. This process of cleaning, seeing, touching, and respecting the machine is essential to increase reliability. As a result of the questions and observations made by people cleaning, the operation and maintenance of the machine can be improved.

Part of the cleaning process is looking for ways to make cleaning easier, or for maintenance avoidance. Perhaps the source of dirt should be isolated to reduce the need for cleaning. In other cases, the machine should be moved or rotated to facilitate access. The book *TPM Development* lists seven steps to a cleaning program:

1. cleaning main body of machine, checking and tightening bolts
2. cleaning ancillary equipment, checking, and tightening bolts
3. cleaning lubrication areas before performing lubrication
4. cleaning around equipment
5. treating the causes of dirt, dust, leaks, contamination
6. improving assess to hard-to-reach areas
7. developing cleaning standards.

Keep Area Clean

Keeping it clean is not only a PM issue. Cleanliness is important in rebuilds, major repairs, and even small repairs. Any mechanic in the business for a length of time can remember a perfect repair gone bad because of dirt.

With all of the attention being paid to dirt and cleaning, one would imagine organizations would take extra steps to exclude dirt when they do major repairs. How many professional maintenance organizations take control of the physical environment with work tents, plastic drapes, or other measures to exclude dirt and contamination.

Even keeping the maintenance shop clean should be a goal of the maintenance program. Issue a periodic work order to clean up the shop. Also look at eliminating the sources of dirt and clutter such as misplaced trash containers, lack of proper storage, broken tools, bad ventilation, inadequate lighting, too small benches.

Bolting

Loose or missing bolts are a major source of breakdown. Even a single missing or loose bolt might cause a failure. In most cases, the looseness contributes to vibration which increases looseness. In electrical joints connected by the pressure of a bolt, looseness is usually the result of thermal expansion and contraction. The space that looseness creates promotes oxidation which increases resistance which expands and contracts the joint which causes more looseness. In other words, loose bolts beget loose bolts.

The misapplication of one nut cost an air charter company $9,600,000. In 1990 there was a plane crash in the Grand Canyon. The nut holding the propeller on a small tour plane came loose causing the propeller to fly off. The jury awarded $9,600,000 for negligence in maintenance practices. The tour operator had a $10,000,000 deductible insurance policy. If a main nut holding a propeller can be missed, what is the chance that you have nuts working their way loose right now as you read this section!

The people cleaning the machines have the best chance of detecting this failure. As they touch and look at the machine, loose bolts should shout to them. The easiest technique is to scribe a line on the nut and the machine frame when the nut is tightened correctly. This scribed line will stay intact (a single line) as long as the nut doesn't move.

When equipment is engineered, the rules of good bolting should have been followed. Much of the process of maintenance is correcting mistakes or deviations from good engineering practices. Many rules concern the size, pattern, torque, and type of fastener. Other rules include head location (nut is accessible), use of lock washers, use of flat washers, and bolt length. In most facilities there are no well-known standards for tightness for task lists, with tasks like check base bolts and tighten if loose.

Bringing equipment to specification is sometimes a lengthy job. In a fleet, vehicles are brought in after 1000 miles to tighten everything up. This short run-in period gives the bolts a chance to seat. This same strategy is not well followed after factory rebuilds or when doing work in buildings.

Good bolting practice takes a while to teach and is not necessarily intuitive.

Lubrication

Lubrication is the "Rodney Dangerfield" of the maintenance field—it gets no respect. It is assumed by people peripherally associated with maintenance that anyone who can find a zerk fitting and squeeze a handle can be a lubricator. Maintenance experts know that tribology is a field that you can get a Ph.D. in. They also know that a good person in the lubricator's role can save a plant, building, or fleet thousands of dollars in breakdowns and potentially millions in downtime and accident prevention.

Failure to lubricate results from several factors. A leading factor is poorly designed or installed equipment where the points are too hard to get to, or there are just too many points. Other factors include too many different lubricants used, not enough time allowed, lack of standards, and a lack of motivation of the worker. The lack of motivation can usually be traced back to a lack of knowledge to know how important lubrication is to reliability, poor self-esteem resulting from the job being a bottom-of-the-barrel type job, and a lack of training and feedback of how the job was done.

Entry level operators take weeks to learn basic lubrication. We assume that journeymen mechanics are experts in lubrication. Frequently they know only what they've seen and tried. This might be only a small subset of the possibilities, and might also be wrong.

A very expensive lubrication mistake almost caused millions of dollars worth of damage on the drawbridges that cross the St. Lawrence Seaway. These drawbridges were activated by two cables which rode on 35′ diameter pulleys. These pulleys were mounted on steel shafts. Partial cracks were found in the shafts. The engineers determined that the cracks were caused by corrosion. A review of the PM work orders for the last 20 years showed that the lubrication was being done at the specified frequency with the correct lubricant.

A tribologist was brought in to review the whole application. He found that the original drawings and specifications called for a lubricant that was inappropriate for a marine environment. The problem took 35 years to manifest itself. Ask yourself this: if an engineer and the people who checked the drawings made a mistake about the functional qualities of a lubricant, what is the probability that the lubricants you've been using are still the best ones today?

One issue is that many plants use too many different lubricants. In some cases you can standardize on the "better" product and save money through larger buys. The cost of the lubricant itself is usually the smallest element of the whole picture. If changes are made, document them and their reasons. In most facilities the lubricants were chosen a long time ago, and the reasoning is lost in time.

For lubrication to be successful, the people involved need to understand why they are doing the lubrication, how to do it, where to do it, and with what. Drawings, charts, diagrams, and annotated photographs are useful in the process. The lubricator must also understand the implications of overlubrication.

One of the biggest areas where cleaning and lubrication overlap is in the cleaning and examination of the lubrication points. Clogged, dirty, or broken lubrication fittings compromise the whole effort. Initial cleaning should highlight these issues and correct them.

Questions (partially adapted from *TPM Development Program*):

1. Are lubricant containers always capped?
2. Are the same containers used for the same lubricants every time; are they properly labeled?
3. Is the lubrication storage area clean?
4. Are adequate stocks maintained?
5. Is the stock area adequate in size, lighting, handling equipment for the amount stored?
6. Is there an excellent long-term relationship with the lubricant vendor?
7. Does the vendor's salesforce know enough about tribology to solve problems, and do they periodically tour the facility and make suggestions?
8. Is there an adequate specification for frequency and amount of lubricant?
9. Are there *pictures* on all equipment to show how, with what, and where to lubricate and clean?
10. Are all zerk fittings, cups, and reservoirs filled, clean, and in good working order?
11. Are all automated lubrication systems in good working order right now?
12. Are all automated lubrication systems on PM task lists for cleaning, refilling, inspection?
13. Do you have evidence that the lubrication frequency and quantity is correct as specified (oil film on moving parts, freedom from excess lubricant)?
14. Is oil analysis used where appropriate?

Automated Lubrication Equipment

One way to improve the lubrication program is the judicious use of automation. There has been significant improvement in the reliability of auto lube systems. These systems can now inexpensively be retrofitted to existing equipment on a one- or multiple-point basis. They provide a level of repeatability and reliability unmatched by most manual systems.

The biggest mistake in the use of automated equipment is that organizations forget to add the automated lubrication equipment to the PM task list. These systems have to filled, inspected, repaired, and cared for.

In most major industrial centers, service companies have been established to do your inspections for a fee. These firms use the latest technology and have highly skilled inspectors. Some of these firms also sell hardware with training. *One good method is to try some service companies and settle on one to do inspections, help you choose equipment, and do training.*

In *Maintenance Management,* by Jay Butler, the advantages of automated lubrication are listed and include reduction in the number of people needed to perform the lubrication, improvement in the amount of lubrication dispensed, reduction in the amount of contamination, insurance against missed cycles due to sickness or reassignment, reduction in the number of interruptions to the equipment. The end result is lower downtime, reduced breakdowns, and reduced cost of operation.

In the transportation field, lubrication is critical. A seized "s" cam or slack adjuster in a trailer axle can fail either actuated or unactuated. When it hangs in the actuated position, the driver can lose control of the rig causing jack knifing and a potential accident. Since the early 1980's, Lubriquip has been providing single-point (semi-automatic) systems. They pipe every lube point to a central location. The mechanic uses the grease gun at the central point. This semi-automated mode saves 25 minutes per trailer per month . Other savings include reduced contamination, reduced missed points, and savings in lubricant. The system costs about $250 per trailer. The system will report if a point is clogged and will count the number of lubrication cycles.

To properly incorporate automated systems, several issues have to be faced. The first is that the automated equipment will create the need for maintenance effort for checking, filling, and repair. Other procedures, worked out over years, will also have to be radically changed to take advantage of the new equipment. Since we will not be in certain areas as often, we will have to enlist other groups such as operators, housekeepers, and security to keep their eyes and ears open.

Planned Component Replacement

PCR is an option on the PM task list. The novelty of this option is the elimination of failure because *components are re-*

moved and replaced after so many hours or cycles but before failure. Depending on the substrategy, some of the components are then returned for inspection, rebuilding, remanufacturing, and others are discarded. The result of this strategy is controlled maintenance costs and low downtime. The strategy does *not* work when the new component experiences high initial "burn-in" type failures.

For example, fleets with time-sensitive loads realized that breakdown costs with downtime are sufficiently high to justify PCR. It is standard procedure in some fleets to replace hoses, tires, belts, filters, and some hard components well before failure on a scheduled basis.

PCR is an expensive option. Even in the aircraft industry, significant effort has gone into improving reliability so that fewer components would be in the periodic rebuild program. According to John Moubray in *RCM II*, after an extensive RCM analysis the number of overhaul items (planned rebuild items) went from 339 on the Douglas DC 8 to just 7 items on the larger and more complex DC 10. While the number has dropped dramatically, PCR is still an important tool to the maintenance professional. PCR is divided into two substrategies called planned discard (where you discard the component like belts) and planned rebuild (for rebuildable components like truck engines).

Planned discard is where a component is removed before failure and discarded. Common examples would include belts, filters, small bearings, inexpensive wear parts, etc. One fleet replaces hoses every two years during its major rebuild cycle to reduce the number of unscheduled hose failures.

Planned rebuild is for major components that are rebuildable such as engines, transmissions, gear boxes, pumps, compressors, etc. Components on aircraft are the best examples of this strategy. The items are removed after a fixed number of operating hours or take-off/landing cycles. They are sent to a certified rebuilder and brought back to specification and returned to the stock to be put on another aircraft.

Since the component is replaced before failure on a scheduled basis, PCR offers the following advantages (partially adapted from Butler's Maintenance Management text).

1. The component doesn't fail. Some of the possibility of core damage is eliminated on planned rebuild parts. The value of the core is preserved. The core is the rebuildable item such as the alternator, pump, etc.

2. Replacement is scheduled so that you can avoid downtime and replace the component when the unit is not needed and reduce overtime (from emergency repairs of breakdowns).

3. Tools can be made available on a scheduled basis to reduce conflicts and reduce costs.

4. Manufacturer's revisions, enhancements, and improvements can be incorporated more easily.

5. Rebuilds in controlled environments by specialists are always better than the same rebuilds "on the floor" by general mechanics.

6. Since it is scheduled, the rebuild can be used for training of newer technicians.

7. The PCR activity is great training for newer or second tier mechanics. Since all work is done on operating equipment, the mechanic gets to see how the equipment should look.

8. Spare components can be made available on a scheduled basis which can minimize inventory (rather than waiting for breakdowns which are known to clump together).

9. Since the component is replaced, breakdowns become infrequent, availability goes up, and the atmosphere becomes more regular.

10. In a successful PCR plan and to maximize the return from the investments, one assumption is that management will take the time to look at any failures that do occur and seek ways to avoid failures of this type in the future. Some options that can be looked into are better quality lubricants, better skill in repairs, design review, OEM specification changes.

How this can be applied: By combining the techniques of failure analysis with the concept of PCR, the manager can choose the most economical situation.

Case Study Comparing Breakdown with PCR Costs

Tom Duvane has hired your consulting company to evaluate several methods of providing certain maintenance services to a produce hauler. They currently use P&M Truck Leasing for full-service leasing on their power units. They are interested in having the Springfield Central Garage provide maintenance services for their refers (refrigerated trailers). Tom feels that the outside work will improve his productivity and provide profit to the company.

Facts for the Case:

50 TK Refer units mounted on Great Dane Refer Vans

Utilization 2500 hours per year per refer

Belt failure rate mean 575 hours, SD 175 hr. (Mean minus 1 SD = 400 hrs)

Failure rate 2500 hours/ 575 hours per failure = 4.35 per unit or 218 failures for fleet/ yr.

Cost per non-scheduled (emergency) failure $285.00

Cost per Scheduled replacement $85.00

Administrative cost per repair incident of any type $20.00

1. Cost for breakdown mode:

failures ×	(cost per failure + cost of admin.)	= total cost of breakdown program
218 ×	($285.00 + 20.00)	= $66,490

2. Cost for PM using PCR: Use 400 hours (mean minus 1 SD from example) to pick up 84.9% of failures:

(Utilization/PCR Interval) × units=PCR Incidents   Failures × 15.9% = Emergency Incidents

(2500 hrs / 400 hrs) × 50 = 312.5 use 313   218 × 15.9% = 34.66 use 35

PCR incidents ×	(cost per incident + admin. cost)	= PCR cost
313 ×	($85.00 + $20.00)	= $32,865
emergency incidents ×	(emergency costs per incident + admin.)	= emergency cost
35 ×	($285.00 + $20.00) =	$10,675

PCR costs +	emergency costs =	total cost of PCR program
$32,865 +	$10,675 =	$43,530

Cost Breakdown: $66,490.00
Cost PCR: $43,530.00

In this case, the PCR alternative saves over $20,000.00. The mathematics of PCR can be tricky, and the analysis can be time consuming. At least one firm offers software help. Their package, RELCODE, helps the maintenance professional by digesting the MTBF and spitting out PCR frequencies. The vendor is Oliver Marketing in Montreal, Canada.

27
Predictive Maintenance

The ideal situation in maintenance is to be able to peer inside your components and replace them right before they fail. Technology has been improving significantly in this area. Tools are available that can predict corrosion failure on a transformer, examine and videotape boiler tubes, or detect a bearing failure weeks before it happens.

"Scientific application of proven predictive techniques increases equipment reliability and decreases the costs of unexpected failures." Predictive maintenance is a maintenance activity geared to indicating where a piece of equipment is on the critical wear curve, and predicting its useful life. Any inspection activity on the PM task list is predictive.

The way predictive maintenance improves reliability is to detect deterioration earlier than it could be detected by manual means. This earlier detection gives the maintenance people more time to intervene. With the longer lead time there is less likelihood of an unscheduled event catching you unawares.

In the appendix to *RCM II*, John Moubray lists over 50 techniques for predictive maintenance. Every year some smart scientist, engineer, or maintenance professional comes up with one or two more. Any technique could be the one that will really help to detect your modes of failure.

Condition-based maintenance is related to predictive maintenance. In condition-based maintenance the equipment is inspected, and based on a condition further work or inspections are done. For example, in a traditional PM program, a filter might be scheduled for changeout monthly. In condition-based maintenance, the filter is changed when the differential pressure (taken before and after the filter) reading exceeds certain readings.

All of the predictive techniques we are going to discuss should be on a task list and controlled by the PM system. In this chapter we will use the common trade press definition of predictive maintenance, which generally refers to technologically driven inspections, in other words, inspections for which you have to buy something (an instrument, scanner, etc.) to perform. Predictive inspections can also come from existing equipment used in new ways, including volt/ohm meters, meggers, measuring instruments, etc.

Maintenance has borrowed tools from other fields such as medicine, chemistry, physics, auto racing, aerospace, and others. These advanced techniques include all types of oil analysis, ferragraphy, chemical analysis, infrared temperature scanning, magna-flux, vibration analysis, motor testing, ultrasonic imaging, ultrasonic thickness gauging, shock pulse meters, and advanced visual inspection. Most metropolitan locations have service companies to perform these services or rental companies to try some techniques in your facility.

Other instruments not discussed here, but which should be considered part of your predictive maintenance tool box, are meggers, pyrometers, VOM meters, strain gauges, temperature sensitive tapes, and chalk.

Almost all techniques depend on baselines to be most useful. The baseline is the readings when the asset is operating normally with no significant critical wear going on. In older facilities getting the baseline is a significant problem. In many fields (such as air handlers, or motors) baseline data can be obtained from the manufacturer.

Many techniques and instruments can also be useful in a predictive way because predictive maintenance is an attitude, not a technology! Before you start a predictive maintenance program, consider the following questions.

1. What is our objective for a predictive maintenance program? Do we want to reduce downtime, maintenance costs or the stock level in storerooms? What is the most important objective?

2. Are we, as an organization, ready for predictive maintenance?
   A. Do we have piles of data that we already don't have time to look at?

B. If one of the PM mechanics comes to us asking for a machine to be rebuilt, do we have time to rebuild a machine that is not already broken?
C. Could we get downtime on a critical machine on the basis that it might break down?
D. Are we willing to invest significant time and money to training? Do we have the patience to wait out a long learning curve?

3. Is (are) the specific technique(s) the right technique(s)?
   A. Does the return justify the extra expense?
   B. Do you have existing information systems to handle, store, and act on the reports?
   C. Is it easy and convenient to integrate the predictive activity and information flow with the rest of the PM system?
   D. Is there a less costly technique to get the same information?
   E. Will the technique minimize interference to our users?
   F. Exactly what critical wear are we trying to locate?

4. Is this the right vendor?
   A. Will they train you and your staff?
   B. Do they have an existing relationship with your organization?
   C. Is the equivalent equipment available elsewhere?
   D. In the case of a service company, are they accurate?
   E. How do their prices compare with the value received, to the marketplace?
   F. Can the vendor provide rental equipment (to try before you buy), can they provide a turnkey service giving you reports and hotline service for urgent problems?

5. Is there any other way to handle this instead of purchase?
   A. Can we rent the equipment?
   B. Can we use an outside vendor for the service?

An excellent treatment of the whole field of predictive maintenance can be found in John Moubray's book *RCMII*, published by Industrial Press. Some material in this chapter is partially adapted from that book. The technologies are grouped around detecting deterioration in the six effects: dynamic (vibration), particle (ferragraphy), chemical (water analysis), physical (crack detection), temperature (infrared), electrical (ampere monitoring).

Chemical Analysis

One of the most popular family of techniques to predict current internal condition and impending failures is chemical analysis. There are 7 basic types of chemical analysis. The first two are related to particle size and composition:

	Type	Material
1.	Atomic Emission (AE) spectrometry	all materials
2.	Atomic Absorption (AA) spectrometry	all materials
3.	Gas chromatography	gases emitted by faults
4.	Liquid chromatography	lubricant degradation
5.	Infrared spectroscopy	similar to AE
6.	Fluorescence spectroscopy	assessment of oxidation products
7.	Thin layer activation	uses radioactivity to measure wear

Oil analysis is a significant subset of all of the chemical analysis that is used for maintenance. The two spectrographic techniques are commonly used to look at the whole oil picture. They report all metals and contamination. This is based on the fact that different materials give off different characteristic spectra when burned. The results are expressed in PPT or PPM (PPT—parts per thousand, PPM—parts per million, PPB—parts per billion).

The lab or oil vendor usually has baseline data for types of equipment that it frequently analyzes. The concept is to track trace chemicals and elements over time and determine where in the machine they come from. At a particular level, experience will dictate an intervention is required. Oil analysis costs $10 to $25 per analysis. It is frequently included at no charge (or low charge) from your supplier of oil.

You are usually given a computer-printed report with a reading of all the materials in the oil and the "normal" readings for those materials. In some cases the lab might call the results in so that you can finish a unit, or capture a unit before more damage is done.

For example, if silicon is found in the oil then a breach has occurred between the outside and the lubricating systems (silicon contamination usually comes from sand and dirt). Another example would be an increase from 4 PPT to 6 PPT for bronze which probably indicates increasing normal bearing wear. This would be tracked and could be noted and checked on the regular inspections.

Oil analysis includes an analysis of the suspended or dissolved non-oil materials including Babbitt, Chromium, Copper, Iron, Lead, Tin, Aluminum, Cadmium, Molybdenum, Nickel, Silicon, Silver, and Titanium. In addition to these materials, the analysis will show contamination from acids, dirt/sand, bacteria, fuel, water, plastic, and even leather.

The other aspect of oil analysis is a view of the oil itself. Questions answered include: has the oil broken down? what is the viscosity? are the additives for corrosion protection or cleaning still active? Consider oil analysis a part of your normal PM cycle. Since oil analysis is relatively inexpensive, also consider doing it:

1. following any overload or unusual stress
2. if sabotage is suspected
3. just before purchasing a used unit
4. after a bulk delivery of lubricant to determine quality, specification, and if bacteria is present
5. following a rebuild, to baseline the new equipment, and for quality assurance
6. after service with severe weather such as flood, hurricane, or sandstorm.

Other tests are carried out on power transformer oil which show the condition of the dielectric, and breakdown products.

The first place to begin looking at oil analysis is from your lubricant vendor. If your local distributor is not aware of any programs, contact any of the major oil companies. If you are a very large user of oils and are shopping for a yearly requirement, you might ask for analysis as part of the service. Some vendors will give analysis services to their larger customers at little or no cost.

Labs that are unaffiliated with oil companies exist in most major cities, especially cities that serve as manufacturing or transportation centers. Look for a vendor with hotline service who will call or fax you in case of an imminent breakdown. These firms will prepare a printout of all of the attributes of your hydraulic, engine, cutting oils, or power transmission lubricants. The firm should be able to help you set sampling intervals and train your people in proper techniques of taking the samples.

Tip: Send samples taken at the same time on the same unit to several oil analysis labs. See who agrees, who's the fastest, who has the least cost. Pick a lab that maintains your data on computer, and be sure you can get the data on diskette or by modem for analysis.

Wear Particle Analysis, Ferragraphy, and Chip Detection

Particle techniques	Description
1. Ferragraphy	20–100 microns, ferrous only
2. Chip detection	40 microns up, metals only
3. X-ray fluorescence	after radiation materials that emit characteristc x-rays
4. Blot testing	blot highlights size and type of particles
5. Light detection and ranging	analyze smoke from smoke stacks

These techniques examine the wear particles to see what properties they have. Many of the particles in oil are not wear particles. Wear particle analysis separates the wear particles out and trends them. When the trend shows abnormal wear, then ferragraphy (microscopic examination of wear particles) is initiated.

Several factors contribute to the usefulness of these techniques. When wear surfaces rub against each other they generate particles. This rubbing creates benign particles that are small (under 10 microns) and round (like grains of beach sand). Abnormal wear creates large particles that are irregularly shaped with sharp edges. All particles generated are divided by size into two groups: small <10 microns, and large particles >10 microns.

When abnormal wear occurs the large particle count dramatically increases. This is the first indication of abnormal wear. After abnormal wear is detected the particles are examined (ferragraphy) for metallurgy, type, and shape. These contribute to the analysis of what is wearing and how much life is left.

The most obvious chip detection technology is a magnetic plug in the sump of an engine. You examine the plug to see if dangerous amounts of chips are in the oil. Chip detection is a pass-fail method of large particle analysis. Too many large particles set off an alarm. Several vendors market different types of detectors. One type allows the oil to flow past a low power electrical matrix of fine wires. A large particle will touch two wires and complete the circuit to set off the alarm.

Vibration Analysis

This is a widely used method in plant/machinery maintenance. A recent study in the city of Houston's waste water treatment department showed $3.50 return on investment for every $1.00 spent on vibration monitoring. The same study showed that a private company might get as much as $5.00 return per dollar spent. The study and the vibration monitoring project were done by the engineering firm of Turner, Collie and Braden of Houston, TX.

Each element of a rotating asset vibrates at characteristic frequencies. A bent shaft will always peak at twice the frequency of the rotation speed. A ball bearing, on the other hand, might vibrate at 20 times the frequency of rotation.

There are over nine different types of vibration analysis. Each individual technique focuses on one aspect of the way assets deteriorate that is detectable by vibration. Techniques include octave band analysis, narrow band frequency analysis, real time analysis, proximity analysis, shock pulse monitoring, kurtosis, acoustic emission, and others.

The most popular is broad band analysis. This analysis measures the changes in amplitude of the vibration by frequency over time. This amplitude by frequency is plotted on an XY-axis chart for a given service load and is called a signature. Changes to the vibration signature of a unit means that one of the rotating elements has changed characteristics. These elements include all rotating parts such as shafts, bearings, motors, power transmission components. Also included are anchors, resonating structures, and indirectly connected equipment.

Many large engines, turbines, and other large equipment have vibration transducers built in. The vibration information is fed to the control system which can shut down the unit or set off an alarm from vibration that exceeds predetermined limits. The system also has computer outputs. This allows transfer of the real-time data to the maintenance information system.

Frequency	Cause	Amplitude	Phase	Comments
½ rotational speed	Oil whip or oil whirl	often very severe	erratic	high speed machines where pressurized bearings are used
$1 \times$ RPM	unbalance	proportional to imbalance	single reference mark	check loose mounting if in vertical direction
$2 \times$ RPM	looseness	erratic	two marks	usually high in vertical direction
$2 \times$ RPM	bent shaft misalignment	large in axial direction	two marks	use dial indicator for positive diagnosis
1,2,3,4 $\times$ RPM	bad drive belts	erratic	1,2,3,4 unsteady	use strobe light to freeze faulty belt
synchronous or $2 \times$ synchronous	electrical	usually low	single or double rotating mark	if vibration drops out as soon as electricity is turned off it is electrical
Many $\times$ RPM	bad bearing	erratic	many reference marks	If amplitude exceeds .25 mils suspect bearings
RPM $\times$ number of gear teeth	gear noise	usually low	many	
RPM $\times$ number of blades on fan or pump	aerodynamic or hydraulic			rare

courtesy of August Kallemeyer's *Maintenance Management*

Quick setup of a vibration monitoring program:

1. Buy or rent a portable vibration meter.

2. Train mechanics in its use, make it a regular task on a task list assigned to the same person.

3. Record readings at frequent intervals. Transfer readings to a chart (or use a spreadsheet program and have it do the charting for you).

4. Take readings after installation of new equipment.

5. Compare periodic readings and review charts to help predict repairs.

6. Do repairs when indicated, do not defer. Note condition of all rotating elements, determine what caused the increase in vibration.

7. Before and after you overhaul a unit, review all vibration readings.

8. As you build a file of success stories move into more sophisticated full spectrum analysis. Train more widely and trust your conclusions.

Temperature Measurement

Since the beginning of the Industrial Age, temperature sensing has been an important issue. Friction (or electrical resistance) creates heat. Temperature is the single greatest enemy for lubrication oils and for the power transmission components. Advanced technologies in detection, imaging, and chemistry allow us to use temperature as a diagnostic tool.

Today, there is technology to photograph by heat rather than reflected light. Hotter parts show up as redder (or darker). Changes in heat will graphically display problem areas where wear is taking place or where there is excessive resistance in an electrical circuit. Infrared is unique since it is almost entirely non-interruptive. Most inspections can be safely completed from 10 or more feet away and out of danger.

Readings are taken as part of the PM routine and tracked over time. Failure shows up as a change in temperature. Temperature detection can be achieved by infrared scanning (video technology), still film, pyrometers, thermocouple, fiber loop thermometry, other transducers, and heat sensitive tapes and chalks.

On large engines, air handlers, boilers, turbines, etc., temperature transducers are included for all major bearings. Some packages include shut-down circuits and alarms if temperature gets above certain limits.

The hardware for infrared is becoming more and more powerful. A typical specification for a top of the line, handheld imager with full accurate absolute temperature measurement capabilities might weigh 5 lbs., with 4× zoom, color viewfinder, self-calibration, 30,000 hr MTBF, thermoelectric cooling, snap-in battery pack, output include to VCR, RS-232, inputs include voice notations of each video image, 500 image storage (and in 1997 it costs $50,000).

Harry Devlin, an Agema representative and infrared survey engineer, explains that the extra money spent gets you high accuracy, where the actual temperature is needed to detect the severity of the problem, and repeatability where you can count on the reading. He says that the high-end equipment is most appropriate for large facilities where there is a need to prioritize the findings (using the temperature gradient to identify immediate and high priority problems) or for industries where calibration and repeatability is an issue (such as nuclear power generation).

An infrared imager only (no temperature reading capability, it can only detect degrees above ambient) might drop the cost below $25,000. An infrared gun that can take spot temperatures (without imaging capability) would set you back around $500–$1500.

Possible uses for infrared inspection	Look for
Bearings	overheating
Boilers	wall deterioration
Cutting tool	sharpness
Die casting/injection molding equipment	temperature distribution
Distribution panels	overheating
Dust atmospheres (coal, sawdust)	spontaneous combustion indications
Furnace tubes	heating patterns
Heat exchanger	proper operation
Kilns and furnaces	refractory breakdown
Motors	hot bearings
Paper processing	uneven drying
Piping	locating underground leaks
Polluted waters	sources of dumping in rivers
Power transmission equipment	bad connections

Power factor capacitors	overheating
Presses	mechanical wear
Steam lines	clogs or leaks
Switchgear, breakers	loose or corroded connections
Three phase equipment	unbalanced load
Thermal sealing, welding, induction heating equipment	even heating

Examples of areas where savings are possible from application of infrared:

- Hot spot on transformer was detected. Repair was scheduled off-shift when the load was not needed, avoiding costly and disruptive downtime.

- A percentage of new steam traps which remove air or condensate from steam lines will clog or fail in the first year. Non-functioning steam traps can be readily detected and corrected during inspection scans of the steam distribution system. Breakdowns in insulation and small pipe/joint leaks can also be detected during these inspections.

- Hot bearings were isolated in a production line before deterioration had taken place. Replacement was not necessary. Repairs to relieve the condition were scheduled without downtime.

- Roofs with water under the membrane retain heat after the sun goes down. A scan of a leaking roof will show the extent of the pool of water. Sometimes a small repair will secure the roof and extend its life.

- Infrared is an excellent tool for energy conservation. Small leaks, breaches in insulation, defects in structure are apparent when a building is scanned. The best time to scan a building is during extremes of temperatures (greatest variance between the inside temperature and the outside temperature).

- Furnaces are excellent places to apply infrared because of the cost involved in creating the heat and the cost of keeping in place. Unnecessary heat losses from breaches to insulation can be easily detected by periodic scans. Instant pictures are available to detect changes to refractory that could be precursors to wall failures.

- Temperature measurement in electrical closets is one area where experience can take the place of baselines. For example, in high voltage distribution if the legs vary by $40°C$ or more then immediate action is required. If the legs vary by $10–15°C$, then correcting actions can come at the next scheduled shutdown.

Ultrasonic Inspection

One of the most exciting types of technologies is based on ultrasonics. It is widely used in medicine, and has moved to factory inspection and maintenance. There are four or five techniques that make up this family.

In one of the most common and inexpensive techniques, an ultrasonic transducer transmits high-frequency sound waves and picks up the echo (pulse-echo). Echoes are caused by changes in the density of the material tested. The echo is timed and the processor of the scanner converts the pulses to useful information such as density changes and distance.

Ultrasonics can determine the thickness of paint, metal, piping, corrosion, and almost any homogeneous material. New thickness gauges (using continuous transmission techniques) will show both a digital thickness and a time-based scope trace. The trace will identify corrosion or erosion with a broken trace showing the full thickness and an irregular back wall. A multiple echo trace shows any internal pits, voids, and occlusions (which cause the multiple echoes).

Another excellent application of ultrasonic inspection is Bandag's casing analyzer. It is used in truck tire retreading. Ultrasonics is used to detect invisible problems in the casings that could result in failures and blowouts after retreading. Since ultrasonic waves bounce around from changes in density, imperfections in casings (like holes, cord damage, cuts) are immediately obvious. The transmitter is located inside the casing and 16 ultrasonic pickups feed into a monitor. The monitor immediately alerts the operator of flaws in the casing.

A different application of ultrasonics is in the area of ultrasonic detection. Many flows, leaks, bearing noise, air infiltration, and mechanical systems give off ultrasonic sound waves. These waves are highly directional. Portable detectors worn like stereo headphones translate high-frequency sound into sound we can hear and can quickly locate the source of these noises and increase the efficiency of the diagnosis.

Some organizations enhance this application with ultrasonic generators. You insert the generator inside a closed system such as refrigeration piping or vacuum chamber, and listen all around for the ultrasonic noise. The noise denotes a leak, loose fitting, or other escape route.

Advanced Visual Techniques

The first applications of advanced visual technology used fiber optics in bore scopes. In fiber optics, fibers of highly pure glass are bundled together. The smallest fiber optic instruments have diameters of .9 mm (.035"). Some of the instruments can articulate to see the walls of a boiler tube. The focus on some of the advanced models is 1/3" to infinity. The limitation of fiber optics is length. The longest is about 12' (getting longer as the technology improves). The advantages are cost (about 50% or less of equivalent video technology) and difficulty to support (they don't require large amounts of training to support).

Another visual technology gaining acceptance is ultra-small video cameras. These are used for inspection of the interior of large equipment, boiler tubes, and pipelines. These CCD (Charge Coupled Display) devices can be attached to a color monitor through cables (some models used on pipelines can go to 1000'). It uses a miniature television camera smaller than a pencil (about 1/4" in diameter and 1" long) with a built-in light source. Some models allow small tools to be manipulated at the end, others can snake around obstacles. It is extensively used to inspect pipes and boiler tubes.

The major disadvantages are cost (currently $10,000 to $20,000) and level of support (they require training to adjust and use). The major advantage lies in flexibility. You can replace the heads or cables and end up with several scopes for the price of one.

In most major industrial centers, service companies have been established to do your inspections for a fee. These firms use the latest technology and have highly skilled inspectors. Some of these firms also sell hardware with training. One good method is to try some service companies and settle on one to do inspections, help you choose equipment, and do training. You can also rent most of the equipment.

Other related visual equipment include rigid bore scopes, cold light rigid probes (up to 1.5 m), deep probe endoscope (up to 20 m), pan-view fiberscopes.

Other Methods of Predictive Maintenance
Magnetic Particle Techniques (called Eddy Current Testing or Magna-flux)

Magna-flux is borrowed from racing and racing engine rebuilding and has begun to be used in industry. This technique induces very high currents into a steel part (frequently used in the automotive field on crank and cam shafts). While the current is being applied the part is washed by fine, dark-colored magnetic particles (there are both dry and wet systems).

The test shows cracks that are too small to ordinarily be seen by the naked eye and cracks that end below the surface of the material. Magnetic fields change around cracks and the particles outline the areas. The test was originally used when rebuilding racing engines (to avoid putting a cracked crank shaft back into the engine). The high cost of parts and failure can frequently justify the test. The OEM's who built the cranks and cams also use the test as part of their quality assurance process.

Penetrating Dye Testing

Penetrating dye testing is visually similar to magna-flux. The dye gets drawn into cracks in welded, machined, or fabricated parts. The process was developed to inspect welds. The penetrating dye is drawn into cracks by capillary action. Only cracks that come to the surface are highlighted by this method.

28
Projects that Put You in the Driver's Seat

Before any system of improvements (including both computerized and manual systems) can be installed, an infrastructure must be built. This chapter is designed to show what structures and preplanning should be done, and how and why to do each project.

Maintenance Technical Library (MTL)

Start with the library because all of the steps that follow will help build the library. The MTL will also give you a place to put the data you collect.

Effective maintenance is increasingly dependant on having access to specific information at a specific time. Imagine servicing a complex system without drawings, program listings, I/O assignments, wiring diagrams, or timing charts. Detailed technical information is just the start of the contents of the MTL. Much of what goes wrong in maintenance is the result of the lack of information or incorrect information.

The MTL is a place where maintenance technicians can have access to repair histories, equipment manuals, PM lists, economic analysis sheets, parts lists, and assembly drawings. The MTL should have a large reference table, shelving for books and catalogs, drawing files, and legal-sized file cabinets.

In addition, the MTL should be the location of a copy of plant drawings, site drawings, vendor catalogs, handbooks, engineering textbooks, etc. If you have computerized maintenance, stores, manufacturing, or purchasing, CADD, CAM, CIM, then a terminal (workstation) could be located in the MTL.

261

Considerations

- Protect from fire, flood, theft (consider fireproof file cabinets).
- Use some kind of sign-out system if material must be removed.
- Make it someone's responsibility to keep it up to date (yearly update meeting?).
- Manage the revisions so that all copies are updated (coordinate with ISO 9000).

When the MTL is set up you will have: a ready reference for make-versus-buy decisions, repair history, repair parts reference with history, repair methods referral, planning information source, time standard development, data bank for continuous improvement efforts, maintenance improvement team headquarters.

The MTL would also house information for research on parts including specifications, past research, past vendors, repair-versus-rebuild-versus-buy decisions, part cross reference, etc. One element of the MTL is the asset file containing all repairs.

Repair History

The information in the asset jacket, book, or folder can be kept on the computer. Some organizations also keep the folder in paper form because of all of the little sketches and as-built doodles. The physical location is less important than the ease with which you can get to the data. The file could include:

1. Survey sheet
2. Equipment record card or contents of the equipment master file from the CMMS
3. Printouts of all work orders (actual working copies with notes if possible)
4. Copy of all engineering data and drawings relating to unit
5. Warranty data
6. Wiring diagrams, as-built drawings, commissioning notes, documented modifications
7. Planned work, shutdown work lists, past shutdown lists
8. PM list

9. PM justification (why should the unit be on PM)
10. Planning guides for various repairs, feedback on results of repairs
11. An ongoing wish list of things you would like to change about the asset
12. Justification for each item on the PM list based on history, experience and economics
13. Copies of any economic models completed on asset or class of asset
14. Copy of recap sheets for that asset
15. Bill of material—parts list
16. This might consist of several physical files

Short Comments on the MTL (partially adapted from *Maintenance Management*, by Jay Butler)

- The tendency of maintenance people is to give the MTL away. Do not lose control of the MTL to engineering, plant engineering, stores, purchasing, or anyone else.
- Make sure the little scraps of paper and notes done by the technicians are filed into the asset jackets—at least.
- Do not allow materials to leave the MTL without a sign-out, 3×5 card, or some method of tracking where copies are.
- Consider a Rolodex card system (or a database) for each asset showing the files, information, manuals, and books available.
- Document management systems (and all of the ancillary support hardware such as scanners, CD ROM, giant disk drives, fast processors, etc.) are becoming more reasonably priced—consider this a possible application.

Maintenance Scrap Book

Keep copies of magazine articles about maintenance and maintenance people. People that are not sophisticated about maintenance realities may be swayed by what has happened in other companies that made it into the newspapers. In one article (mentioned earlier in the book), a misapplied nut cost a company $10 million. This could be impressive for a manager who was on the fence about a proposed improvement in a PM system.

Tools

One of the five elements of maintenance planning is tools. Tools are a significant motivator/de-motivator for maintenance professionals. It is important, in the words of Jay Butler in *Maintenance Management*, to create an attitude of value. He goes on to enumerate some of the steps you can take.

1. Inventory all tools.

2. Indicate the purpose of the major tools (which asset or plant capability they support).

3. Indicate which technician is checked out to use the tool accurately and safely. This is particularly important in an ISO 900X environment with test or measuring gear.

4. Chart and display value of each tool.

5. Provide safe and secure storage.

6. Provide for cleaning, service, and calibration.

7. Create clear policy on personal tools, borrowing tools for home use, insurance for personal tools kept on premises.

8. Forms are available from Snap-on and Proto.

9. Provide a budget for tool upgrades.

Technology is infiltrating every aspect of maintenance. Remstar International (Westbrook, ME) is advertising a card access tool crib. To obtain a tool, the mechanic swipes his/her card through the reader and keys in the tool requested. The crib indexes to the bin of the selected tool and releases the requested tool. To return the tool, just reverse the process. This gives you the security advantages of an attended crib without the extra labor expense.

Building, Equipment, and Vehicle Inventory

One of the first elements of the MTL is the complete list of maintainable units. It is very difficult to get a handle on maintenance activity without a complete list. The maintainable unit inventory procedure takes two passes.

On the first pass through the list, create a catalog of the following:

- all buildings—location, size, usage, cost account numbers, date in service

- all systems within buildings—description
- all stationary equipment—description, manufacturer, model, serial number, location, account #,
- all mobile units—manufacturer, model, description, year, serial number, license
- all of your major tools, compressors

On the second pass:

- add details of subcomponents, ultimately bill of material for asset
- identify special skills required to support units
- insert approximate replacement, current market values
- date in service, warranty data
- vendor, service vendor, service contract information

After a complete list is compiled and checked by your internal people, consider publishing the listing by department, cost center, or area of responsibility. Give copies of the list to the responsible party in each area to check and correct. One maintenance manager added an incentive—they would not maintain any asset not on the list!

Component Numbering

The issue of component numbering rears its head in any effort to catalog or categorize maintainable units. There are many good ways to think about numbering which depend on your specific operating conditions. When equipment is frequently moved, rebuilt, and returned to service (in other asset types), or frequently reconfigured, the numbering system has to be particularly flexible.

Jay Butler, in his seminar text *Maintenance Management,* recommends that components be numbered by the level in the assembly tree they occur. He thinks that a hierarchy from the complete asset to the individual parts or components is the most useful. For a detailed explanation of this strategy, contact the ATA for a copy of the VMRS for vehicle maintenance where this issue has been worked out.

XXX-YYY-ZZZ-PPP

XXX: asset category such as air handler, tractor, elevator

YYY: system or large component such as transmission, door assembly or fan assembly

ZZZ: component such as a starter in the cranking system, motor in a refrigerator compressor

PPP: down to an individual part level such as a linkage pin on a manual shifter

Where an asset number like XXX would be the category of asset like compressor, and the systems would add YYY for the systems within the asset such as motor, then ZZZ would be added for the assemblies such as the front bearing assembly. In some cases, the parts for the asset would be concatenated onto the end of this number.

Repair acts use the same numbering system with the work accomplished instead of the part number. The advantage of this approach comes in data analysis. When the computer sorts repairs, then all repairs on the same system and component would come out together. Analysis is then expedited.

Physical Inventory

Physical inventory is the first step to control usage of and expenditures on parts. Part inventory is a major expense of maintenance and lack of efficient handling of parts is the single major contributor to maintenance inefficiency.

Procedure:

1. List all of the parts.
2. Include quantities and last costs.
3. Identify where each part is used.
4. Segregate any parts that are for assets that are no longer owned, used, or maintained. Investigate ways of converting these parts into cash or usable inventory.
5. Segregate large expensive parts that have been in inventory for a number of years where you have an excessive quantity in relation to your use. Investigate ways of converting these parts into usable inventory. *Caution:* If the part is unusually difficult to get, or downtime on that equipment is very expensive, *keep the part.* Consult the inventory chapter for specific rules about inventory evaluation.

Paperwork Study

We are indebted to George Gross of J. George Gross Assoc. and Abe Fineman of ICC for this procedure. All organizations still use paper to control some aspect of maintenance or initiate their repair activities. Over the years these systems become more and more complex and have documents added and seemingly never removed. This is an excellent study to undertake for newer maintenance managers to help learn their operation in detail. The actual execution of the study, when managed by the maintenance manager, is ideal as a senior project for business students. The purposes of this study are:

1. learn all of the paperwork activity
2. investigate what forms can be eliminated
3. determine which forms can be consolidated
4. streamline paperwork waiting time
5. plug any holes in the control of your operation.

Work Rules

Make a list (or get a copy of the bargaining agreement) of the current work rules. Include start and ending times, make ready, clean-up times, and breaks. Are you multi-skilled or craft skilled? Observe if craft roles are rigidly defined or loosely defined.

Are there any unique work assignments, crewing requirements, extra safety rules, special work rules? Examples would include clean rooms, level 2-3-4 virus labs, top secret facilities, explosive environments, radioactive areas, etc. Are there informal rules (such as all people coming back to the shop for lunch)? Are there special rules for workers in certain categories such as union stewards, light duty people, etc.?

Each rule that has an impact on your ability to deliver service should be noted. The importance of this can easily be overlooked. In a cheese processor they had the bulk of their maintenance crew on day shift, with skeleton crews on 2nd and 3rd. Many problems occurred when they tried to start up the lines after sanitation on 3rd shift. If a line didn't have time to fill, then production would run out of cheese to pack and shipments were lost. If the situation was really bad, production people would be sent home. The usual response was to call someone in early, put up with their protests, and pay the overtime. At the monthly meeting management would complain about the overtime.

Their solution was simple, they shifted half of the day time crew to start at 4 and 5 am (and work 8 standard time hours). By having a full crew working early, the lines were up and running when production came to work and overtime was reduced. If several repairs were still underway at 1 pm, then overtime could be authorized and was usually easily accommodated.

Cost Reduction Effort

Is there money to cut from the maintenance effort? The best maintenance operations look very much like production facilities, with thought being put on reducing travel, getting information, obtaining parts, availability of tools, etc. Cutting costs with intelligence is a goal of the management of maintenance.

While cost reduction is not the true goal, it is often the stated goal of management changes. Many maintenance departments have been subject to the unthinking reductions (without regard to the asset being supported) from management upheaval, reengineering and wholesale downsizing. These changes have affected and will continue to affect maintenance effectiveness, pride in a job well done, quality of life, and safety. As the leaders and experts in what is important in maintenance, we are in the best position to cut costs.

The cost cutting effort is an evolutionary continuous improvement activity. Each element of maintenance is viewed with an eye toward reducing the resources consumed. There are several places to look for reductions.

The rule for cost reduction (as well as crime investigation and mysteries) is to follow the money. Cost reduction should look at all money line items sorted by amount. Follow consultant Ed Feldman's acronym TIDY, which stands for *T*ime *I*nvested should equal *D*ollar *Y*ield!

Following up on TIDY check out the project in the purchasing section to list all the SKU's by dollar yield. Use the highest dollar yield parts to run your cost reduction effort.

One place to look for cost reductions is the chain of events that leads up to a job being completed. As you build the chain look for the value added by each step. Look particularly hard where there is a handoff of the work requested to another function. Of all the functions performed, only fixing the problem and telling the customer the job is complete adds any value to the customer. Follow the flowcharts developed in the paperwork exercise to see what activity actually added value for the customer and what can be eliminated.

1. Any repetitive maintenance efforts such as PM, start-up are subject to revised procedures, reduced frequencies, or job engineering to reduce time or materials.

2. Reduce wasted effort such as walking to storeroom (build a cart with the parts inside), put the parts in a locked cabinet next to the machine.

3. Equip and train an emergency squad. These responders are given interruptible jobs and are the ones interrupted. Your regular crews are not interrupted with every little breakdown.

4. Make sure the handoff to the mechanic is accurate, quick, and complete. How can you improve it? What about better technology?

5. Cut travel by assigning people into geographical zones. Dispatch by fax, or cellular phone fax. In a big facility designate home areas, set up sub-storerooms.

Disaster Planning

It is the responsibility of the maintenance department in all but the largest organizations to plan for catastrophe and disaster. There are several dimensions to this planning effort. Some organizations use their PM systems to develop special events lists for different types of disasters. It goes without saying that paper copies should be kept in case the disaster takes out your electricity or computer.

The federal government has excellent resources to help your planning process. The Federal Emergency Management Agency (FEMA) is the primary responder to disaster in the U.S. They have materials for individual organizations in different parts of the country.

Disaster planning is a logical process that you go through and document before the disaster so that during the disaster you maximize the things that you can do (much of what will ultimately happen is out of your hands).

The first step is to identify the likely disasters for your part of the country and topography. In the mid-1990s we saw floods in the midwest and northwest; earthquakes, mud slides and forest fires, in California; hurricanes in Florida and on the Gulf coast; hundred-year snows in the northeast; and airplane crashes and train crashes in various locations. Once the likely disasters are identified, the planning process can begin for each disaster.

The planning process starts with the lists available from FEMA and your state, province, or local governments. In addition to this generic information, a team should walk around and survey the site in view of the catastrophe considered. The last step is to get a knowledgeable group from operations, maintenance, engineering, and financial management together and brainstorm possibilities for the disaster being considered.

The information developed from government agencies, the walk-around, and the brainstorming session should be edited and then divided into work group responsibilities. The list of items for each work group should be estimated, and tasks could be traded off between work groups until the estimated time is reasonable given the crewing and the notice given for that type of disaster.

The last step is to schedule a drill and time out each work group's task list. Also each work group would list problems to see if they can be addressed in the planning document (such as locations of critical tools, utility shut-offs, critical phone numbers, etc.). More likely events should have follow-up drills annually.

If disaster is part of your business, then the protocols are usually well worked out and practiced. A large airport in the northeast has several potential airplane emergencies a day (when a pilot calls in with a problem, such as no light, indicating that the landing gear is locked down). The airport goes through a well-oiled routine where they call the fire companies, hospitals, police, etc., and everyone goes to a well-defined alert status. The fire company actually mounts a limited response to these calls four or five times a day. Internally, personnel are moved to forward positions, and equipment is prepared for a response.

In the same airport, a hurricane response is less worked out (keeping in mind that this is not Florida but the northeast) and much less tested. Their list, which is generated by the CMMS, has things like:

1. Check and clear all airfield, roof, and pavement drains.

2. Check all sump pumps.

3. Fuel all vehicles, fuel all emergency equipment.

4. Charge all radio batteries and check all units.

5. Relocate emergency electric generators from the tool room to connection points.

6. Test all emergency generators and top-off fuel tanks.

7. Move trash storage containers inside.

8. Review apron areas and remove any small, loose equipment that could cause damage.

9. Provide auxiliary fuel tanks in case electricity goes out (fill gravity tanks).

10. Prepare for glass damage.

11. Execute storm manning plan.

12. Prepare emergency contact list.

13. Deploy wet vac units, prepare for water management inside the building.

14. Secure computer, make back-ups, and remove data to safety.

The maintenance department is responsible for ensuring the capacity of the organization to provide its product or service. Disasters are part of the risks that the maintenance department plans for and puts into action when needed. Be sure your department is never caught without an effective plan.

Summary

After these steps are complete, you will have the basis to re-engineer or improve your current maintenance function.

1. *Standard operating procedure book (SOP).* The paper flow diagrams will show how information already flows through your organization. The new organization charts show both users and resources. The work rules section has all of the information about the definition of a work day.

2. *Complete system sizing and scope.* The paperwork analysis can be designed to yield the number of work orders per month, the asset list will show the number of units under maintenance control, and the organizational plan will show numbers of mechanics. The physical inventory will determine the number of line items that the computerized inventory system will have to store, process, and cope.

3. If you are planning to computerize, then these steps will define the masterfiles for your organization. All of the work so far can directly or indirectly be used in computerization. If you are in the purchase cycle for a system, then a summary of these steps will make an excellent system specification.

4. You have started people thinking about maintenance in new ways. Cost cutting programs are in place, information for new topics at contract negotiations are noted, and the scope of the effort is better defined.

5. You have completed a disaster plan that will give your organization the best protection given the resources and time available.

29
Use of Statistics in Maintenance

Statistics are a powerful set of tools that can help improve the way maintenance is delivered to the customer. While there are some advanced statistical concepts that are beyond the scope of this book, there also are some useful ideas that can be put to immediate use.

The simplest idea is the mean. The mean is also called the sample mean and is shown as $\bar{X}$. For our purposes the mean is the same as the average. To calculate the mean, add up all of the readings (the data) and simply divide by the number of readings.

It is valuable to evaluate a number of readings or measurements and compare them over time. The mean can be used to determine maintenance function effectiveness in a wide variety of ways, as follows.

Mean hours to respond (average response) to an emergency	1.5
Mean work order hours per person per week	37.2
Mean backlog hours per person	67
Mean hours between stoppages on the giant press	790
Mean hours to repair when the giant press does break	12

The mean is tied to and used with the standard deviation (SD). The SD measures the variability of the measurements. For example, the mean of the three readings (1, 10, 250) is 87. The mean of a second group of three readings (79, 89, 93) is also 87. As you can see, the mean doesn't express a sense of the variability of the readings. The SD of the first distribution is 115 while the SD of the second is 5.9.

In any of the 5 statistical measurements above, the SD is increasingly important as the variability of the readings gets larger.

273

If the backlog by craft or skill set has a mean of 67 hours per person but it also has an SD of 52, then we must look deeper into the number. If the SD was 10 then the mean would be all we would need.

How to calculate the standard deviation (SD)

1. Subtract each reading from $\overline{X}$: Difference = $\overline{X}$ – Reading
2. Multiply each difference by itself (difference)^2
3. Add the difference for each reading and divide by the number of readings Variance = Σ (D)/# readings

(readings-1 is the weighted and just # of readings is not weighted)

4. Take the Square root of the Variance SD = $\sqrt{}$(Variance)

What is the normal curve, and what aspects are important to us? There are hundreds of curves or distributions in use to represent events in the world. The most useful one that is easy to understand is the normal distribution. The normal distribution curve can help us identify the "bad actors" in our buildings, equipment, and fleet, among our mechanics, and of the parts used.

The normal (or bell-shaped) curve is a graphic representation of a large number of similar readings. The readings tend to be more frequent the closer you get to the sample mean (the $\overline{X}$). This sample mean is also the average reading. One of the several useful properties of the normal curve is that it is symmetrical on both sides of the $\overline{X}$.

The readings can be maintenance hours per ton of steel, hours per automobile, miles per gallon, flat rate hours completed per week, or any numerical measure. In the real world, when you plot readings from any measure that is a normal distribution, your data will only approximate the normal curve. Usually a greater number of readings will smooth the curve and make it more bell shaped.

You can divide a normal distribution into partitions that are extremely useful to help you analyze a problem. The size of the partition is called one standard deviation (SD). The useful property of the SD is that 68.27% of your readings will be within 1 SD of $\bar{X}$ (that is, $\bar{X} \pm SD$) and 95.45% of your readings will be within 2 SD of $\bar{X}$ ($\bar{X} \pm 2SD$).

How can the mathematics of the normal curve help manage our assets? The normal distribution curve is related to the important and commonly known management rule known as the *80/20 rule*. This rule states that 80% of your problems come from 20% of your units, people, parts, etc. We call these 20% the "bad actors." Our job as maintenance managers is to identify and manage the "bad actors."

In terms of the bell-shaped curve, the "bad actors" group we are referring to is the one more than 1 SD on the weak side of the mean. This group represents 15.9% of the whole population. If we manage the weak 15.9%, we will be managing the whole group. Also bringing them up to average results in the greatest impact per unit of work.

The following is an example using the normal distribution—fuel consumption.

Actual MPG Readings	Mean	SD	Mean - SD = lower boundary (LB)	Vehicles with fuel consumption below the LB
5.63, 5.62, 5.57, 5.13, 5.12, 5.11, 5.11, 4.93, 4.93, 4.86, 4.80, 4.55, 4.40, 4.39, 4.36, 4.20, 4.00, *3.81, 3.58, 3.10, 3.05.*	*4.566* MPG	*.7512*	4.566 - .7512 = *3.81*	3.81, 3.58, 3.10, 3.05

Improvements to the performance of these four units will have maximal impact on your fleet operation costs. Using averages, let's calculate the gallons saved by improving only these 4 units to the mean.

Miles traveled for this group (1 year):		100,000
Gallons currently used: Average MPG	= 3.385	29,542
Gallons potentially used: Average MPG	= 4.566	21,901
Gallons saved		7641

No other group of vehicles will give you as good a return on your time and money investment as those 1 SD or more below the mean.

Another use for statistics is failure analysis. Failure analysis reviews the failures, and using statistics comes to some conclusions about their frequency. The technique of failure analysis is to determine the elapsed utilization between incidents of failure (MTBF) and the time it takes to put the asset back in service (MTTR).

Detailed failure analysis that is statistically valid is not for the faint-hearted. Accuracy dictates large populations of failures. Of course, the better the PM system, the more unlikely you would have an adequate number of failures to analyze.

Good engineering practices dictate tracking each mode of failure separately. These details generate enormous amounts of data and take significant resources. Some CMMS's have primitive statistical capabilities. In the best cases, the CMMS will collect the data needed for statistical analysis and export it to a spreadsheet or specialized statistical package.

How to use failure analysis:

1. Failure analysis can be a major tool in the establishment and updating of PM systems. If failures are too frequent (in relation to the frequency of the PM's) then increases in the depth of the task list or in the inspection frequency would be required. If failures dip too low, then the reverse may be true and too much money is being spent on the PM activity for that component. Note that this is an economic analysis. To set the frequency of a task, an analysis of the P/F curve is necessary. The interval between inspections should be less than the amount of time it takes for the system to fail after the defect could be detected by the task.

2. Use the information to set up a PCR program. If the failure of the component follows a normal distribution, then PCR might help. Planned component replacement frequencies should be chosen to program an allowable failure rate. If you have 1000 failures per year under current conditions and change out the component at 1 SD before the mean time between failures (MTBF), then the new failure rate will be 15.9% of the old rate. In the new scenario you will have 159 failures next year. The aircraft industry uses 3 SD (3 sigma's) which gives them less than 1% of the original failure rate.

3. Use the information to compare two makes of components. You might want to compare two makes of bearings to choose one over the other for a particular application. Look at the MTBF for each component, and factor in the cost.

4. Failure analysis can be used to interpret the results of experiments and provide data for efficient decision making. An example is looking at compressor failures for synthetic versus natural oils.

5. Maintenance departments constantly evaluate their specification. Failure analysis can help improve specification. While there are many factors to the choice of a component or system, MTBF and MTTR (mean time to repair) should be among them.

6. Divide assets by lifetime utilization for greater accuracy (older assets are more like each other in failure frequencies). Division of the units into lifetime utilization leads to higher levels of accuracy and usefulness in the outcome of the analysis. In this case, we break up the incidents by the total lifetime utilization on the unit at the time of the incident. There is an indication that failures of new components such as belts (or bearings, switches, circuit boards, almost anything) have different failure curves based on the age of the unit they are going into.

30
Planning

Planning is the job of looking into the future and anticipating the resource needs of a project or repair. The planning process uncovers needs that are not obvious in the hustle and bustle of the everyday maintenance department.

Planning is related to the issue of maintenance leadership. *The Maintenance Supervisor's Standard Manual* sums up half of the equation in the statement: your people depend on you for leadership so that they have the drive and confidence to carry out your plans. The other half of the picture is: adequate planning is a powerful expression of leadership itself.

There is some confusion between the functions of scheduling and the functions of planning. Part of the problem is that the two functions are frequently compressed together, particularly on smaller jobs. The easiest way to differentiate between scheduling and planning is to think of a military battle. Military planners look at several probable scopes of the battle, predict the duration of the engagement, number of troops needed, decide on the best equipment needed, and calculate the logistical support for the troops and weapons. The military planner will also look at the command and control issues (battlefield communications and supervision). The plans developed might be put on the shelf as a package awaiting a "go" from the commanders.

The plan is independent of execution. Nothing has been ordered, deployed, or moved in the real world. The maintenance plan is the same as the battle plan. It lists all of the resources needed, approvals, material/part/supply lists, tooling requirements, access requirements, time, skills, any other special information or requirements.

In the military, getting a "go" from the commanders begins the execution phase of the plan. Troops, ships, ordnance, aircraft, support units, and logistical units are scheduled into a theater of operations. Food and supplies are ordered. Spare parts and reserve troops are put in place. Orders might be placed for parts, ordnance, fuel, and food to fill up the pipeline. The battle plan is revised based on changes from intelligence gathered. It is turned over to battlefield managers. The success of the battle depends on the quality of the initial plan. Many battles were saved by individual valor and genius, but no one would argue that the right plan makes it easier.

The same things are true in the maintenance world. A bad plan might be saved by individual heroism or genius. It is still easier to work from an even adequate plan. When we are ready to start the maintenance job (the battle), we would say that the planning phase is over and the scheduling, coordinating, and management phases have begun.

Maintenance Planning

On larger jobs you can save 3–5 hours of execution time for every hour of advanced planning. Advanced planning is a proactive skill. Few users scream at you to plan repair jobs. This section shows the steps for successful maintenance planning. The plan consists of a detailed list of the work to be done and an evaluation of each of the five elements of a successful maintenance job. The work to be done is the first part of the planning process. The goal of the plan is enumerate all of the resources needed for a job to avoid all of the avoidable collisions of these resources. The example below shows the average ratio of planning to executuion in various crafts in heavy industry:

Electricians	1 to 20
Welders	1 to 11
Machinists	1 to 14
Riggers	1 to 24
Pipefitters	1 to 17
Laborers	1 to 30

Very few organizations now have full planning staffs. When organizations still have planners, the job has been expanded to include computer "feeding," permits/safety duties, and other staff functions.

These numbers might be a guide as to the amount of time that should be spent planning by the people in the craft themselves. In addition to the normal time the mechanic spends doing routine planning, the chart shows an additional 5% of high level planning for electricians, 7% for machinists, and almost 6% for pipefitters. Use these as guides if your tradespeople are expected to plan their own jobs.

Pre-planning: There are questions to be answered before planning begins. What job is to be done? What is the scope of work? What is the priority of this job? What are the work steps? What is the preliminary budget? What is the likely elapsed time? Plans must be considered about the probability that the job is significantly larger then first thought. Is engineering required? If engineering is required, order drawings and specifications.

Planning: Once the scope of the job is complete, then details have to be ironed out about each of the six elements.

Mechanic(s), techs, helper: What skills, what licenses needed, how much craft coordination, time per step, what craft needed when, total and individual craft crew size, contractor needed, back-up plan if the scope of work isn't adequate and job doubles or triples in size.

Tools: What tools, where to procure, how to ensure availability, lead time, vendor for rental or purchase.

Materials/parts/supplies: What parts, how many, availability, in stock, lead time, vendor.

Information: What drawings, spec sheets, wiring diagrams, pin-outs, are needed to complete job.

Availability of the unit to be serviced: Best time to do, check production schedule for likely times. Look for the best time in the business cycle for a window to repair an item. Determine the effect of this repair on related units.

Authorizations/permits/statutory permissions: Hot permit, open line permit, tank entry, lock-out/tag-out, EPA involvement, Safety Dept.

Planning Tasks

Inspect job and discuss the scope of work, priority with customer.

Determine the scope of the job. Make sketches if necessary.

Look into history files to see if this or a similar job was done.

Maintain packages on all jobs planned but not started.

Prepare work order. Get authorization if it exceeded your authority.

Break job into smaller projects or job steps.

Write out work plan (steps to complete project).

Locate all materials in stores.

Determine lead times and vendors for non-stock materials.

Determine special tools and equipment needed for job.

The planner (or the planning function) is a key to the effective daily operation of the maintenance department. Competent planning requires a wide range of skills.

Qualifications for a Planner

1. Ability to think of jobs as having both mechanical aspects and abstract (time, tools, space requirement) aspects.
2. Ability to express themselves both verbally and in writing.
3. Gets along with others.
4. Can work with different types of people from all levels in the organization.
5. Can represent the organization's interests in discussions with outside firms.
6. Has knowledge in crafts to be planned (not necessarily skill, although skill would help).
7. Has respect of craft workers.
8. Has a positive attitude toward company, supervisors, and managers.
9. Has the ability to plan work and foresee problems (doesn't like surprises).
10. Understands or can be taught job planning and scheduling.
11. Computer literacy (or can be taught, has positive attitude).
12. Some training in budgeting.
13. Understands the issues of the user department.

31
Project Management

On larger maintenance jobs, the techniques of project management can be used to control the planning and execution to good effect. These techniques efficiently manage large repairs but are inappropriate for smaller jobs. Some CMMS software deals with all of the jobs that the maintenance department is involved in as a single project, and use project management techniques to manage the whole mass of activity. The maintenance professional should be conversant with the set-up of large scale repairs on one of the popular project management packages.

"A project is a unique venture with a beginning and an end, carried out by people to meet a specific objective within parameters of schedule, cost and quality. Project management is a set of processes, systems and techniques for effective planning and control of projects and programs." This is a quote from one of the project management experts, Project Management Mentors of San Francisco. They go on to say that there are two distinct skill domains needed to successfully manage projects.

Project management has a set of technical skills and systems that help the manager to handle the complexity of tasks, timing, and resource allocation. The second set of skills is just as important—it is people management for recruitment to the project and job assignment within the project.

Project management was one of the first areas to be implemented on computers, and then on microcomputers, so there is a wide variety of packages at all price ranges. In a nutshell the project management packages require you to list all of the steps (called subprojects), resources, and dependencies. From this information, the software builds a model of the project and prints it out for visual reference. They also can track projects real-time with alerts when you fall behind or when there are resource conflicts.

History

In 1917, Harry L. Gannt of the Frankford Arsenal in Philadelphia developed a systematic technique for tracking and scheduling projects called the Gannt chart. The Gannt chart is one of the oldest planning tools available to maintenance managers. Since then, dramatic improvements were developed to manage larger and more complex projects. The CPM (critical path method) improved the ability to determine the critical events that could hold up a large project. The PERT (Project Evaluation and Review Technique) system, a refinement of CPM—including most optimistic, most probable and most pessimistic—was adopted by the military.

One of the first such projects where PERT was tested was the design and assembly of the first nuclear submarine, *Nautilus*. The U.S. Navy and Electric Boat in Groton, Connecticut, had to schedule 250,000 major activities of 250 contractors and 9000 subcontractors in the multi-year project. At any given time, delay in any one of hundreds of activities could throw the whole project off schedule. They used the PERT project charting and management method. Because of the complexity of maintaining the critical path with the project changing, activities being complete on schedule, ahead of schedule, and behind schedule, the entire system was eventually programmed in FORTRAN and run on the then powerful IBM 360 mainframe. The project was completed close to schedule and budget.

The concepts are extremely powerful, and are as follows.

1. Collisions of labor/material/tooling/machine/order are substantially easier, cheaper, faster to resolve on paper than in the field. This is also a golden rule of planning.

2. There exists a group of activities within the project, the sum of whose times regulates the length of the project. The longest path through the project (that includes these activities) is called the critical path.

3. We also know that time estimates are more likely to err by being too short rather than too long. On the *Polaris* program, they used a distribution called the Beta distribution which is not symmetrical. The more pessimistic (overdue) estimate had the greater probability.

4. If you keep the ever-changing critical path on schedule, the project will run on schedule.

5. Conversely, if an item on the critical path falls behind schedule early in the project, you know: the whole project is in trouble, and only an intervention (more labor, etc.) can bring the project back on track.

How to Set Up a Project Management Chart

1. Define the project. This is the same as the scope and specification of work in maintenance planning. This project planning is actually a more systematic and complete methodology for larger jobs than the static model introduced in the previous chapter.

2. If the project is large enough, choose the project team. The team should work through all of the steps of project management together in the beginning unless they are very experienced at project management.

3. Establish responsibilities for each project management stage. Assignments for requirement writing scope of work, develop planning system including tasking (biggest job to start), design and execute control structures (who will control), close-out and analyze what happened.

4. List all activities for the project. Some methods of project planning create a structure with several levels of subprojects. This is also known as the work breakdown structure. Tom Belanger (Sterling Planning Group of Sterling, MA) uses the model from architecture; his example is building a house, as follows.

> 1.5 Woods and plastics (this is an objective or a major milestone, depends on a set of tasks being complete)
>> 1.5.1 Rough carpentry (this is an activity made up of tasks)
>>> 1.5.1.1 Framing (task)
>>>> 1.5.1.1.1 Exterior framing (subtask)
>>>> 1.5.1.2.2 Interior partitions (subtask)

1.5.1.2 Sheeting (task)
1.5.1.3 Flooring (task)
1.5.1. ... next tasks

The rules: A task has a defined beginning and ending. No other task has to start in the middle of the task, if one does, then split the one task into two subtasks. Determine time for each task, optionally determine effort level (number of people). You might list three times: optimistic, probable, and pessimistic.

5. Determine what tasks must be complete for each task to start. Once that is known, start to analyze the critical path through the project. Add up each path, the longest one is the critical path because it determines the total length of the project.

6. All of the activities along the longest path are critical path activities. Slippage in any critical path activity will result in the project being late. Develop contingency plans for problems. This might mean talking to some contractors who can step in with increased people or getting overtime preauthorized.

7. There is a certain point that, if delayed, non-critical path activities become critical. The amount of time is called float or slack.

8. Using Gannt-type charts or software, start adding objectives, activities, tasks, and block in the times. Start the objectives in the appropriate time block.

9. On the first planning pass, use relative time (elapsed days, weeks from day one). When the actual beginning date is known, redo the chart to absolute (date) time.

10. Be alert for conflicts in labor/materials/tooling/machine/safety/permits/order. Any person, tool, or material is known as a resource for that task. If you added effort level then you can sum the columns by craft to determine number of people required for the project. Studying and manipulating the resources for the desired outcome is called resource leveling.

11. After the project is started, mark off the projects that are completed. Estimate the durations of the uncompleted projects and shift them on the chart. Recalculate the critical path. If anything is out of bounds then plan your intervention. Keep notes of all decisions and rationales for your post-project walkthrough.

12. After the project is complete, gather all information, notes, and data and plot how the project actually went. Compare this plot to the original plot, and look for areas where the actual project substantially deviated from the plan. If possible, have the team replay the project and see what mistakes were made. This is not the time for finger-pointing but for learning what happened.

32
Estimating Job Duration

There are three major reasons to develop, use, and track work standards. The first reason to use standards is to develop a scheduling system. Scheduling repairs and PM activity requires estimates of time to be effective. The most effective ways to schedule a maintenance facility require some idea about the amount of time a job is going to take. The era where shop labor cost $5 to $6 per hour is gone. Labor is still the major component of the maintenance dollar.

The second major reason for time standards is to be able to accurately predict when a unit will be returned to your users. Accurate estimates improve your customer service and increase the value of the service you perform for your users.

The last reason is to be able to evaluate individual mechanics. Comparisons of individual work to work standards over time will uncover and, perhaps more importantly, prove who needs training, re-assignment, or to look for another job.

Reasonable returns should be expected from investments in job estimating. A contractor wouldn't last very long if they couldn't accurately estimate job duration. On the other hand, firms used to major estimating efforts don't survive where the estimating is more informal. A maintenance shop should balance its investment with the potential returns available. Certainly non-repetitive jobs of short or moderate duration do not require elaborate estimates unless the timing of the return of the asset to service is critical. Repetitive jobs such as PM tasks, PCR jobs, and repetitive corrective actions will respond well to estimating.

Work standards should be based on:

1. class of equipment (like equipment in like service is grouped together)

287

2. concise description of repair or PM tasks performed (for example, R&R left front brake shoe)

3. work accomplished (was this an adjustment, replacement, inspection, what was work done)

4. labor hours (should reflect crew size)

5. location of the unit repaired (conditions may vary from shop to shop and repairs in the field take significantly longer)

6. factors like mechanic, repair reason, and labor or parts cost are not considered in labor standards.

Estimate

By far the most common type of time standards are estimates made by supervisors, craftspeople, and planners. Anyone involved in maintenance has been asked how long it will take until this bathroom is fixed or this building is cool. The answer is an estimate.

Slotting Method

Compare the job to be evaluated to a group of known and studied jobs. Determine where the job fits in. Is it bigger than slot 2 but smaller than slot 3? The time would be between the two.

Slot	Job Description	Duration
Slot 1	Replace electrical outlet or switch	15 minutes
Slot 2	Remove and replace toilet	60 minutes
Slot 3	Install door with trim	240 minutes
Slot 4	Remove and replace hot air furnace where there is very little modification needed	420 minutes

Flat Rates

A flat rate is a standard published by either the manufacturer or a third-party publisher. It is developed by comparison to other related jobs (similar to slotting) and by timing skilled mechanics. Other skilled mechanics should be able to meet or exceed flat rate work standards. Activities such as diagnosis are separated from the actual repair. The flat rate for a repair is sometimes too fast if the vehicle is old or has had very rough service and the components are rusted in place or the mechanic is not fully trained. If rusted bolts or studs break during the repair, then even a highly skilled mechanic is likely to miss the flat rate with good reason.

Flat rates are often used in incentive schemes. Some automobile dealers charge the customer by the flat rate for the job and then pay the mechanic a percentage of the flat rate. Repeat repairs are handled on the mechanic's own time (or other arrangement).

Flat rates are a good place to start as long as they are compared to the other standard techniques. If you start using flat rates to estimate repair times, then 100% would be the average productivity rate for skilled people. Schedules based on flat rates should be de-rated by a percentage based on the relationship between the flat rate and the experience of your crew (use flat rate with reasonable expectancy below).

Examples of commonly used flat rates include *Chilton's Professional Truck and Van Service Manual* and Means *Maintenance Standards.*

Historical Work Standards

Just about all computer systems track the time it takes to do every job. Some systems can evaluate all of the repairs of a particular type on a particular class of equipment. These standards can be useful since (unlike the flat rate) they factor in important variables such as the actual condition of your equipment, the skill level of your work force, shop practices, the layout of your shop, and your tools and equipment.

The disadvantage of historical standards are the accuracy of the data collection and of the definition of the tasks completed. For example, the work order might read "Fix Steering." The mechanic might have to check tires, alignment, frame damage before accepting that there is even a problem with the steering. All of these activities are lumped together in the historical standard.

Schedules based on historical standards include a full amount of lost and wasted time. When an employee is talking, eating, or not at his/her job, the time is still charged to the repair and will find its way into the historical standard. Historical standards are a less adequate place to start a performance evaluation scheme because of the factors discussed earlier. Flat rates don't usually include lost time but don't take into account conditions and skills of your crew.

Reasonable Expectancy

The reasonable expectancy (RE) is a work standard based on observation. The definition of RE is a reasonable, observable amount of work in a given amount of time. The reasonable expectancy is the most accurate and most expensive way to schedule a repair facility or evaluate a mechanic.

The concept of reasonable is important. By observing several actual people doing the repair, you have a good idea of how long it should reasonably take. No speed-up is needed to improve productivity. We will improve productivity by recapturing time lost in non-productive and marginally productive activities.

Case Study: Example of Comparison and Use of Three Standards Systems

Tom Duvane wanted a method to measure the productivity of his mechanics. He obtained the flat rates from the Chilton manual. The historical standard was calculated from work orders for the same repairs over the last year by Ann Moore, a co-op student from the local engineering school. Reasonable expectancies were taken by Ann Moore in the month before the study.

Tom chose Joe Dillion, a responsible but slow employee, to evaluate. Joe didn't like paperwork at all and constantly complained about it. Tom figured if the standards worked with Joe they would be a breeze with the others.

The week of October 1–7, Joe Dillion had 32 regular time repair order hours and was paid for 40 regular time hours. He completed 8 chargeable jobs. The remaining 8 hours was spent on non-chargeable activities such as cleaning up and fixing the grounds department lawn mower.

	Hours	Performance Std hr./ R/O hr.	Effectiveness Std hr/pay Hr
Flat rate hours completed*	28	28/32 = 87.5%	28/40 = 70%
Historical standard hours completed*	31	31/32 = 96.8%	31/40 = 77.5%
RE hours completed*	25	25/32 = 78%	25/40 = 62.5%

*To determine the standard hours, add the individual standards for each of the 8 jobs Joe Dillion completed.

Tom noticed several things from the study. The flat rates were somewhat faster than the historical performance at his main garage. He also looked at the reasonable expectancies and couldn't see why anyone couldn't make them. As he observed Joe Dillion over the next few days, he saw that Joe worked at a pace which would easily make the reasonable expectancies. When Tom looked closer, he saw that Joe spent an excessive amount of time filling out the work orders. Tom discovered that Joe could barely read and write English. Remedial classes solved the problem and made a happier employee.

Tom concluded that the reasonable expectancies presented the greatest opportunity for improvements. They seemed to point out where problems were and also seemed intuitively fair. These advantages outweighed the higher costs associated with them. He also realized that in the future reasonable expectancies and flat rates used together might be the easiest to use on a day-to-day basis. He could see adjusting the flat rates by the reasonable expectancies to minimize the number of observations.

33
Maintenance Scheduling

You can schedule your people, or you can let breakdowns or irate, bored, or lonely maintenance customers schedule them. One way or another, your people will be sent to jobs every morning.

Scheduling is the execution step in the planning process. It looks at today's demand for service and matches that to today's available resources. Scheduling takes into account people in training, on vacation, with lack of particular skills, even (when you are not too busy) individual fears and preferences. Good scheduling will match people with complementary personalities and skills. (See Figure 33-1.)

The process of scheduling is first and foremost an evaluation and a prioritization of the backlog, pending jobs, and emergent urgent requests. That mass of (frequently) conflicting demands is where the scheduling process starts. The second element is the available hours, skills, preferences, and experience of your maintenance people and contractors.

In a production environment, scheduling means (above all), conform to the windows where uptime is essential, and service equipment when the equipment is not in demand. Where this is impossible, then schedule the downtime to minimize disruption to the production schedule. Scheduling in buildings such as hospitals is similar in that servicing certain systems has to wait until it is convenient for the patient (who might be in the ICU or might be having an operation) or staff.

The advantages of active scheduling are in the areas of efficiency (you can optimize travel time, stores trips, special tool usage, etc.), customer service (being able to tell a customer when a service person will visit), training (give people experience they

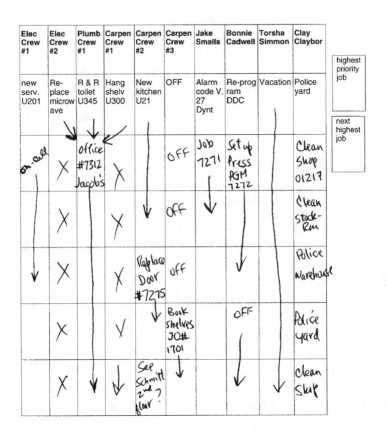

Figure 33-1. Maintenance scheduling board.

need to develop, pick crews with a trainer and a trainee), and morale (assign constant level of work, smooth out the feasts and famines).

Scheduling means:

1. Bringing together in the precise timing the six elements of a successful maintenance job—the mechanic(s), the tools, the materials/parts/supplies, the availability of the unit to be serviced, information needed to complete job, and the authorizations/permits/statutory permissions. These elements were listed, considered, and researched by the maintenance planning process.

2. Review the work plan to see if the situation has changed in any material way. An accurate plan is necessary to know where and when to manage the individual job steps.

3. Keep ahead of jobs that are going to start, and check in on jobs in progress. Materials and parts are common culprits for a well-planned job to go awry. Parts require constant attention and chasing, scrounging, persuading, pleading, and sometimes yelling.

4. If the scheduler finds that a large repair is falling behind schedule, they should investigate interventions (adding overtime, people, contractor, etc.) so that the situation can be corrected while they can still have an impact on that job. Waiting until a monthly meeting to find out that a job went overtime is too late.

5. Work is quantified in the job plan. A reasonable amount of work is expected each day. Workers are freed from a hurry atmosphere one day and a kill-time atmosphere the next. Back-up jobs are given so an employee can jump to another job if they get stuck by a lack of parts or tools.

Jobs of the Scheduler

When the scheduling job is separated from other jobs such as supervision, coordination, or planning, they pick up responsibility for several functions including preparation of the schedule itself, tracking jobs, checking parts and materials, and reporting schedule status.

Preparation and Maintenance of the Schedule

Review all jobs in backlog, review incomplete jobs from last shift.

Assign priority or review existing priority. Put jobs in priority order.

If there is a supervisor, coordinate the next steps with them.

List all of the people you have to complete the jobs, with their available hours.

Create a list of work for each person or crew, assign high priority first.

Look for opportunities where the assignment can take advantage of jobs in the same location, jobs using the same tools or materials, jobs needing the same skills, other ways to improve efficiency.

Prepare dispatch sheets in consultation with supervisor.

Keep schedule updated as new information comes in, jobs get completed.

Job Tracking

Secure permits and ensure safety instructions are in the hands of the craftspeople or their supervisors.

Maintain a tickler file on all projects that have been started and are pending.

Notify and consult with customers about any pending interruptions or disruptions.

Get information about status of all repairs on a regular basis.

Review all projects for their adherence to the plan.

Secure special tools and equipment as defined by the plan before job is to start.

Attend weekly meetings to discuss work progress, progress on projects, and updates on materials.

Be alert for continuing schedule miss conditions.

Detect when a job runs into trouble before it misses a milestone.

Material/Parts/Supplies Tracking

Follow up on availability and delivery of parts for planned work orders.

Prepare requisitions for new jobs.

Call outside vendors before jobs start to get parts that might slow down or stop a scheduled job.

Get purchasing department involved when a vendor is being unreasonable or is suspected of lying.

Verify material is on hand. Keep check on most used and critical items in stores. If appropriate, place parts in protected area.

Work on outage to assure more accurate usage information on parts.

Update the work plan with actual part usage.

Reporting, Paperwork, and Tracking Schedule Status

Clean up paperwork at the end of the job.

Verify that job was done according to the plan. When a job deviates, learn why.

Update the planning package.

Complete daily, add to weekly reports for higher management.

Problems When a Schedule is First Installed

An inevitable byproduct of this approach is the uncovering of many hidden operations problems. Scheduling highlights areas where mechanics cannot do their job due to a problem outside their control and other problems with maintenance procedures and systems. These previously hidden (or unpublicized) problems suddenly come into the foreground.

- Mechanics being pulled off scheduled jobs to work on non-productive activities without consultation or communication with maintenance leadership or dispatch. This is a common problem where the maintenance department is also used in personal service to the plant higher-ups. It is also a problem where mechanics are traditionally assigned by being collared by users needing service.

- Mechanics running to the airport, picking up documents or setting up for picnics cannot stay on schedule. From a scheduling point of view, this is only a problem if the scheduler is not informed (in advance) of the borrowing of the person.
- The initial planning package does not reflect reality. Parts are incorrect, job steps are outdated or wrong, lockouts and new regulations are not included. If the original planning does not reflect the job planned, then the scheduling function is going to be scrambling to keep on top the situation. This is like keeping your footing on a pitching ship's deck in a hurricane.
- Stock room regularly contributes to problem conditions. Items shown on the inventory are not on the shelf, quantities are wrong, stockouts are frequent on regularly used items, and excessive time is spent waiting for clerks.
- Failure to put equipment, buildings, trucks back into service when promised. This builds an attitude of distrust. Your user never wants to give up a machine because they never get it back when promised. This attitude stands in the way of good communications and contributes to the problem of an inability to get custody of assets when scheduled.
- Lack of cross training causes clashes of resources. You might have enough people but are in chronic short of specific sets of skills. The cause is lack of flexibility because your people's skills do not match the skills needed to maintain your assets.

34
CMMS—Computerized Maintenance Management Systems

"Put the most powerful technology to work . . . your mind" is what Einstein said decades ago. Let's not ever forget that the computer-based solutions available today are almost unbelievably powerful and they are not even close to the power of the human brain. Computerization of maintenance is the use of a powerful tool. It is just a tool!

The reason why we computerize is the same reason we manage maintenance in the first place. We computerize to lower or avoid costs, improve service, control costs, ensure uptime, improve quality, etc. The computer allows us easier access to masses of data and to levels of analysis not readily available manually.

The computer also has the capability to perform tasks too time consuming for people. This enhances the ability of a lean maintenance department to have top quality analysis. The newest hardware/software has capabilities that cannot be duplicated manually with any number of people such as real time monitoring. The newest systems now can have an impact on the shop floor while the repair is in process. There is tremendous excitement in the field because we have passed the threshold of automation of existing manual systems. We are in the realm of process management only possible with the computer.

The decision to computerize is actually a surface decision for a much deeper decision. A decision to computerize is also a decision to treat maintenance as a serious profession. The decision to computerize is also a decision to impose discipline on a group of mechanics (which are traditionally independent and hard to control). The computer is a tool that maintenance managers imagine will allow them to predict, affect, analyze, and eventually control what goes on in maintenance. This computerization decision and the deeper decisions that it represents go to the core of the culture of maintenance in your facility.

The reason that most CMMS installations go astray and never realize their promise is because the firms thought they were only computerizing, and only asked hardware, software, and database type questions. They never asked themselves the deeper questions of what we are about, how do we view maintenance, and what is our role in future incarnations of the organization.

In the old model of maintenance, mechanics are looked at as fixers/maintainers—not as thinkers. It is logical in the old model to assume that the tradespeople have no reason to use the computer, and that additional people are needed to enter the work orders and other data.

Maintenance leadership time is clogged up with paperwork, cracking the whip, complying with new regulations, quality inspections, meetings, and the problems of scarce resources. They are already working long hours just to keep the ship afloat. There is no time to even read a report let alone deeply study an issue. Using other resources is a joke because the maintenance engineers, planners, analysts are mostly gone in the downsizing, and the ones left are buried in regulatory/safety issues and endless meetings.

Herein is the dilemma of the computerization effort. How we grapple with this problem will in a large part regulate how well the system is used. There is only one solution to this problem in the new paradigm of maintenance management. The solution is distasteful for many organizations. The solution is as follows.

- Mechanics and tradespeople enter their own data. We should help them in all ways possible, with bar code scanning, scanning, voice recognition, rational data structures, and streamlined data entry.
- Mechanics and tradespeople know how to—and are encouraged to—analyze data to uncover and solve problems. We should help them by standardizing types of analysis and teaching predesigned models.

Our mechanics have far more mental capabilities than we give them credit for. One of the precepts of reengineering is to use the technology as an enabler of new techniques that could not even be contemplated before the technology was available.

Computerization of maintenance is a complex job when measured against other computerization efforts. The reason it is so complex is in the nature of the data collected. Maintenance data flow in copious detail, have inconsistent nomenclature, and come into maintenance in a variety of channels.

Some of the data:

Who called in, time called in, time complete, elapsed downtime, what they reported, why do the repair, where the event came from, who authorized it, what priority, where to deliver the parts, when was the last time this happened, what to service, parts used, supplies, trade, technician, hours, crews, bench time, rebuild effort, crafts, assets, components, subcomponents worked on, what was done, downtime, what the mechanic saw or recommends, who to charge, comments from the mechanic, comments from the operator, what cost center, etc.

Communications channels to the maintenance department:

Telephone, E-mail, face to face, fax, computer networks, engineering drawings, specifications, gauges, MAP systems, PLC networks, cycle counters, storeroom communication, time clocks, authorizations, locations, machine tags, old history cards, written communication from mechanic/from operator, verbal communication from mechanic/ from operator, evidence from the broken parts, outside laboratories, etc.

The second reason that computerization of maintenance is so tough is that maintenance information systems are among the most complicated packages that are commonly found in industry. To compound the problem, knowledge about good maintenance practices is not well distributed throughout the organization.

People from accounting, data processing, or even production do not have enough expertise in maintenance to be of much help in choosing software. At the same time, the maintenance department people are usually not knowledgeable about computerization in general and the other business systems to which maintenance must interface in particular.

Maintenance is low priority when it comes to attention from data processing. Work on maintenance systems is not usually viewed as mission critical. This leads to a common situation where the maintenance system is delayed years and eventually the maintenance department bootlegs a small standalone system that does not talk to the organization's other systems. Data being generated by the maintenance system are not available to other departments and the decisions made reflect this ignorance.

Most organizations focus on picking the right system vendor and the right package. Successful implementations of computerization also focus on the readiness of the organization to accept a new culture. This new culture needs to be sold to all parties involved and interested in maintenance. Without adequate preparation, the system, at best, will only enjoy a superficial acceptance. To improve acceptance, the preparatory training effort must start well before the system is selected.

Return on Investment from Computer Systems

Overwhelmingly, the return from computer systems comes from the discipline imposed by the "outside" force represented by the computer. If you entered all the data on your old card system, if you summarized all the numbers, if you poured over the jackets, you would already be running a tight operation and might not get all of the advantages of the computer system.

The computer system can have an eventual significant impact on the cost areas. The costs areas differ in ease of impact. Operating costs are usually the easiest to impact followed by maintenance costs, downtime, then ownership and overhead costs.

The most important fact to keep in mind is that computers don't maintain assets (yet). Any impact from the system will come from action taken as a result of information provided by the system. This is the concept mentioned earlier. The computer system will identify the bad actors early enough for you to have a positive impact. If you only "feed garbage to the pig," as one maintenance manager referred to the process of using his CMMS, and the reports and screens don't impact your decisions, then you will not get benefit from the CMMS.

Don't expect to save in clerical labor by having a system; in fact, plan on an increase in clerical support (especially in the beginning). Later you may find that the same support staff can handle a much larger asset base.

In addition to the specific areas mentioned on the next few pages, returns flow from improvements to the following general areas.

1. Improvements to labor scheduling. Only repairs where materials are available are scheduled, information on jobs done before are available to planner, labor requirements to complete schedule are easily calculated and recalculated. Savings range 3 to 8% labor.

2. Improved parts availability. Less time is spent locating parts, parts can easily be kitted for jobs, parts and kits can be prepared ahead of time. Savings range 1 to 2% labor.

3. Increased availability of equipment. PM system (when followed) will reduce emergency downtime, jobs will not start as often without adequate labor/parts to complete. Savings 0.5 to 2% improvement to availability.

4. Reduction to inventory. Through parts cross reference (reduce the number and quantity of parts), automatic reorder, min/max, EOQ, exchanging old inventory for usable inventory, ease of physical inventory. Savings of 10 to 20% of inventory level.

Early Areas to Look at for ROI
1. Within a class, sort units by energy usage (like MPG on vehicles). Take action on the worst 15% in each class starting with the class with the highest gallons used. Action taken could be: PM unit looking for fuel consumption items, retrain operator, move unit to lower usage service, etc.

2. On equipment that is still under warranty, collect all warranty claims and analyze the component systems that were hit. Where one component system has had a high frequency of failures, apply for warranty adjustment for all of the assets that were purchased together for that system (such as ballasts). You may be able to get either additional warranty recovery or some other special consideration (extended terms). Note that many manufacturers reserve a percentage of the value of the sale for warranty (typically 1.5 to 2%).

3. If you have added your history when you started the system and you feel that the $ information is accurate, generate a report that shows total cost for each type of asset. Target the worst 15% for action such as: retirement, rebuild, move the unit to less strenuous service (standby status).

4. Review inventory parts as you set up the masterfiles. Look for parts for assets no longer in service. Note the parts or put them aside for sale or trade. This cannot only generate funds for reinvestment in needed parts but also free up space.

5. Review all assets during the first few months of the system looking for units with no activity. Some systems have no activity reports. Many organizations have assets in reserve, like trucks parked against the fence for emergencies. Unfortunately, when they are needed they don't quite run. Retire, trade, scrap those units and lease emergency units when required.

6. Review the history of any unit requiring a major repair. Decide (before you start the repair) where on the critical wear curve you are. You may want to consider scrapping the unit. The questions to ask concern frequency of repairs, emergency breakdowns in spite of PM activity, and increases in the cost to maintain over a period of time. Also ask if this is likely to be the last major repair for awhile or if another component system is showing signs of failure.

Secondary Returns

After the system has been installed for at least a year, and preferably two years, other returns are available. Analysis of your "mass" of data will uncover the specific facts of your situation and environment. All areas of cost can then be looked at, analyzed, manipulated, and compared.

1. Adjust PM task lists by failure experience. This will maximize your effect per PM hour.

2. Review failures to improve specifications for new equipment.

3. Create experiments with additives, synthetic lubes, auto lubrication systems, bypass filtration, heavy-duty bearings or components, etc. Be able to actually know numerically the result of your experiments.

4. Tracking productivity of mechanics will help pinpoint people who need training, rewards, discipline, or who need to be removed from work force.

5. Look at inventory not used in 1 or 2 years. Is it for assets that were retired?

6. Isolate costs of satellite operations; are they cost effective? Would satellite facilities cut the overall cost of operation?

Cost of Installation

To accurately calculate the ROI of a CMMS, you have to be able to determine the total cost of the installation. If you are also planning to install a comprehensive PM system at the same time, the costs outlined in the chapter on installing PM systems have to be added to these figures.

Note in the chart on the next page that the software is only 1 of 15 items. A software vendor estimated that the true cost of his CMMS was over 5 times the software cost (his was a $15,000 package). Most firms underestimate the costs and end up running out of resources midstream. A leading consultant rough estimates 1 hour per asset to gather nameplate data, enter into the system, and audit the results.

How Software is Priced

Pricing for software is complicated. You can spend $2000–$200,000 for similar looking software. The introduction of the IBM PC in 1981 changed the face of software pricing and marketing forever. There are several factors in pricing software such as development, support costs, sales costs, current operating costs, and profit. These costs are spread over the sales of that product for that accounting period.

Compare two organizations with maintenance systems of high quality, whose costs are the same but whose marketing assumptions are different.

Cost Item	First Year	Ongoing Costs
Software	X	
Yearly upgrade costs		X
Supplies—diskettes, paper, tapes, etc.	X	X
New computers	X	X (pursuit of power and speed—always!)
New printers	X	
Network hardware and software, installation	X	
Communications line costs for WAN, LAN	X	X
Site wiring	X	
Consulting services	X	
Installation services	X	
Time to enter database	X	
Time to update database	X	X
Time to audit database	X	X
Training for every level: managers, supervisors, and support staff mechanics	X	X
Benefit training (how to use the system to provide the return on investment that was promised)—this is beyond which keys to press		X
Other?	X	X

A typical package shows a dramatic swing in prices.

Per Project Costs	Company A	Company B
Development costs	$200,000	$200,000
Company operation cost	$1,000,000	$2,000,000
Assumption # of units sold	200	5000
Project cost per package	$6000	$440
Per Package Costs		
Support cost per package	$1000	$250
Sales cost per package	$5,000	$2,000
Profit per package	$5,000	$2,000
Minimum Selling price	$17,000	$4690

You can see from these examples that the same quality package can vary in cost all over the place. In reality the situation is worse because some packages were developed for particular large companies so there aren't any development costs. Other arrangements call for a percent of sales royalty or a fixed dollar per package royalty.

Rule of Diminishing Returns and Software

If you are searching for (or designing) software for maintenance, there is a rule that states that you will get 80% of the benefit of a software package from the first 20% of the investment. In other words, if you expect to pay $50,000 for software you can get 80% of the benefit from an investment of $10,000.

In the maintenance field, getting the first benefit is usually easy. Usually the discipline of using any system will give you benefits. The last 20% of the benefit is very hard to capture. It requires knowledge of your operation and sophisticated knowledge of maintenance.

Large firms can usually justify the extra expense of going for the last percent of return because of the large number of assets. In a 1000 unit department, $100 per unit saved each year will fund a quite large system. The extra savings per unit are usually quite substantial. (See Figure 34-1.)

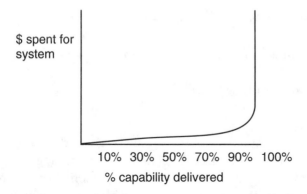

Figure 34-1. Chart showing diminished returns from increases in computer system expenditures, development cost.

Dealing with Computerphobia

As you begin to use computers in your maintenance operation, you may run into resistance that may not be associated with just the change to the new system. Some of the resistance might be due to fear of computers. This fear can show up at any level in the organization, and if not dealt with can cripple your computerization effort. After conducting seminars with psychotherapists on computerphobia, we have come up with some ideas about this condition.

1. Some fear of computers is normal for people who don't come in regular contact with them.

2. There are degrees of computerphobia, from mild anxiety (very common) to full panic attacks with heart palpitations, hyperventilation, and other symptoms (very rare). Psychotherapy can and should be used in cases of severe computerphobia. The mild forms can be dealt with successfully in most job settings with a little understanding.

3. The antidote for the fear is systematic contact. Plan to gradually introduce the person to the computer system, increasing the amount of contact over time. Basic to this technique is to support the computerphobic person, especially early in the process.

A. In the early stages, these people need very well-defined tasks.
B. The tasks should be chosen so that the person can be successful.
C. Training and computer time should be kept short in the beginning and gradually increased.
D. Tasks should gradually get more complex along with training so that mastery can be achieved.

4. Reasons for computerphobia can only be guessed at; some may be: fear of looking stupid or bad, dislike of technology, fear of change, fear of being replaced by a computer, or some combination. These fears are very real for the person and must be dealt with.

5. If resistance of several key people is strong and the project is important, a psychologist can be brought in to help people deal with their fear, anger, and whatever is blocking them.

6. In our experience, once the initial fear is worked through these same people become the system's greatest boosters.

In-House System Design Option

As you look at the market you might want to consider a home-brewed system. Some of the in-house systems designed by maintenance or data processing departments are so good that they end up being sold on the market and can compete effectively with existing packages. Unfortunately there are more horror stories than success stories. In *Maintenance Management*, Butler reviews some of the pros and cons of in-house development.

Pros
1. You get to participate in design.
2. System can be designed around special regulatory or business constraints.
3. Information Services (IS) department (also known as Management Information Services, or MIS) might be looking for work.
4. System can be designed to communicate with other business systems.
5. High level of support, help desk support.

Cons

1. Staff might be reassigned to a higher priority project.
2. Your IS department might not understand maintenance issues and refuse to listen to you.
3. Your IS department might listen only to accounting about maintenance needs.
4. The design might be narrow and not be up to date with the industry.
5. After it is done, it might be impossible to get revisions or updates.
6. Lack of documentation, training, hand holding.
7. There might be no available expertise to handle a project of this size or type.
8. It might take a very long time to get your system.

How to Look for a System

Shopping for a system is a daunting undertaking. There are 200 or more vendors of software for maintenance. There are an additional 250 vendors in specialized areas such as fleet maintenance, building maintenance, etc.

The salespeople learn that no sale takes place unless someone in your organization gets excited about their offering. To create excitement the salesperson shows you how to solve real problems with the system's inquiries and reports. You complain about a PM problem, so they show you the PM screen that solves the problem. You would like to track the costs going into a machine you are building, so they show you cost accumulation reports. There is a problem for you with this approach.

The reason that systems fail in their implementation rarely has to do with the lack of reports or inability to get information out, but rather from a basic misfit between the system and the existing maintenance culture or organizational requirements. If your culture requires mechanics to keep log books and the system you chose doesn't support that format, then they might rebel against the duplicated effort. If you use a 16-digit asset number for accounting purposes, then the 10-digit field will not do. These pitfalls can be avoided by looking at any prospective system using a technique described in the next section. Look at the parts of all of the prospective systems in the same sequence.

Selecting and installing maintenance software is a major effort that will require the time and energy of key maintenance players. The best implementations have the support of the mechanics, supervisors, and other maintenance staffers before the system is turned on. To facilitate this, involve as many levels of the maintenance department as practical in the search for a system.

After choosing a system, with the guidance of the vendor and both the maintenance and data processing management teams, encourage the supervisors and lead mechanics to do the research and help type in the system's masterfiles. Kent Edwards, Vice President for Four Rivers Software Systems (the organization that markets the TMS package), has a rule of thumb that no more than 90% of the system setup should be done by either the vendor or the customer. That ensures the expertise of the two groups is commingled. For a more complete discussion, see the section on setting up your PM system. That section also includes a detailed discussion of the steps necessary to set up the CMMS.

Before you decide on a system, answer the following questions.

1. Is there enough time, money, and interest to involve all levels within the maintenance department and other stakeholders in the decision process to buy CMMS? Is there support from top management to see you through the inevitable ups and downs of the entire installation process? Management support is important but maintenance staff support is essential.

2. Sufficient resources for a complete installation are also essential. The resources include training dollars, time replaced on the shop floor, and computer access. If necessary, can you get typing and basic computer skills training for your mechanics? Will management tolerate the initial research and keying of files by your mechanics and staffers? Can you get the budget authorization to replace the mechanic's slot on the shop floor by overtime or by a contract worker?

3. After the maintenance system is in operation, will mechanics and supervisors have the training, knowledge, positive attitude, and access into the CMMS to investigate a problem? Is there continuing training in advanced concepts beyond "which key strokes to get which reports" type classes? Is there regular time set aside for thinking and using the system for research into problem areas? Do mechanics and supervisors have easy access to terminals or PC's? Are these devices hardened against the shop environment?

4. Is there organizational willpower to ensure that garbage and faked data will be kept out of the system? Another way to put that is, is falsifying a work order to fill 8 hours viewed as a minor offense or a crime? Will the data coming out of the system be commonly held by management and the workers to be accurate and useful? Are maintenance records treated as seriously as payroll or other accounting records?

5. Does anyone (including mechanics) have the time to investigate repair history to detect repeat repairs, trends, and new problems? Related to #3 above, do they have the training to use the system to answer the questions that they have? Most maintenance people enjoy solving problems, do yours?

6. Can you and your staff spend enough time designing the system's categories to make meaningful comparisons between like machines, buildings, and cost centers? This is a two-step process. The first step is to have the vendor's trainer conduct a class in the category and codes model of that system and how things are commonly handled. The second step is to actually fight out the categories and codes that you want to use. It is *critical* to *understand* and *wrestle* with the decisions that you make at the early steps in the setup of a system.

7. If you have 100 pumps, probably 20 of them create the most maintenance load. This rule of management is called the Pareto principle. Has the Pareto principle (the 80/20 rule) been taught and used to isolate the "bad actors" (that is, to identify the problem machines, craftspeople, or parts). Do you understand how to generate these Pareto analysis or exception reports in the system you chose.

8. Will you have the support of a responsive data processing department (or a very responsive vendor)? You will want changes, fixes, enhancements. In fact, your ability to handle technology and sophisticated systems will dramatically improve after the first 6 months. Many organizations outgrow their first systems in a year or two.

9. Does the longer-range plan include CMMS integration with stores, MRP, purchasing, payroll, CAD/Engineering? The trend is toward company-wide networks. Organizations want everyone discussing a problem to be working from the same data. This means linkages of the maintenance information system to the corporate information systems, with all the links and hooks that implies. Increasingly information systems are viewed as strategic advantages. Access to information makes a major difference.

Design of the Maintenance Management Application

All business software packages, maintenance management included, have four logical components. Together these components are the "system." A maintenance system might consist of several to more than 500 programs. All of the programs are linked together to form what you see as a seamless system. The completeness and quality of the system depends on the care, knowledge, and goals of the designers of the system in the four areas. When you are choosing a system, designing a new system, or revamping an older system, consider these components separately.

Part 1—Look at First

Daily Transactions: Includes all data entry such as work orders, packing slips/receipts of parts, payroll information, energy logs, physical inventory information. A defect in this section of the package is usually fatal. It is usually very difficult to repair or reprogram this section for the vendor. The main reason that problems here are fatal is the amount of time your staff will spend facing these screens. The second reason is that defects here will adversely impact all other parts of the system and may limit the usefulness of the system.

Look for: Completeness, quick data entry, logical format, consistent format, alternate data entry paths. Alternate data entry paths include work order, log sheet, standing work order, string PM entry screens. The more flexible, the better. You will spend more time entering data into a maintenance system than anything else that you do with it. Most of the garbage that will plague you will enter at this point. Data elements not collected at this point tend to become problems later. For example, a coal to coke manufacturer did not collect the data element "who performed the work." As a result, they could not look at rework or callbacks to determine who needed additional training.

The second issue is consistency. If three people enter the same repair differently (and the system allows that), the computer will have limited ability to analyze that data. An example would be:

#1 Replaced bearing A2,
#2 Thrust bearing on shaft 'A'- R & R
#3 I unscrewed the housing D1 put my hands around the end of the shaft holder and...

These could all be the same repair. A person rather than the computer would have to decide if they were the same.

The other issue of consistency is system editing (discussed at length in the chapter on work orders). Better systems test data coming in against the masterfiles or with logic to determine that it is valid. For example, equipment ID numbers are checked to see if they are valid (already in the file), subsystems are checked to see if they exist for that asset, etc. Some systems will not allow you to log a repair that isn't on the master repair list for that asset, such as changing the tires of a pump. Other checks might be that you can't close a work order before you open it, or meter readings go up unless the meter was replaced.

A good system might go through 50 different edits to keep garbage out of the database. Total vigilance is required to keep garbage out of your database. Garbage in your database undermines all of your work. One of the rules of data processing is to move the data entry chore as close to the generator of data as possible. Any firm that decides to go that way should be aware that their training bill will be high for the first year since 20 people or more will have to be trained.

Lack of speed is the greatest killer of systems. Clumsy data entry design and slow networks will cause a revolt among the mechanics or supervisors (whoever enters the information). To test the speed of a system, develop a set of 4–5 typical work orders. Enter them into the target systems and count the key strokes or time the process. This would give you one way to evaluate competitive systems. Be careful that you are making a fair comparison between like levels of detail capture. A fast system might fail because it doesn't collect enough details. Some of the speed issue might be related to your network configuration and loading. Consider network traffic as a component of the system speed. On a slow network, even a speed demon system will operate at a snail's pace. The best way to look at CMMS is to load a working copy or demo on your network and test it. The speed of the sorts, screen changes, reports, and inquiries is likely to be the fastest you'll ever see. The system will gradually slow down as you add files.

Indications of a speedy design includes on-line look-up tables to ease the way when the work order is missing data, copy function for common repairs, speed typing where you can type the first few letters and have the computer fill in the rest, field duplication keys and programmable defaults (with the most common choices as defaults). The other thing to look for are speed keys (such as Alt + Function keys) to move around the system to avoid the slow but friendly menu structure.

Some of the alternate paths for data entry include bar code scanning, modem, radio, or LAN communications from other data systems (such as time card information from the time keeping system), umbilical connections to vehicle on-board computers, direct feeds from factory floor communications network and radio transmission from handheld terminals. Systems are in use to read handwriting, typing and check boxes of work orders directly with retyping information. Verbal data entry is starting to show up in selected venues.

Part 2—Look at Second

Masterfiles: The masterfiles are the fixed information about the assets, parts, mechanics, and organization. The masterfile structure reflects the designer's biases more powerfully than any other part of the system.

Look for: Completeness is the big issue because it is very difficult to add any fields to a masterfile after it's in use. Not having space in a masterfile for information that you want to store (perhaps after system is in use) is a common major difficulty.

If a data element is not in a master file and it is not collected in data entry, then it is not in the database. If it is not in the database, then you cannot easily analyze it with the CMMS. The easiest example of this problem is when you want to look at maintenance activity related to some outside facts such as downtime, shift change, or even contract negotiation. If those outside data items are not on the database, then analysis will likely be manual.

In large networks or mainframe computers, the data might be in another system. In that case, the analysis could be done with special programming on the computer or in a spreadsheet package. The usual process to determine the contents of the major masterfiles is a two-step process.

The first step is to look at your data needs. Make a list of the elements necessary to produce the information you need to manage the operation. The second step is to survey several systems' master file layouts to look for good ideas and additions. The first step could be repeated after the system survey.

One system has 105 64-character fields in the equipment masterfile. The first 10 fields are reserved by the system for their canned reports and inquiries. You have 95 large fields to store information. Systems such as this are very flexible. The price you pay for flexibility is the setup is more complex and requires more effort. You have to design the system! On another system there was a field of 35,000 characters (that's about 8 pages of information) for comments!

The following is an example of the typical data required for the asset file in a building maintenance package.

Typical record for a building that has a water pumping system from a cistern (supplied by city water)—information is taken off the survey sheets.

Asset number:	MWI #4
Asset description:	Water pump system
Location:	South basement main building
Manager responsible:	V. Santiago
Manufacturer:	Pacific Pumping Company, Oakland, CA
Model number:	Series III
S/N:	51B39721

Specs, Elect.: 3 pump units tied together through control panel 208V, 3Ph, 178 amps max.

Connection to which asset: Cistern, potable water system, electrical system panel #4

Condition: One of the three pumps out of service, the other two alternate weekly

Work to be done: Rebuild or replace bad unit

Estimate: Central pump quoted $12,000 to rebuild

Prob. of replacement: Immediate on 1 unit, 5–10 years on other two

From Asset Detail Sheet:

Vendor: Universal Plumbing Contractors

Installer: Universal Plumbing Contractors

Date Installed: 1972

Date in service: 1973

Original Cost: unknown (part of general contract not detailed)

Warranty: 1 year

Control Panel Pacific Pump

Model Series III

Below are examples of masterfiles from a large vehicle maintenance system (from Petro-ICC fleet maintenance system). Each of these masterfiles is an actual file in the computer. The information in these files comprises the system. Since the masterfile data are different for each organization, each "system" is unique.

Unit master
Product class master
Technician master
Computer system master
Class master
Product master
PM master

Reason down master
Trouble ticket master
Exception master
G/L master
Division master
Company master
Vendor master
VMRS
Influence master
System master
Permit master
Tire master
Meter master
Critical message master
Area code master
Cost allocation code master
Facility master
Location master
Product group master
Cross reference master
Inventory master
Type code master
Operator master
Organization master
Manufacturer master

An example of masterfiles from a small PC-type system (masterfiles from VEMS, a PC vehicle maintenance system, compliments of Metzco) follows:

Equipment master
Employee master
PM master
VMRS master (supplied with system)
Work accomplished code master
Fixed company information master

Part 3

Processing: The daily transactions are processed either in a traditional batch mode or on-line. Processing updates the PM schedule, summarizes detailed repair data for reports, machine histories, and keeps all financial accounts current.

Look for: Does it work? Process some data through the full cycle and see if all the accounts, schedules, and master files are updated correctly. Accuracy and completeness are the difficult areas in this item. Most of your bugs will occur during unusual processing conditions.

Reset the system's clock for the year 2000 and run some work orders. Print some reports that include 1990's data with 2000 data. Many bugs will show up in software that uses the last 2 digits of the year (instead of all 4 digits) to calculate elapsed time. Test if possible by printing a report on a machine. Then add a work order on that machine. The test is to print a second report on the machine which should be different from the first by exactly the amount of the work order that you entered.

On one mainframe system, the processing program represented 4 or 5 months of intensive work. After the original programmer left the company, none of the other programmers wanted to venture into the code. Instead of fixing the problems, they insisted they wanted to rewrite the entire sequence from scratch.

Part 4

Demands, Reports, Inquiry Screens: The demands on a maintenance system include reports and screens. There should be reports where there is a large amount of data or where analysis is required. Inquiries should not have to require going to print. Imagine how you expect to use the system and then see how the system will behave.

Look for: Many different ways to look at the data, complete basic set of reports and screens, future ability to alter or add reports/inquiries to suit your changing needs and growing expertise.

The reports and screens are the reason that you bought the system. You've been feeding data and maintaining master files for this payoff. The reports should be useful, not too detailed (with the ability to go to a detail level), include the information you need (not results from 10 other divisions), and be easy to read.

An example of useful levels of detail can be taken from the popular financial program Quicken® (Intuit Corp.). You can ask for a summary of the totals of all money spent by category. If you highlight one total and double click, the detailed transactions that make up that number pop up in a window. Hitting the escape key returns you to the original report. In this way the details don't overwhelm you and are easily available.

A Canadian maintenance manager complained that his system gave him too much data. Every week he was treated to a 1400-page report of all maintenance information with detailed comparisons to other divisions. He referred to three or four pages and occasionally skimmed the report to see what his buddies were doing in his old division. This is a waste of resources and paper. He asked to have a summary weekly, and the big report come monthly or quarterly. Information services reported that they couldn't find the time to make the change.

Role of the Manufacturer

OEM equipment manufacturers have a critical role in the ease and efficacy of the CMMS. As of this writing, the manufacturers, for the most part, hide details of what is happening in the field from the maintenance departments. By opening up the books they might suffer increased warranty costs or increased liability from admitting that a problem exists. Have you ever called a manufacturer help-line to find out that no one has ever complained about that problem before?

The OEM's don't want to disclose details about parts because that detailed information might cost them parts business or parts markup. The markup on parts is 300% more profitable than the markup on OEM equipment. Manufacturers will go to great lengths to preserve their parts business.

Call to Arms

Buyers should demand the following (or you won't buy). The format should be on diskette or CD ROM (or downloadable from the Internet).

1. Bill of material with part numbers, descriptions, basic specifications.
2. Lists of repair acts with labor estimates, parts lists and tool lists.
3. Drawings of machine, installation drawings.
4. Operators manual, detailed TLC (tighten, lubricate, clean) instructions and sketches.
5. Drawings of parts with a 15-year lock. These drawings cannot be viewed until the manufacturer will not support the machine, then the files unlock (I can dream, can't I).

Evaluating CMMS

You need to ask yourself—and your vendors—how to avoid the most common pitfalls of choosing, purchasing, and installing computer control and information systems. A full-function CMMS should be able to help in many areas. Many organizations purchase systems to solve specific problems. They don't need other functions or don't consider them important at the time of purchase. The following 50 questions will help you focus your attention in the various areas. They are not in priority order. Also consult the maintenance fitness questionnaire for additional ideas.

Fifty Questions to Help Your CMMS Search
Work Order

1. Produces an easy-to-use work order that allows future conversion to bar codes and other improvements to technology.

2. Work order classifies all work by some kind of repair reason code: PM, corrective, breakdown, management decision, etc.

3. Provides an easy way for a single person or designated group in maintenance to screen work orders entered by customers before authorization that work can begin.

4. Prints up-to-date lockout procedure on all work orders automatically.

5. Automatically costs work orders.

6. Provides status of all outstanding work orders.

7. Records service calls (who, what, when, where, how) which can be printed in a log format with automated time/date stamping.

8. Allows operations people, tenants, or facility users to have access to the system to find out what happened to their work request.

9. Records backlog of work and displays it by craft.

10. Work orders can be displayed or printed very easily.

11. The system facilitates labor scheduling with labor standards by task, ability to sort and resort the open work orders by location of work, craft, other ways.

12. Records changes to inventory (receipts, chargeouts, physical inventories).

13. Does the storeroom part of the system have part location to help the mechanic or store keeper find infrequently used parts.

14. Can the system generate a parts catalog by type of part, vendor with yearly usage to facilitate blanket contract negotiation.

15. Does the system recommend stock levels, order points, order quantities.

Maintenance History and Reporting
16. Maintains maintenance history that is detailed enough to tell what happened.

17. Provides information to track the service request–maintenance work order issue–work complete–customer satisfied cycle.

18. Provides reports for budgets, staffing analysis, program evaluation, performance.

19. Provides information for work planning, scheduling, and job assignment.

20. Is able to isolate all work done (sort, arrange, analyze, select, or list) by work order, mechanic, asset, building, floor, room, type of equipment or asset.

21. Provides the ability to easily structure ad hoc (on the spur of the moment) reports to answer questions that come up. This is sometimes called a report writer.

22. Has the ability to generate equipment/asset history from birth (installation, construction, or connection) with all major repairs and summaries of smaller repairs.

23. System reports are designed around Pareto principles where the system helps identify the few important factors and helps you manage the important few versus the trivial many.

24. System reports on contractor versus in-house work.

25. Provides reports charging back maintenance cost to department or cost center.

26. Has reports with mean time between failures that show how often the unit has been worked on, how many days (or machine hours) lapsed between failures, and the duration of each repair.

27. Will the system highlight repeat repairs when a technician needs some help.

PM System
28. Allows mechanics to easily write up deficiencies found on PM inspection tours as planned work to be done. System then automatically generates a planned maintenance work order.

29. Automatically produces PM work orders on the right day, right meter reading, etc.

30. Is able to display work load for PM for a future period such as a year by week or month by trade.

31. Is able to record short repairs done by PM mechanic and actual time spent.

32. Does the system support multiple levels of PM on the same asset, does it reset the clock if the high level is done (if you do a yearly rebuild, does the monthly PM clock get reset).

33. PM's are generated by location by trade to facilitate efficient use of people and minimize travel.

34. Allows the input of data from Predictive Maintenance sub-systems.

35. Highlights situations where the PM activity is more expensive than the breakdown.

36. Are there simple reports that relate the PM hours/materials to the corrective hours/materials to the emergency hours/materials. This will show the effectiveness of the PM program.

General
37. Can the system handle 3–4 times more assets than you imagine having.

38. System has a logical location system to locate assets and where work is done.

39. System tracks the warranty for components and flags warranty work.

40. Is easy to use for novices, and quick to use for power users.

41. System is integrated or can be integrated to purchasing, engineering, payroll/accounting.

42. Can the system easily handle a string PM such as a lube route, filter change route.

43. System runs on standard computer hardware, not some special hardware incompatible with everything else. Is the system compatible with Local Area Networks if it is a PC product.

44. System vendor has filled out vendor information sheet and has the financial strength to complete the contract (and stay in business for several years).

45. Does the vendor have software support people, can you easily get through to a person. Is there an 800 number. Once you get through, do the people know the product and something about maintenance. Is there an Internet site with technical support, user discussion groups, updates available for downloading, and other useful information.

46. Can the vendor provide economical, necessary customization. Is this capability in-house.

47. Does the vendor have a local installation organization.

48. Are they experienced in management of installation projects of the size of your facility. Do they have start-up experience with projects this size.

49. Are the vendor's technical people well cross trained (software, hardware, and reality wear, like how a real building works). It's important that the computer people have experience with building/facility maintenance.

50. Has the vendor been in business 5 years or more.

35
Maintenance Parts and Part Vendors

Part analysis is fundamental to maintenance control. Most maintenance jobs require some parts so that reducing part usage will automatically reduce labor. Follow your dollars of part usage to uncover areas where your time would be well invested. Parts are the pointer to the maintenance problems. Parts consume 40% to 70% of every maintenance dollar. Continuous improvement will follow from this investment.

The issue of parts and materials is a multifaceted one. This and the next several chapters concern some aspects of the parts and materials issue: maintenance stockroom, inventory control and part vendors, the Internet and maintenance, how maintenance interfaces to other departments, and accounting issues of maintenance.

The other issue is false savings. Saving money on parts in a vacuum (without looking at the other factors) may cost the organization 100 or more times the savings in additional costs. The actual part cost might be the smallest factor in the entire equation. For example, downtime costs per minute on an automotive assembly line ($4000–$6000/minute) might cost more than all of the bearings in one section of the line. Having a vendor who will drop everything to deliver the bearings needed might be worth hundreds of thousands of dollars more than the value of the parts themselves.

One of the strangest circumstances in maintenance is the way parts are accounted for. In many organizations, when parts are purchased for maintenance they are expensed immediately. Unlike raw material inventory, they are never assets of the organization. Raw material is an asset and is on the asset side of the ledger

325

(same side as cash or receivables). Even if maintenance parts are kept on the shelf for future use they do not exist as far as the books are concerned. Keep this in mind when discussing maintenance inventory with accounting-oriented top management people. For a more complete discussion of this issue, see the section on accounting.

Concept of "Insurance Policy" Inventory Items

Some items for critical units are stocked because the lead time on the part would create unacceptable downtime. These parts are an insurance policy against excessive downtime. In the electric utility industry these parts are called capital spares. They are capitalized and depreciated like a machine or other asset (unlike regular inventory items that are expensed).

These parts have to be considered differently from other parts. They have long lead times and high downtime costs (for the associated equipment). These parts can be considered insurance policy parts. They are kept in stock *even if they are never used*. Consider them like your organization considers liability insurance. You never want to use it but if you have to you're glad it's there.

An excellent example concerned the inventorying of a spare rotor by a southern power generating utility company. The part is for a peak load shaving turbine. The rotor cost a few hundred thousand dollars and had a lead time of 24 months. The cost of downtime for this turbine during peak load was almost $150,000 per hour (cost of purchasing power from the grid). This insurance policy was well justified. The goal is to carry as few insurance policy parts as is prudent—and to carry all that are necessary.

	Insurance Policy
Premium:	Few hundred thousand dollars for rotor plus storage space and upkeep
Risk Insured:	Downtime @ $150,000/hour
Beneficiary:	Power Production

Do you have specialized, hard-to-get parts? Are these considered insurance or do they affect your performance measures? When looking into insurance policy parts, always investigate alternative stocking strategy such as shared parts (several companies share expensive spares), vendor stocking, or, as an alternative, set up backup systems so the asset is not critical.

Analysis of Parts Usage

Dividing your parts into categories will facilitate analysis to lower your parts costs. We are using analysis to uncover the parts which consume your dollars from your inventory. These "bad actors" are indicators of situations where an investment of your effort could pay off handsomely.

The way to analyze inventory is to sort each part by dollar volume (price times yearly usage). In one study it was determined that the top 7.3% of the line items represented 76% of the yearly volume (see Figure 35-1). These high-moving items are called the "A" items. In the same study, the "B" and "C" categories were the remainder of the line items (92.3%) and only comprised 24% of the dollar volume.

Inventory Technique of Analysis

1. Divide your inventory into "A" items and other items.

2. Apply rigorous purchasing and negotiating techniques to "A" items to lower costs. This is an excellent application for the skills

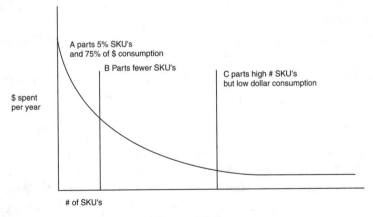

Figure 35-1

of the purchasing department. Use catalogs of your purchases, blanket orders, long-term commitments, etc., to reduce costs.

3. Review specifications to get better parts at the same costs or equivalent parts at lower costs. This is the situation where some level of engineering can pay off in big returns. Areas can include lubricants, will-fits, belts, fasteners, wiring/electrical.

4. Apply sophisticated standards to setting reorder point and economic order quantity. Factory production inventory control experts have long studied inventory strategies. The "A" level items respond best to these techniques.

5. Consider creative new vendors, purchasing modes and approaches such as: instead of purchasing high usage fasteners from automotive distributors go directly to screw jobbers or manufacturers. Another significant savings is to install tanks for oil and antifreeze and purchase truckloads.

Example: Change in purchase strategy—installing tanks for motor oil in a fleet maintenance operation that uses 25,000 gallons per year:

Purchase Units		Cost per Gallon	Cost per Year
Cases (24 Q)		$3.50	$87,500
55 gal drums		$2.50	$62,500
5000 gal bulk deliv		$1.90	$47,500
Savings if you change to:		55 gal	5000 gal bulk
If you now use:	Cases	$25,000	$40,000
	55 gal	-0-	$15,000

The current cost of above-ground indoor tankage is $1.00 to $1.25 per storage gallon. Payback on a 6000-gallon tank could be as little as 7 months ($8000 installation costs being funded by $15,000 return or $8000/$15,000 = 6.4 months). Additional investment in piping shop for overhead dispensing could cut labor costs on add-oil and oil changes. Other savings include reduced contamination, reduced labor, increased efficiency, reduced need to dispose of drums or odd lots of lubricant.

Manufacturer and Vendor Cross Reference

Manufacturers purchase parts (such as filters, switches, axles, transmissions, seals, bearings, etc.) directly from OEM manufacturers of those items. For example, Fuller truck transmissions are available from virtually all truck manufacturers. Many of these manufacturers assign their own part numbers to the Fuller parts. There may be as many as 10 different manufacturer's numbers for the same Fuller part. These are identical parts from the same factory with different part numbers, not after-market parts that are reverse engineered.

There are three reasons why knowing the interchange can give you an edge in parts purchasing and inventory control.

1. *Reduce Your Inventory:* You may be stocking the same physical part under several manufacturers' part numbers. This is especially true if you have many types of equipment. Among the parts of this type, you may be able to reduce your line items by 10–15% and your inventory level of these items by 5–10%.

2. *Reduce Parts Cost:* Once all of the interchanges have been identified, you will find that different manufacturers charge different amounts for the same parts. Each manufacturer has a different volume (pays a different price) and has a different markup. There may be as much as 20% difference between the highest and lowest price on the same part. Your organization may be purchasing higher quantities of, for example, SKF bearings and be getting a better discount than on Timkin bearings.

3. *Reduce Downtime:* You may be out of stock of the primary part, for example, a MACK 235414. By searching the interchanges, you might have the equivalent Ford part C6TZ7101A. These are identical to the Fuller part #14326. If you stock-out on all the interchanges you have many vendors which increases the probability one will have stock. On older or obsolete parts the interchange route may be the only way to easily locate the part.

There are interchange libraries for most factory parts and for heavy duty trucks.

Vendors of Maintenance Parts

Reduction in parts costs is one of the ways to reduce your overall maintenance bill without changing the entire way you do business. Reducing costs means negotiation with your parts vendor to increase your discount. Particular attention should be paid to the items that turn over the most rapidly.

For example, the fleet field has one of the most complex and fluid sets of discounts. In this field, parts are usually purchased with discounts from the published price list. These discounts have names (such as fleet, jobber, etc.) whose definition varies and whose discount varies in different subspecialities and different locations.

The lowest discount is the *fleet* discount. Every fleet should be able to purchase at this level. Only the smallest non-stocking fleets should accept this level. The second tier is *dealer* pricing. This is actually only slightly better than fleet pricing. Even moderate sized fleets should negotiate to the next levels. The discount level is *jobber* level. This is the price small parts houses pay for their parts. Fleets with inventory levels up to $20,000 could save some money at this level. Major fleets with inventory in the $100,000 range and above should be shopping from the *warehouse* distributor. WD's purchase directly from manufacturers and usually offer the best discounts. Since they don't carry all lines, some flexibility is important. In some cases, you can purchase directly from the smaller *manufacturers*. These prices can be very attractive if you are purchasing in large enough quantities.

Vendor Selection

If the contract exceeds $50,000 for the next year and the potential firm is small, then review the financial status of the organization. This would include Dun and Bradstreet, trade references, bank references, financial situation of owner (or annual report if public). Do they have the financial clout to support you?

Review in detail their terms of sale. What is their published return policy, stock balancing policy, warranty policy, shipping/delivery/pick-up policy; are there any incentive plans?

Ask for some non-competitive customer references. Ask the customers about delivery, returns experience, stock situation, accounting problems and accuracy. Ask how the counter/phone people are to work with.

Look at their inventory and decide if it matches your needs. If you visit the warehouse, how does it look. Does it look like you'll get the parts you need when you order? Do they carry the lines you want? If you give them a blanket contract, how will they treat the low volume one-shot type purchases.

Steps in the Review of Parts Vendors
1. In order to negotiate with the parts house, warehouse distributor, or manufacturer, you have to know your current prices and the volumes you currently purchase.

2. Establish a parts book (most computerized systems will print this) with current, last price and average price, monthly usage, and usual purchase quantity.

3. Many parts houses respond with lower prices if blanket quantities are specified, guaranteed monthly dollar volume or some other sales guarantee is offered. The best guarantee for you is to guarantee that when you buy a part it will be bought from them with an estimate of last year's volume. Then there are fewer arguments about committed quantity.

4. For several vendors (and several level vendors), compare their net costs against your parts book costs.

5. Pick the best two vendors for each of your fast moving parts. Review the vendor's history. Reduce list to a few general vendors.

6. By reviewing the prices, swapping parts from list to list, and further negotiation, allocate the purchases to each vendor to support whatever guarantees you made to secure the advantageous prices.

7. Inform winning vendors, update your parts book (or the computer system then generate new parts book). Inform losing vendors what it took to get the order. Inform second place finisher that they will get the first call if the prime source gets into trouble. Secure "next best pricing" from second source.

Specific Criterion for Selecting a Major Component Rebuilder

- Does the rebuilder offer a reciprocal warranty (with rebuilders in other locations)? This is particularly important for fleets as the vehicle might be 1000 miles away when a newly rebuilt component fails.
- Is the rebuilder certified by the OEM? In addition, is the component updated to the latest engineering revision of the manufacturer? Is the component rebuilt to OEM specifications?
- Can the rebuilder give you an analysis of why the component failed and keep records on previous rebuilds of the same component?
- Will the rebuilder stock your requirements on a consignment basis?

Partnership

It is a very competitive market for general parts vendors. These distributors must distinguish themselves in order to build long-term relationships with maintenance departments. The following example is typical of the industry where the goal is bigger than just getting the next order.

ABABA Bolt, Inc. of El Cajon, California, designed the ASAP (Ababa Stock Assurance Plan) which, based on their program description, provides bolt users inventory management, sales automation, applications assistance, and quality assurance. These added services add value to the sale and make it desirable to stick with one vendor. Keep in mind that in maintenance frequently the soft costs (figuring out what bolt, thinking about how many and if it should be stocked, setting it up in the stores system, placing an order, receiving, matching up the packing slip, etc.) of parts exceeds the hard costs.

In their inventory management section, they set up racks and bins for bolts where they are needed, with the sizes and types used. Backup products are stocked in the Ababa warehouse until needed. Ababa personnel restock the bins with the correct products on a JIT basis. A bill is generated by the restocking amounts.

The sales automation uses computerization to automate the buying process. Bar codes on the bins are scanned and bins counted to determine the reorder amounts. The purchasing department just has to approve the entire sheet which serves as a PO, picking list, and packing list.

Since the bolt house deals with bolting issues on a daily basis, they have wide experience. They help the customer look for the lowest installed cost. Ababa also maintains a bolting laboratory for testing and applications engineering.

Finally there is the quality assurance program. It includes mil specs, dimensional reporting, lab testing, and other post-production activities. The entire quality assurance program is documented in a manual available to the customer.

In this example, a company goes out of its way to become part of a customer's organization. The savings from the merger comes from elimination of the duplicated overhead in the customer organization (stock system, applications engineering, automated purchasing, and quality system). The best maintenance vendors are burning the midnight oil thinking of how they can add value to their bearing, bolt, bushing, or whatever they sell to the maintenance market.

Buying Service

A buying service replaces your purchasing department with a group that has significant expertise with the type of items you are buying.

Consortium Buying

With the advent of the giant chains in the retail world, small retailers cannot compete because their volume is smaller which makes their purchase price higher. In certain industries such as drug stores and appliance stores, the retailers got together and negotiated their purchases as a group. The appliance consortium buys for 300 or more small retailers. They can now compete with the chains.

Replace the Stockroom (outsourcing)

Keeping a stockroom is expensive and is not something for amateurs. In a military installation, they bid out the storeroom function. Now when a tradesperson on that base needs a part, they go into the same building, except there is a Graingers sign above the door. Graingers supplies all of the MRO items needed from their catalog, special local lists, and their already powerful stocking system. This is an example of outsourcing a entire function.

Consignment (Rack Jobbing), Guaranteed Stock Level

If you go into a convenience store (like a 7-Eleven) you will notice chips, bread, milk, newspapers, etc. These items are owned by the vendors and are paid for when they are sold and replaced. The same thing happens in maintenance storerooms with bolts, bearings, filters. Dealers will come in, usually buy your existing stock, supply or clean up your existing shelving, survey your equipment to see what should be stocked, and deliver and fill the shelves. Every month they return, clean up the shelves, and restock what's been used and prepare a bill. This cuts your storeroom costs and gives you adequate stock levels without tieing up organization funds.

Blanket Ordering, Prenegotiated Bid Items

Running a parts business is a risky proposition. With adequate volume and even small markups you can make it. A smooth running parts business built on volume can make money on the lower volume buys (at higher markups). One of the ways to build volume is to guarantee low prices if the organization promises all their business to you through a blanket order.

A way that the part vendors can help the purchasing department is to create prenegotiated bid items at the beginning of the year. This arrangement allows the planner, supervisor, or tradesperson to buy selected items without specific action from purchasing. It speeds up and cuts costs of the internal process.

EDI Linkup, Maintenance Networks, Engineering Hookup

Sophisticated electronic linkages between customers and vendors are becoming more and more common. Graingers publishes an electronic catalog on CD ROM that includes software to link into their network. This linkage allows customers to place orders, check stock, check on orders without leaving their computer screens. In some cases, the vendor's engineering data are available on-line. If you need specific data to solve a problem, you can query the database and get a drawing or technical specification.

The next step is dedicated maintenance networks. These networks are unique (in the factory arena) in that they tie a maintenance user to all of their vendors with one interface. You shop for maintenance parts at your own private maintenance mall! The maintenance network ties order entry, stock checking, engineering data, and E-mail together for a nice clean package. Your maintenance users can learn one interface and "talk to" all of your vendor computer systems. Detailed information on maintenance networks can be found in the chapter on the Internet.

36
Maintenance Stockroom and Inventory Control

If you could get the parts you need, at the lowest cost, without downtime (due to parts) and without inventory, then you don't need to maintain much of a maintenance stockroom. The reality usually is that the speediest source is also frequently the most expensive. Many parts are difficult to get on short notice. Even with a fully equipped stockroom, equipment has downtime due to stock-outs of critical parts.

The maintenance inventory is designed to support the performance of the maintenance function. The building, stocking, and upkeep of a maintenance storeroom must reduce the overall cost of providing maintenance. Decisions about inventory *always lead back to an economic justification.* Economic justifications include avoidance of expensive downtime or avoidance of breakdowns that will cause loss of customer goodwill. Both can impact the bottom line. Non-economic justification of high inventory usually covers holes in the control of the maintenance function.

There are three types of inventory. Each type is subject to different types of analysis. The problem for maintenance is that the first two types are well understood and well studied in business schools, and the third type is almost unknown.

Type	Function	Analysis Techniques, Strategies
Resale	To be bought by customers	turnover, sales per square foot, when last used, rack jobbing, consignment, cross docking
Raw materials for production	To be used to make something	JIT and daily or hourly delivery, deliver to machine with no stock, 8 hour window, tie vendor to production line
Maintenance	Used to provide the capacity of an asset	dollars per asset value, cost and amount of downtime, insurance policy stock, have the part in stock or available in 24 hours

20 Symptoms of Inadequate Inventory Control

The following problems indicate that inventory is not under proper control. Review your operation to see if these symptoms are present.

1. Stock-outs on critical parts when they are needed.
2. Inventory for units no longer in service.
3. Inventory on shelf for 1 year or more.
4. Inventory cannot be reconciled (parts on shelf $\neq$ on-hand + receipts – usage).
5. Parts can be added to or taken from inventory without proper paperwork or computer entry.
6. Purchase orders issued after item was received.
7. Routinely items are purchased with petty cash on a rush basis.
8. Little knowledge of location, inventory level, turnover.
9. No established min/max, reorder points, E.O.Q's.
10. No competitive bids or sweetheart deals with certain vendors (not partnerships).
11. Parts purchased but never used; no accountability of parts used.
12. No proper storage, unlimited access to parts room.
13. No physical inventory taken.
14. Excessive hoarding of parts outside of parts room by mechanics.
15. No knowledge of the current value of the inventory.
16. No analysis of equipment to estimate spare part requirements (unit number where each part is used is not known).
17. No knowledge of quantity on hand at the moment.
18. Constant calls to vendors for emergency dropoffs.
19. Field reengineering changes parts needed without changing bill of materials.
20. Bad relationships between stores, maintenance, and purchasing.

The Elements of Inventory Control

1. Building adequate storage with limited access.
2. Requirement that all parts removed are recorded on WO's or equivalent document.

3. All parts must be received, price checked, physically checked, counted, and signed for (unless a partnership method has been designed).
4. Parts are assigned locations.
5. Periodic physical inventory is taken to verify quantity and location.
6. Some means for recording usage, price history, where used, and substitutions.
7. Periodically, parts are shopped and vendors evaluated.
8. Periodically, applications and specifications of parts are reviewed.
9. Periodically, part usage and lead time is reviewed to adjust min/max and economical order quantity.
10. Parts are divided into classes for different treatment.
11. Parts for assets that are retired are reviewed for use elsewhere or disposed of.

Ideas for Layout of the Storeroom

- Self-closing gate with fencing around area (the idea is to show that the organization values the inventory).
- If there is no computer or part room clerk, put usage log sheet where it cannot be missed.
- Number all shelve units, shelves, bins, and drawers.
- Drawer units make more efficient use of space than shelves, and protect items better.
- Give all materials a home. Consider how the part is picked.
- Make space for returns, and for incoming and outgoing rebuilds.
- Have a cage or area for reserved parts or for staging special incoming parts.
- Keep heavy parts down low, fast moving parts near window.
- Keep low value small parts outside room for free access.
- Arrange by family in order of popularity.
- Put items that are used together next to each other.
- Within a shelving unit, set it up like a book—left to right, top to bottom.
- Consider keeping top shelves empty for overstocks and expansion.

- Store in manufacturer's cartons after removing wires, strapping, shrinkwrap, etc.

Computers Can Help

Normally computerized inventory is a section in the CMMS. The two are very closely related. It is not very useful to maintain a maintenance inventory without also automatically costing work orders (parts charge-out cost × quantity) and automatic reduction of inventory when parts are charged to a work order.

The computer excels at applying rules and doing calculations over and over. Inventory is a perfect example of a task that is well suited for computerization. The issue with manual inventory is discipline. Any manual system that is also rigorously followed will limit stock-outs and high obsolescence costs. When computerizing the stockroom, review the elements of control (also in the beginning of this chapter).

Rules of a Computer-Aided Maintenance Stockroom

1. All parts removed are recorded on WO's or equivalent document. The system facilitates this element by entering all items used on WO's, and charging them to units. With more complete systems the stockroom can actually charge the part to the unit and WO at the time it's given to the mechanic.

2. All parts must be received, priced, physically checked, counted, and signed for. Much of the checking, such as exact part number ordered, price, quantity ordered, etc., is very easy to do and will probably get done more often. The exception is when you have a vendor partnership where the vendor performs some of these functions.

3. Parts are assigned locations; physical inventory verifies quantity and location. All parts' locations are logged on the inventory system. Parts are easier to find and count for physical inventory. Systems can usually generate a physical inventory form sorted by location with the part number, description, location, and a place for the physical count. Large stockrooms can benefit from hand-held computers with counting software (similar to supermarket inventory systems).

4. There must be some means for recording usage, price history, where used, and substitutions. Usage and price history are automatically captured by the system (requires no additional steps). Where used and substitutions can usually be entered into the part masterfile. In some cases, the system will capture the asset the WO has been pulled for and associate the part with that asset.

5. Periodically, parts are shopped, specifications reviewed, and vendors evaluated. Since the system can easily generate parts catalogs periodic shopping of higher volume items is a great deal easier. Performance reporting is also possible from some of the systems.

6. Periodically, part usage and lead time is reviewed to adjust min/max and economical order quantity: The real power of computerization lies in its ability to capture and analyze usage data and apply preset formulas to determine EOQ's.

7. Parts are divided into classes for different treatment. Certain systems can set up the ABC classes of parts through analysis of yearly dollar volumes.

8. Parts for units out of service are reviewed for use elsewhere or disposed of. Outside information such as asset retirement and changes in asset makeup usually have to be manually factored in. If the system has "where-used" as a data element in the part master file, it may have the ability to isolate all parts used on the retiring unit for review.

How Computerized Inventory Control Can Reduce Stock-Out and Obsolete Parts

The computer system can apply the check for minimum stock level every time a part is requested. If the parts are actually ordered, then stock-outs can be reduced to the level that you set. Inventory levels can be adjusted up or down by allowing more or less stock-out conditions. Once everything is settled down, fine tuning for seasonal variation and for age of equipment can bring inventory into line.

Many items are purchased for particular jobs and never get added to the inventory. The system can (and should) track these items and periodically report on items that start showing up regularly. You might start to stock these items to reduce costs, downtime, or both.

On the other end of the scale, we are concerned with inventory that hasn't been used. The system can easily print parts that have not been used in 6 months, 1 year, or 2 years. These parts can be investigated (see if they are hard-to-get "insurance" stock). If they are available from outside sources, you can try to sell or trade them for usable stock.

These symptoms indicate holes or voids in your organization's control of inventory. In small fleets, factories, and medium-sized buildings, you may have to put up with some of these situations because of inadequate volume to justify the control structure and people. However, *even the smallest operations can justify some level of control.*

There is a common misconception that computerization can somehow get your inventory situation under control "once and for all." The fact is that without physical and procedural controls, computerization will only make the situation worse. The wonderful advantages of computers flow only to organizations committed to the controls. Once controls are in place, computerization will greatly simplify the clerical work and the analysis.

Reorder Point and Economical Order Quantity

We will review a financial method of determining EOQ and reorder point. In an informal way, keep in mind the cost of issuing a PO, receiving the material, checking the paperwork, approval of the invoice, and issuance of the check. Frequently this number is $20–$100. Figure 36-1 shows the tremendous variability of this number. Consider arranging your purchases so that you have a minimum of $200–$400 on each order.

There are several interrelated numbers to determine the critical control numbers for an inventory system.

Annual requirements: How many of this part do you use per year? Referred to as R in formula.

Carrying cost and downtime cost: You determined the carrying costs under the section of true cost of inventory. The downtime cost has to be determined through discussions with the accounting department. For the example, 25% is used for the sum of both numbers. Referred to as K in formula.

Usage: Average parts used per month.

Lead time: How long does it take to purchase and receive the part? Include time from when you discover it has to be ordered to when it will be received.

Part cost: Unit cost of part or material. In the example, we are using $240.00 ea. Referred to as C in formula.

Reasonable maximum usage: During a period of maximum usage, what would be your maximum average usage?

Restocking cost: Also known as Cost of Acquisition (COA). How much does it cost to order any item from a vendor? These are the costs of purchasing, receiving, and accounting. For the example, we are assuming $7 in the first example, and $250 in the second. Referred to as S in the formula.

"S" is a can of worms. In typical resale inventories, the value of S is well established and can be used across the board. Maintenance is fundamentally different because different parts have vastly different S values. Dr. Mark Goldstein uses the following

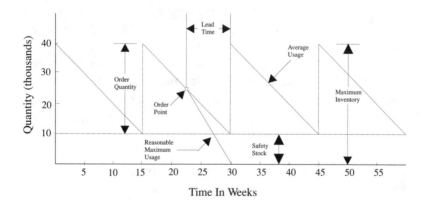

Figure 36-1. Saw tooth curve of inventory levels.

table to explain the different values of S based on what is known when you place the order. Blindly using any particular S value will cause the stock level to be vastly different than optimum.

Cost per line item for S	Description of Situation	Part Known	Price Known	Vendor Known	Delivery Date Known?
$7.00	Stock item from Grainger or McMaster Carr	Y	Y	Y	N
$70.00	Bearing shopped several vendors	Y	Y	N	N
$700–$7000	Formal bids and proposals, writing specification, investigation of vendors, advertising, responding to questions	Y	N	N	N
$7000–$700,000	Engineering bid where system or part is designed, vendors proposals and designs are analyzed	N	N	N	N

Economical Order Quantity (EOQ) = $\sqrt{(2RS/KC)}$

for a "Grainger" type item with low acquisition cost:
7.5 (or order 8) = $\sqrt{(2 \times 100 \times 7/.25 \times \$100)}$

but if the S is $250, then the EOQ increases to
44.7 (use 45) = $\sqrt{(2 \times 100 \times 250/.25 \times \$100)}$

As mentioned in the section on accounting issues for maintenance, the variability of the S factor could be the driving force toward streamlining the MRO supply chain. Anything that simplifies and reduces effort and reduces mistakes will have a great impact on both the cost and effectiveness of maintenance.

If you open any business text on inventory, you will find the S variable as a constant. The reason for this is that it is a constant in a department store or almost any retail inventory. Compare COA costs (or S in the formula) of buying a spare part for a Dayton wet-dry vacuum from Graingers to buying a pinion gear for a several-decades-old German mixer. If it only happened once in a while, there would be no problem. Maintenance requirements call for the "unusual" on a regular basis.

Order point (also called reorder point or minimum stock level): The order point is where the inventory on hand equals the safety stock plus the expected demand for the item during the lead time. Ideally an order is placed for the EOQ at that time (any hold up) could cause a stock-out.

Safety stock: The amount of safety stock equals the difference between the consumption of the item at normal usage levels for the lead time and the consumption levels at reasonable maximum usage during the lead time.

Maximum inventory level: This equals the safety stock plus the EOQ.

Obsolete Parts

One of the problems that the stockroom has to face is obsolete parts. This is a problem because the obsolete parts take up resources. They not only take up money but also space, and also they consume management time for disposal. Many of these parts can be found a home if someone has the time to search.

In *Maintenance Management,* Butler discusses the reasons that inventory becomes obsolete. Parts become obsolete because the asset is retired, engineering changes, field reengineering, asset has moved on to a different repair cycle so the failures change, and items are ordered on emotion rather than calculation.

Seasonal or Business Cycle Demands for Inventory

There is a problem with this fine level of analysis for specific types of parts. In a truck fleet, if we would do this analysis on heater system parts in the early winter or cooling system parts in the summer, we would be overstocked for 9 months of the year and hand-to-mouth or out of stock when we need the parts. In your analysis, consider the effect of seasons on its usage level.

Consider this issue when you are designing or purchasing a computer system. Few systems are "smart" enough to correct the Min/Max and EOQ by the season or time of the year.

How to Reduce the Inventory: Big Ticket Item Analysis

The first type of analysis breaks out big ticket items. In a review of fleet inventories, a major soft drink bottler found that these big ticket items (their definition is $200-up) consist of 15% of the yearly volume and 70% of the total inventory value on the shelf. Review your items that cost over $500. Consider the following ideas in your review.

1. Determine your 90-day usage and sell or trade the rest of the parts. *Note:* if the part is difficult to get and the cost of downtime on the unit is high, leave it on the shelf.

2. If you can get the part from a supplier (even at a small premium), consider selling or trading off your entire stock (or letting it run out). Remember that if you can get the part in 1 or 2 days, that is equivalent to stock because it usually takes that long to prepare the unit for a major repair.

3. The premium that you can afford to pay is related to your organization's carrying charges and the average amount of time the item spends on the shelf. If an engine costing $8500 normally sits for 5 months before use and your carrying charge is 15% per year, then the extra costs are $531.00. You might use 50%–75% of the carrying charge or $300 premium ($8800) to not have to stock the engine. Use 50%–75% because some of the carrying charges are fixed and would not decrease if you eliminated these items.

The resultant reduction to your inventory means you will have cash available for investments that can pay more significant dividends. Your stockroom can be better arranged to take advantage of the extra space. You also increase the number of inventory turns per year, which increases the efficiency of your use of money.

37
The Internet and Maintenance

The Internet has several uses for the maintenance profession. An advantage is that all the capabilities mentioned below are on-line, they are available 24 hours, 7 days a week. Usually the server computers (where the information is located) are up (available) all the time except when they are being backed up or serviced.

Internet and Maintenance Management

Advertisements and catalogs: Companies can make their latest catalogs available as soon as they are complete. It is much less expensive to provide the catalog on-line rather than printing them. It also saves your shelf space and trees, too! Because of the increasing cost of paper, expect to see a push for on-line catalogs.

Technical bulletins: The latest technical problem and fixes can be available minutes after the vendor's engineer's decide to put it on-line. No longer is there a weeks to months lead time to publish and mail the bulletins. Also, having the latest information is a click away.

Drawings, field modifications, manuals: The same way you are updated by technical bulletins, you can view manuals and download drawings. Field modifications can be fed back to the OEM engineering departments if that is appropriate.

FAQ's (frequently asked questions): In every field and on every piece of equipment, there are FAQ's. These are what most of the time of the telephone response department is spent on. These FAQ's can be posted to be read by novice or new customers. Since they are on-line, they too are available 24 hours, 7 days a week.

Technical help: Technical help is one of the greatest growth areas of the Internet. You can ask questions by E-mail and get answers back to solve your problems.

Locating used equipment and parts: There are classified ad sections where companies and individuals buy, sell, and trade equipment.

Software changes: This capability is already used by most major vendors of software. You visit their site and download the latest version of their software. Also, software that you want to try is also available as a download.

Directories of installers, vendors: When you are looking for vendors or installers, you can ask members of a newsgroup of interest, make an electronic query from a home page on the World Wide Web, or use E-mail to the companies' postmaster. In most systems, the postmaster is a person designated to manage incoming E-mail that is not addressed to an individual. They maintain directories of all their addresses and lists of resources to redirect your inquiries.

Access to library: Many university libraries and information databases are available on-line. The Library of Congress is putting their enormous library on-line. A group is making available the complete texts of great books available for downloading.

User groups: Do you own a CMMS and want to talk to others with the same system? Many user groups are going on-line as newsgroups (see Newsgroups). Here you can read other's comments about the software, ask questions of the whole group, get help, and gripe to your heart's content.

Newsgroups: These are groups that are bound by a common love, hate, interest, membership, or whatever. In early 1996, there were almost 20,000 newsgroups.

Example

General Pump of Mendota Heights, Minnesota, has developed a service on the Internet they call "Datalink." This site is up and running 24 hours a day, year around. It features downloadable engineering drawings, bills of materials for their pumps, new product information, discussion groups, training capabilities, and the ability to place electronic orders. This is the near future of the MRO business. Wouldn't you be more comfortable dealing with an organization where you could converse with engineers, other users, and experts, or where data are immediately available when you have a problem (rather than rooting through your files for an obsolete booklet)? You can find out about this service at http://www. generalpump.com.

What is the Internet?

The Internet is a giant network of computers connected together by telephone lines. It has an agreed-upon system of sending and receiving messages. It also has a standardized system of addresses. From your computer you can enter the address of any computer and any user anywhere in the world. The computers that make up the Internet include most large universities, most companies, and most governmental agencies.

The Internet is a direct outgrowth of work by the advanced research arm of the U.S. military called ARPA (originally the Internet was called ARPANET). The function of this network was to keep military installations connected in the event of a nuclear strike or terrorist action. It is a completely decentralized network with "communications backbones" (high-speed communications lines) spread all over the world. It is designed to survive even if major sections are removed or damaged.

The Internet took off in the popular press in 1994. By 1995, hundreds, then thousands, then hundreds of thousands of sites were created that could be visited by entering an address on your net navigator (software to help navigate the Internet). At the end of 1995, 10 million people in the U.S. used access to the Internet. In early 1996, AT&T offered 1 year free access to their 80,000,000 customers. Also by early 1996, one of the listing bureaus that keeps track of sites was receiving over 1000 site descriptions to catalog every day! The Internet is changing so fast that only generalities about it will be true one or two or five years hence.

The Internet is changing the way we communicate. The changes will be radical, like the changes brought to maintenance by the fax machine or the computer itself. The maintenance field is already in flux because of the new capabilities available on the Internet.

Who Pays?

One of the reasons the Internet exploded is the low cost of use. The Internet is funded by the military, universities, and organizations who set up servers. These servers are all linked together and are the nodes of the network. Each server pays for the high-speed communications line to a central backbone that snakes around the world.

How to Get Access

The lowest-cost option is to subscribe through a service provider. Most of the long distance companies and hundreds of smaller providers have cropped up with local phone access to the Internet. They charge $15–$25 per month for mail and complete Internet access. The on-line services such as America On-line also provide Internet access. Their prices are higher but the interface is much easier for casual users and there are capabilities not available on the Internet itself. In all cases you need a computer and a modem.

Some organizations can afford to become an actual part of the Internet. To become a server you have to connect to the nearest node or communications backbone. Individuals and small businesses that can afford the communications fees and hardware (in 1996 the costs were about $100–$500/month for a dedicated communications line, plus $5000 and up for the server and communications equipment). With this setup you could host hundreds of web home pages and provide all kinds of files and space to users. Most larger servers and the core of the Internet is designed using the UNIX operating system. Newer and usually smaller servers are now using Windows NT.

Addresses

Each address is unique. The address has three or four parts and cannot have commas or spaces. By convention only lower case letters are used. On FTP or TELNET (explained later) sites, the name is further followed by the path to the directory where the information is stored. In UNIX, forward slashes are used to show the directory level (rather than backward slashes on PC-type computers running some version of DOS).

Domain: This is the very last part after the period. It puts you into a large category like ".com" for commercial, ".edu" for education, ".gov" for government. Outside the U.S., the domain might be a country such as "uk" for Great Britain or "fr" for France.

Organization or location: Each user is part of a higher level organization. For example, organizations try to get their name or initials such as "osha-slc" for OSHA Salt Lake City. @ is the divider between the user name and the organization name. The organization name follows the @ symbol. This is optional where there is no particular user.

User name: This is the unique name of the specific user. Each user is assigned a name within the organization and domain. A user name (like "johnjones") could be duplicated in another organization.

Major Sections and Capabilities

E-mail is one of the most used and most powerful parts of the Internet. This channel links the entire world together and enables researchers, business people, and even elementary school kids to send messages worldwide. There is no extra charge beyond the local phone call. A recent survey showed that over half of the users of the Internet just used E-mail.

World Wide Web (WWW) is the hot area of the net. The explosive growth mentioned earlier is in this part of the Internet. Every organization that can afford $100/ month can afford a home page on the WWW. All of the addresses that start with http:// are World Wide Web sites. Originally the Internet ran on text only. Many parts still run on text only. The World Wide Web was designed to allow graphic transfers of information. This is also the area of most interest to the maintenance community.

FTP (file transfer protocol) allows you to go to thousands of computers and transfer files to your own computer. These files could be weather maps, programs to solve engineering problems, games, electronic books, bibliographies, or just about anything else. FTP sites allow access and have public directories that you can browse (although you have to use the next capability, Telnet, to browse).

Telnet allows you to go to a remote computer and act like you are directly connected. You can browse the directory, run programs—in short, do almost anything a local person could do. The system administrator set up what Telnet guests can do and where they can go. Usually Telnet sites set up Pub directories for the public. In selected systems you can cruise throughout almost the whole server.

Newsgroups are people interested in particular topics like real estate investing, wine tasting, presidential politics, or just about anything else. There are tens of thousands of Newsgroups.

Search methods are essential in a entity growing as fast as the Internet. There is no yellow pages because it would become obsolete before it is printed (actually there is an Internet Yellow Pages, it is not exhaustive but rather instructive in helping you in finding where to start). The search engines look at their individual areas like Newsgroups, WWW, FTP sits, etc.

Archie search for files for FTP

Gopher menu driven server program to search entire Internet

URL (universal resource locator) will help you locate a WWW address

Veronica program to help you search through gopher sites to find filenames

WAIS Software used to index large files; it finds and retrieves documents using key words

Yahoo (http://www.yahoo.com) searches WWW sites by keyword

Some Sites to Visit

At a recent trade show, 20 major vendors were touting their home pages. By January 1997, Dr. Mark Goldstein predicts the growth will be so explosive that you will have to have Internet EDI (EDI—electronic data interchange—a fancy way to allow buying and selling directly by computer) to be viable in the MRO field.

The venerable Thomas Register is on the world wide web. No maintenance-oriented tour of the web would be complete without visiting http://www.thomasregister.com.

Try http://www.facilitiesnet.com for something different. This is an advertisement for a facilities oriented pay network. The annual subscription is under $200 for an individual (there is also plenty to do for free). Like many sites, this one is sponsored by a magazine, in this case *Maintenance Solutions.*

Another service is http://industrynet.com. All kinds of companies have joined together on this service bureau. Many links to company home pages. Also try http://www.usinternet.com to find links to information or organizations you seek.

For the engineers out there, there are many sites to visit. Start with http://www.csemag.com. This is the site for *Consulting-Specifying Engineer* magazine. CMMS buffs can talk to DPSI software company at their home page at http://www.industry.net/DPSI. If cleaning and housekeeping are interests check out http://www.cleannet.com. It is sponsored by the Cleaning Management Institute and its magazine. You fill out a guest book and then you can look at the entire site.

Magnetic drives feature zero leakage, high efficiency, low maintenance and safe operation—want to hear more? Send an E-mail message to nml@nsrfc.ns.ca and talk to Nova Magnetics. Another pump company has mag drive pumps in Sonoma, CA. Instead of a long distance call sign on at http://www.crl.com/. The pump people really found the world wide web because Cat pump can be found at http://www.usinternet.com/catpumps.

Cutting tool users have access to 24 hour on-line catalog and a section on FAQ (frequently asked questions). Also new product announcements http://darex.com/sharpeners. If you want to visit Trane (the HVAC people) visit http://www.trane.com. My favorite transducer company is Omega. Their catalogs have graced my shelf almost two decades. They can be visited at http://www.omega.com. If you are interested in tool storage and retrieval (with automated charge-out) check out sales@remstar.com

38
How Maintenance Interfaces to Other Departments

In a factory, the maintenance department has to work closely with purchasing, stores, traffic, and accounting, in addition to production. In a school building, the maintenance department interfaces with the principal. Maintenance also interfaces with food service, security, transportation, and housekeeping (if they are separate). In a typical snack distribution fleet (like a chip distributer), the maintenance and sales departments have daily contact; in addition, accounting, warehousing and administration are also interfaced with.

This chapter explains the point of view of some of these departments that have not been discussed before, and their relationship to maintenance, and how to effectively work with them. The goal of each department is to serve the organization in its own unique way. Each department serves the organization with differing assumptions, tools, and world views.

Accounting Department

It does not matter if the organization is a sports stadium, space shuttle assembly building, or a microprocessor manufacturing facility, the accounting department will serve some oversight functions. Additional information on this important relationship can be found in the chapter on accounting issues of maintenance.

Topics of Interest to Accounting

Cost reduction: The maintenance budget was $7,000,000 last year and will be $6,500,000 this year with these changes.

354

Cost avoidance: At the rate of deterioration our costs will go up $500,000 next year unless we take these steps.

Cost control: The maintenance effort cash needs vary by 50% month to month for the last 3 years. With these changes we will be able to predict the cash needs to plus or minus 10%.

Increased production through increasing the availability of assets: We achieved 7950 hours of production last year on the big widget line. With these modifications and changes we will run 8220 hours next year.

Savings acceleration: Maintenance savings flow directly to the bottom line. The profit of a typical organization is $.05 to $.10 for every dollar of revenue. A dollar of maintenance cost avoidance or cost reduction is worth $10–$20 of revenue.

Insurance policy as a model for capital spares: The failure of any of the 30 parts on this list could shut down production for at least 4 weeks and probably as much as 12 weeks. We lose $1,000,000 of direct revenue each week we are out of production. We currently stock only 5 of these parts. For an investment of $375,000 we could stock the rest of the parts and insure the uptime against catastrophe.

Maintenance has impacts throughout the organization. Its impact affects the entire cost structure of your product. When talking to the accounting department specific causes and effects must be mapped. Look for costs outside the normal maintenance area for persuasive arguments.

Cost areas to discuss:
maintenance parts, labor

contract labor, service contracts

energy cost including electricity, gas, oil, coal

water costs

cost of shorter equipment life

cost of downtime

cost of scrap

cost of poor quality (variation in the process)

cost of accidents—medical, lost time, legal

cost of environment fines

cost of lowered productivity

costs of angry customers who go elsewhere

Charge-Back

One of the best ways to manage maintenance, where there are many masters who demand service, is to give each department a maintenance budget and charge them for every job. This is called charge-back. Maintenance becomes a business providing services for departments of the company. See the chapter on accounting issues of maintenance for additional details on charge-back.

Purchasing Department

The purchasing department's mission is to buy materials necessary for the business at the lowest cost, that meet the specifications. In addition, purchasing usually is involved in contractor negotiation, new machine procurement, and sometimes is in control of the storeroom.

Maintenance is a headache for most purchasing departments. Maintenance demands make it very difficult to implement continuous improvement in purchasing. The needs of maintenance are at odds with some of the methods usually used for improvement in purchasing.

Continuous Improvement Technique	Maintenance Reality
Reduce the number of vendors	Hundreds of vendors
Become a big fish for better service	Small buys
Planned purchases	Unplanned events
Consolidate buys, blanket orders	Uneven requirements
Take time to do it right first time	No time, rush, ASAP
Time for low cost shipment	Air freight, courier required
Maintain quiet, efficient atmosphere	Crazy, emergency room atmosphere

The purchasing department is one of the major interfaces to the outside world. It is also charged with the mission to protect the organization from certain types of lawsuits and fraud. In the U.S., all organizations are regulated by a series of commerce laws called the UCC (Uniform Commercial Code) which regulate commerce. Other countries have similar sets of statutes. Additional coordination responsibilities are added when the firm is a JIT (just-in-time) manufacturer.

Public sector purchasing departments are also constrained by statutes that define when items need public bids (possibly over $10,000), competitive bids (over $2500), and when low bid must be followed (almost all of the time). The process time for a major procurement might be 4 or more months (in the case of a city airport buying complex contracted services a year or more).

Every dollar saved by purchasing flows to profit. In some organizations the major portion of funds are spent by purchasing for raw materials and supplies. This direct relationship to profit sensitizes purchasing agents to saving money. Losing control of even dimes and quarters can affect profit by thousands of dollars at the end of the year. The actions (or inactions) by purchasing are sometimes painfully visible to accounting and organization management.

The main technique used to get the best price, specification, and delivery is shopping and contract negotiation. Because of the requirements from the organization's auditors and legal council, the preference is to shop around and get bids from several sources. In public agencies this process is required by law. In most organizations it is strongly recommended by customs and regulations. Fair shopping is the best protection the purchasing agent has against charges of fraud, incompetence, collusion with vendors, and other malfeasance.

As a result of shopping, the purchasing process takes time. Preparation, workload, complexity of the buy, experience, quality of history files impact the time it takes to issue a purchase order (PO).

To a maintenance professional, doggedly shopping each part looks like a waste of time. Shopping is good purchasing practice. Maintenance has good practices that are equally obscure to outsiders. A good mechanic will not necessarily fix the first thing they see until they understand why the failure occurred. That is good maintenance practice—it takes a little longer but in the long run makes sense.

The other problem that seems to occur with regularity is that the maintenance planner or supervisor locates a part needed. To locate the part, they work with a vendor (who might put significant time into the research). The requisition with all of the research and the vendor recommendation are sent to purchasing.

Purchasing seems to ignore the recommendation and buys the part from another vendor that is a few dollars cheaper. This is an unfortunate communication problem. Meetings with the purchasing agent and other managers should discuss and determine policy in these situations. Sometimes the value added is worth the higher price and sometimes it amounts to a sweetheart deal for the vendor.

A purchasing department for a large metal producer issues over 13,000 purchase orders in a typical year. In a recent downsizing their staff was halved. Their reaction, which will be hailed by maintenance departments, was to set up 30 prime maintenance vendors and several thousand commonly used items. They called these items "E" items. These vendors had pre-negotiated blanket contracts with the 30 prime vendors for "E" items. The maintenance planning department was given the authority and responsibility for buying "E" items from these vendors. They had a limit of $500 per line item.

The move in manufacturing toward JIT (just in time) adds to the complexity of the purchasing mission and increases the stakes. Traditionally buffer stocks insulated the factory (and the purchasing department) from costly stock-outs of critical raw materials. JIT techniques eliminated the buffers. Now when delivery problems arise, purchasing is on the front lines to contribute to solving any problems.

When talking to maintenance managers, one constantly hears stories about the purchasing department buying junk to save a few dollars. Purchasing uses specifications prepared by engineering, production, and maintenance to determine what it buys. Specifications are detailed descriptions of the qualities, performance, function of the item purchased. Usually these mistakes are traceable to the specifications being incomplete or incorrect. Purchasing is totally dependent on the completeness and accuracy of the specifications. They cannot independently judge the qualities of the parts or supplies purchased.

Over the years, organizations make mistakes and learn things to avoid. The purchasing department (in conjunction with legal counsel) keeps track of this learning in its "terms of purchase" (which are usually printed on the back of the PO). Some typical terms of importance to maintenance (definitions in glossary): changes, force majeure, design responsibility, warranties, termination, assignments and subcontracts, delivery, price warranty.

Ideas to Improve the Purchasing–Maintenance Relationship

1. After an emergency is over invite the purchasing agent or buyer into the shop to show them what was bought. Frequently a $35 solenoid is very impressive when actuating a $700,000 machine which generates $48,000 a day.

2. Give time when possible. Many emergencies are due to a lack of planning on your part. Work on getting the maintenance act together to reduce the number of unscheduled events.

3. When you or your staff does research into a vendor for a part, pass along your work but do not expect that it will be followed exactly. Have a meeting between emergencies to discuss this and work out how the research will be used. Also work out the position of the company toward vendors that take on some of the work load.

4. Look closely at specifications on items that have been giving you trouble. It is possible that your problem is in the specs and not in the purchasing department. Work with purchasing and engineering to develop good, usable specifications that are also easy to purchase.

5. Humanize the relationship by finding out about the people in purchasing. Find out what makes them tick and what pressures they face. Realize that purchasing serves masters of their own and that they also contribute to the overall success of the organization.

6. In larger departments try to negotiate a specialist to handle the bulk of the maintenance buys. In the best cases the specialist can be moved to the maintenance department.

7. When setting up the stock list of a new asset, get the purchasing department involved at the beginning of the process. Let them help with the stock list, initial vendor selection, etc.

Traffic

The mission of traffic is to move freight into and out of the plant using the most economic method consistent with the delivery timing requirements. Traffic saves the organization money by knowing the best modes, carriers, and strategies for moving freight. Maintenance usually represents a minor but persistent headache to the traffic department. The typical maintenance shipments are LTL (less than truckload) and time sensitive. Modes of shipment include truck, rail, air, and other. Within each mode there are several distinctions.

Truck can include small package delivery, LTL, truckload, special trailers (drop frame, flatbed, refrigerated). Truck also includes permitted loads (too large, heavy or dangerous for conventional moves). Rail cars have similar characteristics but have almost double the size and weight of trucks. Type of rail cars include boxcars, tank, flat, and special cars.

Airfreight includes small package delivery service, regular air freight, regular airlines airport to airport, charter. Occasionally helicopters are used to deliver freight to places that are hard to get to. Other shipment modes include all special moves including barge, ship, courier, and anything not already mentioned.

Costs within each mode are determined by weight, distance, and class of commodity. For example, one of the lowest cost commodity classes is Class #50 iron and steel fittings. The reasons that Class #50 is inexpensive to move by truck are the products are hard to damage and dense (small cube with high weight) and not usually desirable for theft. A high cost commodity class would be computers which are easy to damage and are low density (the trailer would be full before weight limits are exceeded). Theft exposure is also high.

Safety

In the U.S., 8% of the workers will suffer some kind of an accident at work each year (*Maintenance Supervisor's Standard Manual*). Many of these accidents are minor cuts and scrapes and a few involve disabilities or death. Because of the nature of maintenance work, craftspeople are usually overrepresented in the injured group. In some organizations, the maintenance people have the highest injury rates, have most of the hazardous chemicals, and are the hardest to convince to use safety gear.

In addition to being a headache to the safety department, maintenance is usually considered the major resource to abate the safety problems. Maintenance is a key player in the safety equation. Maintenance workers are expected to identify hazards, repair potential safety problems for other workers, and be advocates for increased safety.

Safety has high priority. Savvy maintenance leaders use safety in a link-up with good maintenance practices such as inspection, TLC, RCM. Good maintenance practices provide support for a safe operating environment. Safety becomes the lever to move organizations toward these better maintenance practices.

To protect maintenance workers, safety consciousness must start on the first day on the job. The safety rules must be taught. Safety training for maintenance people has to go well beyond the training given to other workers. It is usually more difficult to get maintenance people to follow the rules. Supervisors, managers, and other people in leadership need to be vigilant and consistent in their enforcement of the safety rules.

Any accident that does occur has to be investigated thoroughly. Your responsibility includes questioning the employee to see if they knew and understood the rules. Your responsibility also includes removing known hazards in the equipment, process or maintenance practices.

Maintenance people are usually deeply involved in safety-related duties such as first aid, fire brigade, and disaster planning/preparation. Maintenance workers make logical candidates for first aid because they know the facility and are used to emergencies that involve their fast action. They are logical fire brigade members for those reasons, and most maintenance personnel are physically strong and they know where the utility shutoffs are likely to be.

Disaster planning is also a possible job for the maintenance team. While they would need support from other departments to know the business consequences of different catastrophes, they would be in a good position to know the needs of the building and its systems. Maintenance would be in an excellent position to make the building safe and secure during the crisis.

Disaster planning starts with a catalog of the types of disasters known in your area for the last 50 or more years. Plans should be drawn up even for the infrequent disasters. Your local office for disaster preparedness is a useful resource for the planning process.

39
Elements of Maintenance Leadership

The managers of maintenance must be effective leaders to improve performance. Within the maintenance community, there are excellent and frequently unheralded examples of leadership. In the words of A.S. Migs Damiani (Jan. 1996, *Facilities Magazine*), a leader accepts the challenge for change, thinks positive and big, knows the business of their business, becomes their boss's teacher, invests in themselves, focuses on training, increases their visibility, involves their people, has financial know-how, values diversity and finally looks for the gold in others. This is an excellent list to help get our hands around the concept of leadership.

Maintenance departments have stories about legends, heros, demons, and devils. These stories from the past frequently reflect the values of the institution. The legends could concern heroic deeds of breakdown repair, saving lives, saving product, or saving machines. The legends could also concern great thefts, hateful chiefs, evil purchasing agents, drinking and partying, high stakes card games, or other illegal activity. These legends reveal insight into the underlying patterns governing the department and, possibly, the whole organization.

Maintenance leadership transforms the legends from great repairers to great reengineers, from stories about great work through the night to great "it never breaks down while we take care of it" people. When we shift the rewards, stories, legends, there is a possibility to run maintenance without repair. True maintenance leadership does not focus on improving the efficiency of the arts of repair, but on looking for other ways to pre-

363

serve the asset and, more importantly, preserve the capability that the asset provides.

In traditional environments, leadership was less important than the supervisor's other capabilities. In the 1981 edition of *How to Manage Maintenance* by AMA (see the Appendix), leadership was the first attribute mentioned but it was never defined. The whole list of selecting a person to be a foreman (the word foreman has been superseded by the gender-neutral word supervisor) is as follows.

1. Who is a leader?
2. Who has demonstrated skills in planning?
3. Who understands the work order system and priorities system?
4. Who is a good communicator?
5. Who can sketch?
6. Who is respected by peers and superiors (note, no mention of respect by their team members)?
7. Good at figures?
8. Who is an innovator?

This list is still valid today. But since many of the tasks mentioned above have been transferred to the tradespeople themselves, leadership is much more important than ever before.

One comment made in the same text was that the supervisor's position is a good training ground for young engineers. Considering the deleterious effect of an inexperienced person in charge of a work group of experienced tradespeople (they might tend to "have him or her for lunch" until the engineer "proves" themselves), it is very true that the engineering profession as a whole needs much more real experience in maintenance before they are let loose as designers.

One of the gurus of management, Stephen Covey, has a story in his seminar "The Seven Habits of Successful People" that demonstrates the difference between leadership and management. A group of people are trying to hack a road through the jungle with machetes. To support them, the managers developed machete sharpeners, apprentice machete wielders, and other systems that made the work smooth and efficient. One of the people climbed a tree and yelled down that they were going the wrong way. The management on the ground told him to shut up because they were making such good progress. In Covey's model, leadership is knowing where to take the maintenance work group and, more importantly, where to take the maintenance function.

Scott King, maintenance supervisor at Forest Home, Inc., is quoted in *The Maintenance Supervisor's Standard Manual* as having four specific areas of focus for improved leadership.

1. Spend time with employees. He spends at least 30 minutes a week one-on-one with each subordinate and believes that this is the most valuable time he spends. When employees complain, he asks them to design their ideal job description in writing. He and they then discuss the job and look for ways to adjust the job to fulfill more needs and desires of the employee.

2. Planning reduces inefficient use of resources. He says that a supervisor should spend one-third or more of their time planning. A well-planned environment improves morale.

3. Getting employees involved starts by having the employees help design their own work environment. When presented with problems needing solutions, King challenges his people to always have a solution when presenting a problem.

4. Keeping a fresh approach is essential to continue enjoying your job. If you can't find a valid reason to keep doing something, try stopping it. If that creates a problem, then look at the whole situation again. His style is to give people enough information with priorities to make their own decisions.

Leadership in a Union Environment

It is essential that the manager has leadership in a union shop. The same leadership skills are important when there is a union involved. In addition to the skills already mentioned, there are some skills that are particularly important when in conflict, and when listening to a grievance. These skills are useful at *all* times, whether there is a union or not—this is called courtesy!

1. Don't rush the discussion.
2. Active listening is important. Try to suspend your judgements and experience and try to see what the problem is like from the

other person's point of view. This is not the same as agreement.

3. Set a time and place for the discussion, and show up on time.
4. Listen to the entire story before discussion.
5. Follow through on the steps outlined in the bargaining agreement.
6. Realize that this is important to the person who brought it up, treat it that way.

What are the Attributes of a Great Maintenance Supervisor?

Leadership is the key to a productive and motivated work group. To identify the elements of maintenance leadership, we discussed the issue with over 100 maintenance managers, maintenance supervisors, maintenance planners, plant engineers, building managers, and production managers throughout the United States and Canada. The organizations ranged from the largest industrial firms, federal and local governments, to small industrial and building management firms. We felt that there already was knowledge about leadership within the maintenance community. We wanted to tap into a rich tradition. The answers fell into three general categories.

People	Management	Technical
Good listener	Is organized	Dedicated to quality
Compassionate	Ability to make decisions	Knows equipment
Can motivate others	Good delegator	Knows job
Fair and consistent	Meets goals of business unit	Knows safety
Is respected	Reanalysis of progress to goal	Can analyze
Honest	Knows what is and isn't	problems
Is an innovator	important	Can evaluate skill
Open-minded	Provides good service to	level
Effective	customer	Understands product
communicator	Loyal to organization	
Is a coach, not	Oriented toward results	
dictator	Good planner	
Good negotiator	High productivity	
Positive outlook	Follows up to see job is done	
Is flexible	Understands importance of	
Treats people as	scheduling	
equals	Can assign and keep priorities	
Can "read" people	Understands and uses budgets	(cont'd on next page)

People	Management
Adaptable to change	Is available
Common sense	Provides a conduit for
Willingness to learn	downward communication
Not afraid to make	Provides intelligence to upper
mistakes	management
Will take control if	
necessary	
Effective trainer,	
communicator	
Can work with	
different types of	
people	
Gives recognition for	
job well done	
Can deal with difficult	
people issues	
Praise in public,	
discipline in private	
Has a cool head, can	
handle pressure	

Notice that most of the comments from these managers centered around people issues and management issues. Many people have an idea about maintenance leadership and manage to exhibit some of the attributes they discussed at least some of the time. In the next question the interviewees passed along some of the wisdom that they have gained from their years on the job.

Gems of Wisdom for New Supervisors

1. Be a good listener.
2. Learn to bend, but don't abdicate.
3. Remember you are in charge, act that way.
4. Strive to be respected, not necessarily liked.
5. Always be available to your people.
6. Cultivate your patience. Know which things can be put off, ignored, and which can't.
7. Pay attention to what your people are saying, and to the back-room talk.
8. Never stop learning.
9. Treat people consistently, fairly, and firmly.
10. Keep your eyes open—don't just look, but **see**.

11. Make time to analyze your problem areas, compile facts before deciding.

12. Those tough humbling experiences are valuable, treasure them.

13. Give clear indications of what a good job is; give praise when it is achieved.

14. Don't be afraid to acknowledge that you don't know.

15. Quality cannot be ordered, it is an attitude.

16. There is a fine line between getting involved and getting in the way.

17. Good supervisors surround themselves with good people and are not afraid of training replacements.

18. Keep a positive attitude; keep company interests at heart.

19. Set goals every day, and review them before leaving. Plan your days.

20. Listen more, talk less. Be able to hear feedback you don't like.

21. Solicit the views of the workers for improvements and problem areas.

22. Use positive, one-on-one techniques with workers.

23. Follow your work plan.

24. "Watch your back."

As you can see, the rich tradition of maintenance leadership has identified 24 items that would promote leadership in any endeavor. The second part of leadership is insight into motivating maintenance workers. Leaders tap into the intrinsic motivational structure of their crews. Motivating maintenance workers is important work.

One thing all maintenance leaders have in common is their ability to get people to want to work for them—the ability to motivate. Motivating maintenance workers is tough and important. Having a motivated workforce is not an accident, but a combination of a deliberate approach with action.

Imagine you are Ron Augustine, a supervisor for Parker Hannfin Corp. in Michigan. He says: "For 10 years my supervisor took credit for the things that we did that went right, but when things went wrong he was the first to point the finger." Imagine your own level of motivation the second or tenth time that happened. Would you be able to maintain a motivated state?

How would you like to go to work every day like David Daugherty for 1615 L Associates in Washington: "It's an environment where everyone does their own thing and doesn't try to work together as a team. The employers don't seem to care what happens. The building systems are neglected and fall apart and we rush to fix them." A department's attitude toward their mission of providing maintenance service can be a motivator or demotivator.

At Eastalco Aluminum in Maryland, Roy Ellison supervises a crew that just went through an 81-day strike. The result is a $.92/hour pay cut after no increases for the last 6 years. He reports "the number one problem is morale and motivation."

We can all identify with these situations. Note that in the first case, the problem was with a person; in the second, with a department; and in the third, with the entire company. Ideally motivation is the result of the three levels working together in concert. You can have a motivated work group in spite of a negative department or adverse company situation.

Let's examine some major themes in successful motivation. There are two types of motivational activities. The first type is a daily way to conduct business, and the second concerns one-time situational (or once a year) interventions. Cases are taken directly from active maintenance supervisors.

Everyday Motivational Techniques

1. *Recognition.* This is the easiest way to increase the motivation level of most maintenance workers. Recognition can come in many forms. For example, Johnny Johnson of Abex Corp. in Virginia reports:

> "We had a press that needed to be converted to run a new product in a short time. I acted as a cheerleader and got all three shifts working together. The job got completed in 1/3 of the normal time. I wrote a letter of appreciation for each person for the personnel file. That motivated them."

In another case, recognition was extended to include the family. Gabriel Saavedra of Baxter Hyland in California once had this motivating experience:

"Upon completion of a remodeling job, the Vice President sent me a note of thanks with copies to my boss and my file. He also sent a basket of fruit home with a note explaining the long hours during the project to my wife."

2. *Treat the maintenance workers like people.* I think this goes without saying. Bernard Rulle, a carpentry shop foreman at George Air Force Base, has some rules for himself:

1. I like to know my people personally. I know their hobbies.
2. I listen to people with my mouth shut.
3. I remember their birthdays and send cards.
4. I praise them in front their peers when they do outstanding work.
5. I ask for donations when loved ones die for plane fares or expenses.
6. I invite people without families to share Thanksgiving.
7. I have a BBQ for the whole shop once a year.

3. *Training and ongoing development.* Train your people and keep them up to date. Factories, vehicles and facilities are changing and our skills are not keeping up. We must train people in new technologies and, for maximum flexibility, cross train in two or more crafts. At Chevron Canada in Burnby British Columbia, Michael Edwards had a problem with the computer process units:

"The computer controlled process units were maintained by instrument technicians and process engineers. The technicians cared for the hardware and the engineers looked after the software. The systems did not work well. There was not good communication between the two groups. We trained the technicians to write and maintain the process software. Since they had complete responsibility they worked harder and now the system runs smoothly. The instrument technicians are confident and highly motivated."

4. *Give the maintenance crew real power.* When you give a crew the power to do the job their way, you'd better be prepared for some improvements. A highly motivated crew with the power to make improvements can be very effective, as Edward Wilanowski from CIL Inc. in Alberta, Canada, relates:

"Reactor rebuilds took 36 hours. We needed to reduce this time to increase productivity. Management tried increasing the number of people but that increased non-productive time while people waited for each other to complete jobs. The extra labor did decrease the turn-around time a small amount. Management then tried pre-planning which also slightly decreased the turn-around. Then the power was given to the trades people. They changed some equipment, added some tooling and now rebuild a reactor in 16 hours with the original crew. They are a highly motivated crew."

In another example, Paul Redding, a supervisor for Purdy Corp., had a problem. When he took over the maintenance and housekeeping department, the group had a terrible reputation. They were lazy and unproductive. Office personnel particularly complained about the janitors. Paul called a meeting which went like this:

"At first there was silence. After a long while someone broke the ice with complaints about the cleaning materials. The flood gates opened and I found out about several problems with the supplies we were purchasing. At the conclusion of the meeting I gave them the power to purchase their own supplies. They were given a monthly budget and the abilities to make their own decisions. Our purchasing department gave a mini-course in buying. The group is now highly motivated and complaints have dropped dramatically."

5. *Ownership.* This is one of the key concepts of the 1980's and 1990's. We assume that someone who owns something will treat it better than just a cog in the wheel of production. The worker as owner has a say from acquisition through maintenance to retirement. At Kimberly-Clark, Patrick McDonough wanted to put this into action:

"We had to purchase a new expensive piece of equipment for the mill. Normally I assigned an engineer to study the need and develop a specification for the new unit. In this case we put together a team with people from engineering, production and maintenance. They worked on a specification together. Unlike other purchases from an 'aloof' engineer this equipment was installed with minimum trouble and was accepted quickly by all parties."

6. *Planning.* Many people enjoy working in a planned environment. A planned environment has fewer surprises, more even work flow, and higher productivity. One of the most important things to plan is your annual shutdown. Keith Brown from Dow Chemical describes the improvement to morale due to planning *with communication of the plan:*

"The first time we did a shut-down with pre-planning and scheduling was an eye opener for the crews. They enjoyed knowing their next jobs and not getting in each other's way. The people liked the visible schedule board."

7. *Use their ideas.* Frequently the best ideas come from the people closest to the action. Skilled maintenance workers have significant experience that, once unleashed, can solve major and minor problems. When maintenance supports production, significant new ideas will flow. Ron Howard works for the Rice Growers Association of California and had a problem of low output from a packer:

"The supervisor got everyone together and stated the problem of low output and suggested a few of his own ideas. He then backed off and let the maintenance people come up with their own ideas. In the end he turned the project over to them and helped the effort by purchasing the parts and materials. The result surprised even him, the packer was hitting production levels 30% higher than factory specs and double the old output."

8. *Communication.* When people understand each other they are more likely to work together with higher morale. High morale translates into smoother functioning with open channels of communication. Jim Wakefield, a Weyerhaeuser maintenance supervisor, had poor communications between shifts. He solved the problem in a straightforward way:

"We set-up an inter-shift meeting to discuss all of the issues that affected the workers. The first few meetings seemed worthless. As time went on more and more things came out. Now if someone is on a job when the shift ends they will stay over a few minutes and bring the new person up to speed."

Interventions to Motivate Maintenance People

These interventions are not everyday techniques, but special events. Interventions are to be used when the circumstances indicate.

1. *The boss is one of us.* In selected settings, letting the people see the reality of the boss as a person can be an excellent intervention. Joe Costa, a lead mechanic at Angus Biotech Inc. in California, felt motivated when one of the bosses took off his suit and got down in the ditch. It sounded like this:

"It was a hot sweaty filthy job that summer. We were working 12 hour shifts 7 days a week to complete the shut down work. All of the upper management would walk by and criticize this or that but never got too close. The chief engineer jumped in and worked with us. He supervised from the inside, took his meals with us, and worked shoulder to shoulder with the crew. Morale skyrocketed. Our whole view of top management shifted. He really showed us he was a regular person and could keep up with us to boot."

2. *Challenge and competition.* People love a game. Challenge (I bet you can't wire the new grinder by Friday) is a great way to complete projects. Competition between groups can be effective and enlivening for the work group. William Castro, the maintenance manager of Ibis Systems, set up the following challenge:

"I gave each person an area of responsibility. We developed a Trouble failure report that indicates the amount and reason for downtime. Each area competed for the lowest downtime statistics. People really started to take an active interest in the areas after the competition started."

Sometimes the challenge is against bringing in an outside contractor. The maintenance workforce is told, in so many words, they are not trusted for this important job. Charles Jones, a first class maintenance mechanic for E-Systems, Melpar Division in Virginia, relates an unintentional challenge that was an effective motivator:

"Early last summer our 450 ton chiller went down because of the motor. The manager was in a panic and wanted to immediately call a contractor. I convinced them that we had the skills to do a good job in less time at a substantial savings. I told our people that management didn't think we could do the job. We rebuilt the motor and had the chiller back on line within one day at significant savings."

3. *Drama.* Sometimes when there is a problem, the supervisor needs to "hit everyone over the head with a 2×4" through the judicious use of drama. Continental Mills in Seattle had such a problem with wasted product. The plant engineer, Dave Sloan, related this story about his president:

"The president called a plant meeting and explained that we were losing $45,000 worth of product each month. He opened a case and proceeded to dump 45,000 one dollar bills into a garbage can while saying 'this is what we are doing with our product.' The impact was amazing. Our waste dropped 50% the next month and has continued to improve since then."

4. *Reassignment.* Motivation within a work group can be greatly undermined by an abrasive personality of one or more of its members. Frequently, reassignment, rotation, or transfer of the discontented parties will restore the work group's motivation level. Ron Vanderpool worked with a shop steward whose nit-picking attitude affected the crew. The organization was also suffering from quality problems. Ron suggested that management:

> "Appoint an hourly employee to the quality program at the finish end of the process (where the shop steward was senior)."

He then sold the program to the steward who really liked the idea. Quality went up sharply while the steward got to nit-pick in a way that *helped* the organization.

5. *Survival.* This is a powerful motivator. When you face a survival situation, people will work very hard to solve the problems. When faced with certain layoff, C. Bostic of Hercules, Inc. went to management to save his people's jobs:

> "We formed a 7 person central day crew. This crew would do construction jobs presently being done by contractors. Management bought the idea and started the project the very next week."

This was a win–win situation. The supervisor solved the layoff problem and management looked good to the workers.

6. *Variety.* Some people get bored. Sometimes low motivation levels are the result of work which underutilizes people's capabilities. Marvin Barry from Ocean Construction Supplies in Maple Ridge, Canada, remembers when:

> "Our mixer men were becoming tired and missing lubrication points and missing inspections. To create more interest we rotated the people through several parts of the plant and through some nearby plants. The rotations seemed to spark enough interest so that the work was done well and the people seemed happier."

7. *Choice.* Maintenance mechanics love tools. Ed Eccles of Whitworth College in Spokane, Washington, gives each mechanic at the college $100 each year for tools. The people can pick any tool they want. He reports motivation and excitement from being able to choose.

40
Craft Training of Maintenance Workers

The skills needed to maintain today's factories, fleets and buildings are changing faster than most people can assimilate. Technological changes disorient even the best-trained worker.

A world class auto manufacturer requires 96 hours of training a year for everyone in the organization. They feel that they will not maintain their world class status unless they continue to invest in their people.

The need for training is driven by people new to the organization, new equipment, new operating requirements, new operating expectations, competition, modernizing old equipment, new diagnostic techniques, new management strategies, organizational reengineering, team building or new financial goals.

In the building maintenance field it used to be enough to hire a licensed tradesperson. You expected that the electrician could handle any electrical problem. Today with digital buildings an off-the-street electrician has months of training before they can effectively serve the maintenance need. A fleet mechanic now spends significant time reading and interpreting a screen of data, typing and running computerized tests. The training requirement is significant and ongoing.

Three Competencies Create Three Training Issues

In training terms, there are three domains of learning: knowledge, skill, and attitude. Many types of training address one or another of these types of learning without regard for the other(s). Maximum effectiveness must come from competence in all three areas.

375

1. *Knowledge.* The *observable behavior* is to be able to describe, diagram, argue, etc. The *performance level* is to be able to answer x of 10 questions correctly.

For example, this domain would include the following questions. What is the process to heat treat steel to a particular specification? Describe the steps to obtain a hot permit in this facility. How would you program a cascade pumping arrangement with an Allen Bradley PLC?

The knowledge domain is what is taught in most schools. Generally it is easy to test in this domain. Sometimes you will find mechanics that can do something, but don't know what or why they are doing it. When people know how to do something without knowing what they are doing, two problems are created. The person might unintentionally create a dangerous situation or a cascade failure in another system. The other problem is that without the knowledge, the person cannot be as creative and think outside the normal way to solve problems.

2. *Skills.* The *observable behavior* is to demonstrate, show, perform, or solve. The *performance level* is do the job in x minutes with no mistakes.

An example of this domain would be to ask the candidate to demonstrate his or her competence by welding two pieces of 308 stainless in a vertical position or make off MI cable connection to at least 3 meg ohm resistance to the casing. Most on-the-job training consists of skill training. In maintenance we admire skilled mechanics. If you have the proper demonstration setup, this is also easy to test.

Many people have the knowledge without the skill. We say they can talk a good game—but can they weld, etc.! A metallurgist might be someone who can explain the details of the welding process without actually knowing how to weld. They could specify the tools, heats, covering gases, wire type of weld—without being able to do the job themselves.

3. *Attitude.* The *observable behavior* is comfort, without hesitation. The *performance level* is to your own satisfaction.

An example of this domain would be to discuss the individual's comfort level with a particular technology or their own work ethic. This is very difficult to test for. Someone could conceal their discomfort. You occasionally run into tradespeople that have the skills and knowledge, but lack the will or the confidence to do the work.

Attitude problems might start with a void in a skill or knowledge that creates frustration and a bad attitude. In other cases, the person might lack the capabilities such as strength or height to do the job. Fear might also play a role.

Steps to Effective Training

Step 1: Job Analysis

Look at the job to be done. Determine what knowledge, skills, and attitudes are needed for the job. Before we can look into teaching anything, we have to see what is needed. Look at the job as it is today, and forecast where the job is going in the short term. The big picture of competencies is called the general learning objective (GLO).

The concrete and specific skills, knowledge, and attitudes required to do the job are called specific learning objectives (SLO's). If properly designed, a person achieving these SLO's would be successful in this job.

Step 2: Candidate Evaluation

Evaluate the trainee's (or group's) current skills, knowledge, and attitudes, and compare them to the results of the job analysis. A direct supervisor might be able to make an educated guess. If the trainee has good insight, they might know where they are weak. Most evaluation situations require some kind of testing (either observation on the job or more formal written or bench tests). The testing should be designed to uncover the skills, knowledge, and attitudes on your job analysis list. The testing should also uncover those people who do not have the aptitude to learn the new information.

It is important to note that success on the test should correspond to success on the job. Testing that does not reflect job requirements is said to be invalid. The Americans with Disabilities Act (ADA) and related legislation are clear that the test must not discriminate against any group, disability, or condition. For example, if the worker must lift 100 lbs. in the test, the job must call for heavy lifts where equipment cannot easily be used.

Step 3: Fill the Voids

Translate the voids in skills, knowledge and attitudes of the potential trainee uncovered above from the required list, to develop a training lesson plan. The training plan should list all of the types of learning that this person/group needs.

Training prescription summarizes the skills, attitudes, and knowledge that the candidate lacks and needs for the job. The estimate of time requirements, costs, and time off for the trainee, and any requirement for supporting staff, would be developed from the training prescription.

You might go through this exercise for all related jobs. The service technician class 1 might have related SLO's to a service technician class 2 that can be economically incorporated into this training.

Step 4: Post-Test

Retest the trainee and determine that the voids in skills, knowledge, and attitudes have been filled. Most training efforts complete three steps and never validate if the learning objectives have been met. Increased rigor in this area will make it easier to choose effective training, and prove the case for additional training resources.

Organizing the Training Effort

Your department (if not the whole company) should set training goals for all people. Look at where they are now, where you need to go, and what is missing. Budget at least 1% (20 hours per year) training for each person per year. Firms in rapid change or with a significant deficit of skills need 5% (100 hours per year) or more to keep skills current.

If your firm does not have a formal training department, you could set one up for your department, area, or even work group. The training director can be one of the people in the department and could rotate among different people every year.

Set up files for each person and each job with the GLO's needed. Review the GLO's with available time, business cycle, and funds. Be sure to act with the full input of the workers and supervisors. Once underway, review every file every six months and be sure everyone gets an opportunity for training.

The horizon for training is usually five years. Your system

should track and highlight training needs by year for the next five years for each person. Budgeting should allow you to replace the skills lost through retirement and turnover. Additional funds are needed to upgrade the existing skill set to the new one needed. Knowing your historic turnover rate and your projected retirement rate is a starting point for the training management exercise. Without a formal program, it will be difficult to replace skilled people as quickly as required when they leave.

Some non-craft topics are: your industry, quality, safety, CPR, fire fighting, toxic material handling, toxic waste regulations, your maintenance information system, statistics, filling out paperwork, PM, scheduling, project management, report writing, shop math, drafting, CAD, computers, engineering, cost accounting, your end products, what it's like to operate your machines, and job cross training. The list is long and growing. Your people will be the better for the attention and the training.

This organizing effort might require the creation of a guidebook. More details can be found in the *Maintenance Supervisor's Standard Manual*. This manual consists of job-related information arranged by topic in a ring binder (or computer file). Topics might include safety, whom to go to for answers to different kinds of questions, description of all major equipment and special tools, glossary of company terms, information on CMMS, information on how the PM system works, uses of idle time, parts procedures, work site rules, and many other areas.

ISO 900X and the Certified Operator, Mechanic Training

One of the items usually covered by ISO 900X is the concept of a certified operator or mechanic. It is impossible to assume quality if the people interacting with the asset don't know their job. Donald Overfelt of Lucent Technologies explained his company's methodology for a certified mechanic under ISO 900X.

They took each machine and broke it up into components. For each component they thought of all of the action words (such as clean, adjust, lubricate, replace, rebuild, align, etc.) and created a matrix first for each machine, then for each mechanic.

Machine matrix: All the things that can be done on this asset

Component	Clean	Adjust	Calibrate	Rebuild	Replace...
Proximity assembly	X	X		X	X
Limit switch	X	X			X
Main shaft and bearings	X	X			X
Component x,y,z	X	X	X	X	X

Mechanic matrix: What activity is this individual certified or checked out on

Component	Clean	Adjust	Calibrate	Rebuild	Replace...
Proximity assembly	C	C		X	C
Limit switch	C	C			C
Main shaft and bearings	C	C			C
Component x,y,z	C	C	X	X	C

C—certified
X—not certified yet

Multi-Skilling and Cross Training

There is no way to perform at world class levels if the maintenance personnel don't have the knowledge, skills, and positive attitude in more than one craft. The writing is on the wall—everyone will have to know the basics of computerization, mechanics, electrical hook-up, etc. In today's slimmed down workforces, the first person who responds must have the skills to deal with the problem. The craft advocates argue (rightfully) that the quality will suffer if unqualified people perform the craft. The whole environment becomes less safe, less reliable, and less secure. The truth is that people will have to be qualified before we let them loose on our multi-million dollar assets.

Even the strictest "craftist" (my own word, meaning a supporter of the craft division of labors) will admit that the base of knowledge for all crafts overlaps so that picking up the second craft will be much quicker than the first. In fact, most experienced craftspeople know quite a bit about the crafts they work around. The most vigorous craftists can be found doing other crafts with competence and safety around their own homes.

The radical change that requires multi-skilling is the changing nature of our equipment. Equipment is designed without regard for craft lines. Digital circuits, pneumatics, mechanical systems, electrical power, and hydraulics are frequently mixed up within a four-square-foot machine. Troubleshooting requires working knowledge of all these systems.

The last factor that seems to confuse the issue is the need for specialization. If you read the want ads for maintenance jobs, you will find organizations looking for specialists in very specific fields, such as a Symax programmer or bench tech for disk drives. This need for specialization is happening at the same time as the drive for multi-skilling.

Many vendors are looking closely at the training issues of multi-skilling and have developed training and can recommend structures for multi-skilled shops. Robert Williamson lays out (in *Facilities Magazine*, Jan. 1996) a possible scheme for multi-skilling with areas of needed expertise.

Another big issue is how the unions are going to react to these changes. The unions and the guilds before them protect their turf to keep the compensation high (if any one could practice medicine, put in new electrical services, or do plumbing jobs, the prices would drop). The move toward multi-skilling poses a real dilemma for the union.

Running a craft shop is a real competitive disadvantage. Slim differences in the ability to deliver maintenance service might help decide who wins the business war. In general, unions have been increasingly flexible when survival issues are on the table.

Supervision

One of the ripe areas of training is in supervision. While there are generic classes in supervision available internationally (by all means send your people to these sessions too) there are some intensive maintenance supervisory training classes offered by the author through several universities (UOA, PLU, Clemson, UW, Concordia—see the Appendix).

Having an assistant is another way of gradually training someone to become a supervisor. An assistant would replace you during vacations, alternate holidays, and whenever you are out for any extended period. Having an assistant, a successor also gives you upward mobility and a trusted person into whose hands you could confidently put your department.

Needs of New Employees

The Maintenance Supervisor's Standard Manual mentions an important point that it is critical to attend to the emotional needs of new hires. These people, even if they are already skilled in the trade, need to feel competent. This is a vulnerable time for most people.

One way to start new employees off is to give them jobs that will be easy to master and that will take them throughout the facility. It can be explained that every new hire goes through this activity for a week to a month or more.

Sources of Training and Training Materials

Training is big business for a large number of organizations. Specific sources follow in the Appendix. Here are some ideas.

1. Your staff is your first choice of potential trainers. In excess of 80% of all maintenance training takes place on the job. You are already in the training business in a big way. I recommend investing energy to manage the effort. Within your staff there are several possible people to consider as trainers. Please note that being a trainer should be viewed as a job-enhancing project. Time should be given for preparation of materials. During preparation of materials and delivery of the training the trainer should be relieved of other duties.

    a. Tap your soon-to-retire people as trainers. This group has significant experience that should be channeled into the next generation. In some organizations people who have already retired are recruited to return as part-time teachers. Stay away from people with negative attitudes toward the organization or they might infect their students.

    b. Use the internal guest instructor concept. In this concept, a staff member would be treated as a guest (lunch, clerical support, off-site for longer training, no interruptions, relief from other duties). Most of all, the internal guest instructor would get respect as if he/she were hired experts from the outside. There are several advantages of this technique including improved self-esteem for the trainer, improving the efficient transfer of information within the organization, and increased internal respect for the whole organization.

c. Tour training is an excellent team-building exercise. Once a month you tour a section of your facility, and the most experienced maintenance person plays "show and tell" about the problems and successes in his/her area.

d. Video technology has rocketed ahead so that most firms can afford a quality video camera, recorder, and an editor. New equipment setup, construction documentation, and specific machine training are popular first subjects. Craft training is a more difficult but rewarding area. After expertise is obtained, any topic can be recorded to good effect. One firm hired a producer from a small local cable station to put the training tapes together. The video professional guided the company's employees in how to produce what they had in mind.

e. Look to other parts of the organization, such as human resources, data processing, engineering, production for expertise useful to your upgrading effort. Topics could include your process, your industry, interfacing to other groups, etc.

f. The supervisor, lead person, or leader (where qualified) should be a logical choice for trainer. Consider sending them to "Train the Trainer" classes locally. A small amount of formal training in adult education will significantly improve your training effectiveness.

2. There are many excellent companies that provide craft training. These firms can provide professional instructors, testing rigs, work books, and video/audio tape. The quality and appropriateness to your operation may vary, so check several vendors.

a. In-house courses are available on a wide variety of maintenance topics. These are most appropriate if many people need the same training. Costs are about $1000 to $3000 per day.

b. Public seminars are very useful for training one to three people. Expect seminars to cost $500 to $2000 and last 1 to 5 days. Try to get recommendations of the better seminars from people in your industry. One big advantage is the interaction with maintenance professionals from other organizations. Some seminars (particularly if they are sponsored by vendors) are very good buys because they are subsidized. Avoid the sales pitch seminars for training purposes.

c. Many organizations sell videos. Expect to spend $50 to $500 for video training in a wide variety of areas.

d. At the very top of their video product lines are interactive videos that lead a trainee through a series of lessons which retrain when necessary. They started on laser disks and have proceeded to multi-media CD ROM's. Expect to pay $50–$500 for a rental, and even $5000 and up to purchase a complete course. As with most areas dependent on technology, the prices are dropping rapidly. Custom designed interactive videos, however, are still as expensive as films to produce. Large manufacturers pay $250,000 or more for a custom project. The hardware needed to view the programs are PC's with multi-media with some added goodies and cost $5000 to $10,000. Many firms package the hardware and lessons into a monthly rental charge.

e. Test beds are excellent training tools since the actual equipment can be seen, touched, and worked on. The company can program the test bed to simulate faults and train your people how to uncover them. These are popular with PLC's, hydraulics, pneumatics, etc.

f. The largest organizations in the information business are the library systems. All libraries have books, audio tapes, and videos for loan. Many of them have extensive sections on craft training. Visit the main branch of your county or city library—you might be pleasantly surprised; you can also purchase videos that are missing and donate them. That way they manage the lending aspect of training.

3. Equipment manufacturers have a vested interest in a trained user base. Many of them subsidize training and charge it to marketing expense. Time and time again, it has been shown that trained users are happier users. This is a growing area.

a. Negotiate training into all equipment purchase contracts.

b. Excellent low-cost training is usually available from vendors of predictive maintenance hardware.

c. Immediately call all vendors of equipment you own and ask if they have free videos on maintenance or operations. Develop a scoring system for the quality and applicability to your facility. Have mechanics volunteer to view and rate them.

d. Also ask for free parts disks or CD's for your computer.

4. Trade and professional associations are striving to increase their value to their membership. One of the traditional ways is to provide industry-specific training in either traveling seminars or at workshops during trade shows. If your association does not provide training that you believe is needed in your industry, why don't you volunteer to put a seminar package together for the association. The AFE (formally the AIPE) plant engineering show in Greenville, SC, had 25 seminars (1997) on many important aspects of maintenance management. Could you imagine the positive effects of two or three technicians being rewarded with attendance to a show and exhibition.

5. Some of the computer networks provide phone access training (dial-in training) sold by the hour. These systems train in electricity/electronics, pneumatics, building trades, business subjects, computer subjects, basic science, and many other areas.

6. Distance training is the merger of video, telephony, and computer presentation technologies to bring training to one or more students spread apart. The technology to conduct distance training is getting increasingly affordable and easy to use. Advantages can be short duration (1–3 hours), any time or shift of the day, key people can get training because there is no travel time, online instructor. The providers can be schools, companies, or almost any traditional training provider with a computer and some extra software and hardware.

7. Tech schools are an excellent source for trade training. Get to know the people running your local tech schools. Visit and walk through the facility. Many companies set up specific labs, benches, or workshops with equipment so that the tech school can provide equipment-specific training to train students on. Many tech schools are very open to negotiate training contracts for some or all of your technical training needs.

8. Community colleges, colleges, and universities are frequently looking for new markets. These institutions have significant expertise in teaching more advanced subjects to adults. Many of them have entered into instruction contracts with private industry in areas including computerization, robotics, regulation, automation, business skills, and other areas. Also take advantage of the professional development departments of colleges and universities. Several excellent departments that feature maintenance training by the author are listed in the Appendix.

9. Unions are looking at their traditional roles. Many see that skill needs are shifting, and they have decided to lead the trend by setting up training for their members. This might be an interesting subject to be discussed if your union is not doing this already.

10. Insurance companies can cut claims by conducting certain types of training. Some firms will send risk managers through your facility and provide specific training in areas such as safety, risk management, liability reduction, fire safety, storage and handling of chemicals, record keeping for maintenance, safety, accidents.

11. Governmental agencies provide seminars and workshops on a wide variety of topics, including environmental issues, hazardous materials, access for disabled, waste disposal, safety, record keeping, dealing with overseas vendors, and many others.

12. Use your contractors as trainers. Assign an observer when jobs are going on and you want some training. Tell your people to ask questions (even stupid questions are OK—never forget you are paying the bill!). Some firms even video tape their contractors (get permission first).

Checklist for Selection
1. Is this resource the best available to achieve the learning objectives? Is the scope matched to the size of the training problem? In other words, are we trying to bring down an elephant with a sling shot?

2. Is this resource consistent with the style of the department and organization? A free-wheeling department will not respond well to a formal authoritarian lecturer. A more traditional department will not respond well to some of the more far-out team-building exercises. What is the style and history of successful training endeavors in the past?

3. Will this resource satisfy the expectations of the trainees? The tradespeople have expectations that must be addressed. These expectations are sometimes hidden. Remember some people respond to training with hostility because it might seem that you are saying that they are deficient.

4. Can we afford this resource in terms of both time and out of pocket expense? It is a bad practice to pull people out of training for anything less than an emergency.

5. Are there other benefits beyond this training from this resource?

6. If it is an outside vendor:
   a. How stable is the organization? How long in this business?
   b. How knowledgeable is the actual trainer in areas useful to you? How effective are they in training your level people? Can you meet the actual trainers (as opposed to the sales staff)?
   c. Will they guarantee results? Can the results be tested?

Methods to Consider for Training
1. *Coaching, OJT:* One-on-one training and encouragement (use the technique TSED—tell, show, have them explain, have them do). Technique is suitable for important new skills where the high time investment is justifiable.

2. *Case method:* Analyze a specific incident, problem, situation, or company.

3. *Correspondence:* Home study of a commercially produced course of study. Can be adapted to tailored training within a firm.

4. *Books and reading material:* Low-cost method; good for people already skilled to add a specific skill or knowledge area. Permanent; can be referred to in the future and used by others.

5. *Video and audio tapes:* Advantages of lecture, books, and demonstration combined. Permanent; can be referred to in the future and used by others.

6. *Conference:* Send someone to a public conference with a training program. Trainee can get exposure to many instructors, peers, and vendors at the same time.

7. *Cross training:* Training someone skilled in one area another skill or craft.

8. *Demonstration:* Trainer shows trainee how to do something, clarify, or highlight the best way to do something.

9. *Laboratory:* Experiments designed to teach material by discovery.

10. *Lecture:* Trainer tells trainees material to be learned. Provides basic information on a topic. Can introduce topics to many people at the same time.

11. *Programmed learning:* Trainee goes through material at their own speed. Accommodation for trainees who need more material in some texts.

12. *Role play:* Trainee plays the role and learns from their reaction and the reaction of the other role players.

13. *Simulation:* Trainee is presented with a realistic scenario and the trainee works through problems and situations.

14. *Distance training:* Using a computer and telephone line, conduct training over great distances. Several plants or buildings can share a "great" instructor at the same time.

Running the Training Itself

1. Materials (called educational software) which includes books, work sheets, etc.

2. Staff time scheduling, contingency plans if key players cannot attend. Replacement on shop floor.

3. Outside firms needed for guest speakers, professional trainers, turn-key training, slides, video production, audio taping.

4. Trainees have been sent invitations and have sent in their RSVP. Be sure to avoid down days.

5. Structure of training might include break-out sessions, hands-on bench work, access to computer, classroom.

6. Aids, including overheads, flip charts, voice amplification, slides, videos, satellite link-ups.

7. Facilities are adequate for the number of trainees and are comfortable. Who is the contact person? Who has the keys? How do you turn everything on and off?

8. Accommodations for trainees who are traveling in from another facility.

9. Food and refreshments help make people more receptive. Remember people have different tastes. Only 40% of people drink coffee in the morning, plan for your group.

10. Check dates for conflicts with vacations, holidays, hunting seasons, schedules, bad weather.

11. If people are traveling by air or train, coordinate pick-ups, tickets, vans, etc.

12. Use promotion techniques to sell the program, persuade people to want to go.

13. Timing can make or break a program. A giant reorganization before a training can be a good thing, but one afterwards can be devastating.

Advice for Trainers: 22 Guideposts for Adult Training

There are many "rules of thumb" in the teaching of adults. The more rules the supervisor follows, the more likely the training will be successful. Larry Davis' leading book on adult training and education—*Planning, Conducting, and Evaluating Workshops*—gives 22 rules for teaching adults which sum up the best thinking on the topic.

1. Adults are people who have a good deal of first-hand experience. Effective training taps into the adult's existing store of experience. Use the people's expertise. In this mode be a facilitator rather than a pure teacher.

2. Adults are people with relatively large bodies subject to the stress of gravity. Effective training allows the adults to take breaks, move around, and change pace. Figure breaks every 60–85 minutes (depending on time of day). Active activities in the afternoons are usually effective.

3. Adults are people who have set habits and strong tastes. Effective training is sensitive to adult habits and tastes and tries to accommodate as many as possible. For example, in the southern U.S., soft drinks are served in the morning along with the coffee. If you host people from around the country, inquire into their habits and tastes.

4. Adults have some degree of pride. Successful training is careful with the egos of the participants and helps develop greater abilities and independence in the areas being trained. Acknowledge people with long years on a job. Lean on them to contribute their "hard won" expertise.

5. Adults are people with things to lose. Good training is concerned with gain and not with proving inadequacy. The most effective training has 100% success ratios. Stay away from graded exercises. Let everyone succeed.

6. Adults are people who have developed a reflex toward authority. Good trainers (and good supervisors, too) know that each adult has a different style of dealing with authority, and don't take any of it personally. Ice breakers and humor in the beginning help people get over many negative attitudes toward trainers and "school."

7. Adults are people who have decisions to make and problems to solve. Effective training is problem–solution oriented and entertaining. Mix theory with actual demonstration, concrete application. The why is interesting, and the how is critical.

8. Adults are people who have a great many preoccupations outside of a particular learning situation. Effective training does not hog the adult's time. Training should achieve a balance between tight presentation and time needed for learning integration. It is sometimes necessary (and rarely effective) to run long hours in full day training. Consider 6 to 7 hours contact time to be a full day.

9. Modern adults may be bewildered by all of their options and opportunities. Effective workshops assist them in selecting what is important at this time. Review of the available options is good, and your recommendation for the best choice is even better.

10. Adults are people who have developed group behaviors consistent with their needs. Effective training concerns itself with satisfying these needs and allowing many different behaviors. Some people contribute by being loud, others by being quiet, some talk all the time, others have to be drawn in. Allow for the different styles.

11. Adults are people who have established emotional frameworks consisting of values, attitudes, and tendencies. Training denotes change. Change puts people's framework at risk. Effective training assists adults in making behavior changes. Effective training assists adults in becoming more competent. Build on what is good. Start from where people are and proceed from there. The behavior change desired should be logical and defensible from the trainee's point of view. Arguments could be provided to allow the trainee to sound good in the new change.

12. Adults are people who have developed selective stimuli filters. Effective training is designed to penetrate the filters. Penetrating agents include dramatic presentation, logical arguments, humor, involvement, role playing, games, conversation, and other techniques.

13. Adults are people who respond to reinforcement. Some respond to positive reinforcement. All occasionally need negative reinforcement. Effective training is built on appropriate reinforcement. Exercises give reinforcement, instructors give reinforcement, and the peer group gives reinforcement.

14. Adults are people who are supposed to appear in control and who therefore display restricted emotional response. Sometimes intense training loosens up these restricted responses. Effective training is prepared for emotional release if it occurs. Set up a supportive training situation (in maintenance you can't be too obvious about this).

15. Adults are people who need a vacation or time off from work. Effective training provides some time away from the grind. One of the best outcomes of training is that people have time to think about their working situations. This time away can be very motivating. In our public maintenance sessions, getting away with other maintenance professionals is very useful.

16. Adults are people who have strong feelings about learning situations. Effective training is filled with successes. Everyone can succeed in one way or another in a successful training.

17. Adults are people who secretly fear falling behind and being replaced. Effective training allows people to keep pace with the field and grow with confidence. It is important to show the context of new technology to maintenance. All maintenance professionals are faced with an increasingly difficult task of keeping up with the technology. One way to help is to show how the technology fits in. The technology changes, but the context stays the same.

18. Adults are people who can skip certain basics. Effective training starts with where the adult is today and builds on that. After knowing where the people are in understanding your topics, certainly skip or skim over basics. Check this out before trying.

19. Adults are people who more than once find the foundations of their world stripped away. Effective training reminds them of their ability to learn and start again. Maintenance is changing so rapidly that the trainee might be reluctant to put themselves at risk. Once at risk they feel that they might have to start over.

20. Adults are people who can change. In the last analysis, people change and like it after it's over.

21. Adults are people who have a past. Effective training is concerned with the development of new competencies. The why's of the past are someone else's concern.

22. Adults are people who have ideas to contribute. Effective training leaves room for their contribution. Training that takes people's ideas builds ownership and positive feelings.

41
Special Issues of Factory Maintenance

Maintenance management issues for factories and process industries are considered together in this discussion. While a chemical plant or refinery is vastly different in details from an automobile assembly plant, many of the maintenance issues are very similar. Each has incoming raw materials, a manufacturing process that adds value to the materials, and a product. It is the relationship to the physical product that makes the management of process and factory maintenance closely related. As manufacturing gets more closely linked by robots and computers, the factory is actually becoming a process type manufacturer.

Most factories have more dollars at stake than most building maintenance efforts. In fact, most of the dollars spent in the field of maintenance are spent servicing factories.

One of the major trends of manufacturing is the trend toward tighter scheduling, just-in-time, and closely coupled operations. The level of scheduling sophistication necessary to run a factory today is impressive. For the most part, maintenance has not kept up. Typical maintenance paradigms were invented for long runs and excessive work in process. Even a company like Saturn Automotive, that has made exponential leaps in every facet of manufacturing, has made only incremental gains in maintenance.

There is great inertia in maintenance. The power of the CMMS has not even been started to be tapped as a standard application on the shop floor. This is of great interest and concern to Dr. Mark Goldstein. Dr. Goldstein is the leading speaker, writer, and consultant in linking the shop floor control to the CMMS. Dr. Goldstein writes that the maintenance application must be directly linked to and run by the MRP II program that runs the shop floor. He says that maintenance leadership has to get on the floor "where the big boys play" and not stay locked in the maintenance shop.

The basis of his argument (and this has been copied by most major systems vendors and many consultants) is that the factory manufactures product and MRP II provides the person, machine, material, and tooling to do this in a 52-week master schedule. This is different from maintenance which manufactures capacity, and the CMMS (should, but really doesn't) provides the person, machine, material, tooling to do this also on a 52-week (in this case, PM and scheduled corrective) schedule.

CMMS and MRP II need each other. The MRP II schedule has custody of the machine which maintenance needs for PM and corrective actions. Maintenance provides capacity for MRP to use. It seems so simple! Yet as of 1997, no major CMMS is linked in this way to the MRP II applications.

He further feels that lack of consistency between the CMMS vendors slows the buying cycle. When maintenance adopts a set of generally accepted principles (similar to accounting standards or MRP standards) then the whole field will take off. The result of this situation is that the companies' CMMS are not supported by shop floor management (the big boys) and the systems are only used a small percentage of their possibility.

Enterprise wide software vendors such as SAP have almost solved this problem. SAP, for example, controls everything from the accounting, purchasing, and MRP functions.

Five Factors that Make Factory and Process Industry Maintenance Unique

1. *Greater resources:* Usually the factory, mill or refinery has significantly greater resources than a similar sized fleet or building maintenance department. These resources include machine shops, electronic benches, design and build capability, failure analysis (even if informal), and detailed process knowledge. This resource is usually generalized for all types of problems that might crop up in a factory.

2. *High cost of downtime:* The second area of uniqueness is the generally high cost of downtime. In the downtime area it is more likely that a factory will have considered the issues of costing downtime than a building maintenance department.

An example of the high cost of downtime comes from a cigarette manufacturing facility. This facility has five assembly lines that each roll cigarettes, make the packs, pack the packs, make the cartons, pack the cartons, make the cases, and pack and seal the cases. The production rate per line is typically 100,000,000 cigarettes per 24-hour day. In the United States the manufacturer gets $.03–$.04 per cigarette. That translates to $3–$4,000,000 per day per line.

Properly calculated downtime subtracts the raw material costs because the tobacco, paper, filters, cardboard, and other materials are still in the warehouse when the line is out of service. Some accounting departments also subtract certain overheads, general and administrative or other expenses from the gross revenue figure. In the case of the cigarette line we could subtract about 50% of the cost of a cigarette to arrive at the downtime rate. Even at the reduced rate, the downtime is $1,500,000 to $2,000,000 per 24-hour day, or over $1000 a minute.

A power utility might run up to $5000 or more a minute for outages. An automotive assembly line might cost $4000/minute for downtime if there is a market for that model vehicle.

3. *Diversity of systems and types of maintainable units:* The third area of uniqueness is the diversity of equipment and systems that an even small plant's maintenance department is expected to care for. It is unusual for a factory to have more than three or four of any machine. This makes for a multitude of learning curves. It also makes skilled maintenance workers very valuable. It might take years for a skilled tradesperson to be effectively able to deal with a majority of the systems and components.

A small cream cheese company has 225 major assets. They have 3 plate heat exchangers, 12 standard filling lines of four types (one type is completely custom for a specific product), 7 fork trucks of 2 types, 25 machine tools in the maintenance shop (all different), and a host of other special equipment.

They are not atypical by any means. In many manufacturing plants, the equipment was specially made and doesn't exist anywhere else. In contrast, most fleets have many of the same type of trucks. Even buildings, which are more similar with one of this and two of that, will have many assets duplicated such as package units, thermostats, or doors. Field service organizations specialize in one type of equipment, so there might be only one HVAC unit in a building: they service 200 buildings with the same brand unit.

This diversity has an impact on the tradespeople. An industrial electrician, for example, will see more variety, solve more types of problems, and work with a greater range of products. A building electrician has less variety but has a more strict appearance component to each job. The challenge of building electricians is usually not the electrical problem but rather the cosmetic problems of getting wires to points without tearing the whole room up.

4. *High level of technology:* The fourth area is the high level of technology that the maintenance department is expected to deal with. Even simple plants have high technology in their machines and process systems. A typical process plant such as an oil refinery, chemical plant, or pulp mill is fully automated. In the same day, the maintenance worker might be typing on a console for information, debugging a digital instrument, and hooking up a portable computer to locally query a subsystem.

All of maintenance is getting more technological. Factories and process plants are moving at breakneck speeds because they are being asked to do more with less, and technology is often the only way to fill the gap. Almost all equipment has microprocessor controls, self diagnostics, and interfaces. The standard ways of interfacing equipment to the factory communication backbone are new and are just now being taught in maintenance tech schools.

5. *Market focus:* The fifth area is the tremendous focus that the shop floor has in systems, hardware, and software. The factory is a hot bed of innovation. More so than in fleets, buildings, or field service, the factory floor is the focus of top brass. Maintenance is somewhat in the shadows compared to production control, MRP I and II, JIT, TQM, coordinated logistics, or factory backbone with all points connectivity. Even the reflected light of these efforts is quite bright. Maintenance is the unspoken enabler of corporate polices.

One of the greatest turnarounds of the late 1980's was the transformation of Harley Davidson. Widely written about and frequently pointed to as an example of the resurgence of North American can-do attitudes, Harley Davidson was owned by AMF, a recreation products conglomerate. It was losing money, market share, and big motorcycle niche dominance. Things looked bleak when a management team that believed that the Harley Davidson mystique could ride again engineered a buy out.

The newly independent company had an enormously devoted core customer base but a low quality, high cost, unreliable product. The problems with the product were liquidating the brand loyalty. They had reached the change or die decision. The change was to JIT, liquidation of the work in process (which provided an initial shot of cash when it was sorely needed), short runs, short setups, rapid response manufacturing. They made only a day's usage of components at a time. They were going great guns, quality was up, press was good, the new company was accepted by its customers.

As this change became manifest, the reliability of the equipment became more and more important. If, because of a breakdown, they missed their production window with critical parts, 440 people could run out of work. Production equipment reliability became the most urgent deliverable of the maintenance department.

The second factor was quality. For the resurgence to keep going, the quality had to continuously improve. Shortening the production runs to a day's worth of components improved quality. Now if a problem with quality was discovered, it could be corrected and new parts were on the line in hours instead of months. Maintenance's contribution to quality is providing machinery that can produce the same part over and over. It is up to design to make sure that the engine will work; it is up to maintenance to make sure that the milling machine will make the same cut over and over and will not vary due to internal wear.

Getting Out of the Maintenance Business to Survive

In factory maintenance, there is a great need to get out of the maintenance business. As companies downsize, there are fewer people in the maintenance department. The equipment is becoming more sophisticated, more complex, and more automated. There are fewer production people but a similar or greater need for maintenance services.

There are three major directions that all have to be pursued. Follow these strategies when you are serious about and need to get out of the maintenance business. They will be mentioned here in the context of the factory and process plant floor and treated in more detail elsewhere. They are mechanical reengineering, process reengineering, and in-sourcing.

Build It Right

The easiest way to get out of the maintenance business is to choose equipment that does not break down as often under your service requirements. This is more serious than it sounds. The tendency to build a new machine or entire facility to a dollar value and not a life cost has hurt maintenance greatly. Designs that include a reasonable amount of redundancy, high quality valves and fittings, well-sized and high quality equipment cost more to build. They cost a lot less to operate for their whole life cycle. The first strategy is to do it right the first time, when you buy or build it!

Fix the Specific Problem

Mechanical or electronic reengineering can dramatically increase reliability. Dick DeVogelaeu suggests a simple strategy based on a success in his plant. He had three condensate pumps with standard graphite packing. The pumps had to be looked after daily and constantly worked on to reduce leaks. The leaks represented loss of treatment chemicals, water, and energy. He replaced the packing with a new product called "O Drip." He has been able to reduce his daily visits to weekly and has not had to adjust one in over a year. The key here is that Dick was willing to create an experiment. Some experiments will fail, but if properly constructed you will know more than before. With the new knowledge, perhaps the next experiment will succeed.

Every place where parts, labor, or other resource is being consumed is open for reengineering analysis. This type of reengineering permanently reduces your maintenance work load.

Trends Toward Outsourcing

One of the fastest growing businesses today in factory maintenance is maintenance contracting. The trend is clear. It is likely that your firm will outsource more in the next few years than in the last few years. This trend is increasing. After restructuring or downsizing there is no one left for those projects, catastrophes, unscheduled events or other extras that you used to be able to slip in. If the older tendency was to crew at 85% of the work load and fill in with overtime, the tendency now is to crew 65% of the work load.

To make outsourcing work, the contractor must be made to feel like part of the family. They must be partners. In one Japanese chemical plant in the U.S. the contractor helped them build the plant, stayed through setup and startup, and has stayed on. They feel ownership. There is significant advantage to having a contractor as a strategic partner.

42
Special Issues of Fleet Maintenance

As a subsection of the larger maintenance field, the fleet field is unique for several reasons. The most impressive point is that fleets have large numbers of similar or identical equipment in the same type of service. These large populations lend themselves to statistical analysis.

Another uniqueness to the fleet industry is the wholesale agreement about definitions of maintenance activity called the VMRS (Vehicle Maintenance Reporting Standard). This standard has propelled the software ahead of similar packages in the factory or building maintenance arenas. The large numbers of similar units and maintenance standards made sophisticated analysis much more attainable.

More than most other maintenance fields, fleet maintenance is viewed by top management as essential to the success of the whole effort, rather than a necessary evil. This means that maintenance and maintenance people are taken seriously and are important players in the organization's future.

The ATA VMRS and Fleet Data Interchange

The Vehicle Maintenance Reporting Standards (VMRS) project was initiated by the American Trucking Association (ATA) in 1968. The initial study was commissioned by the Union 76 of California. After 2 years of study and consultation with different fleets and discussions with other committees involved in standards, the first VMRS were published. (The complete manual is available from the ATA Management Systems Committee whose address is in the Appendix.) Since then, VMRS has gone through hundreds of revisions to keep it up to date. In fact, any carrier can apply for extensions or additional codes for special situations. These codes, once approved, will be added to subsequent editions of the VMRS.

The standard was originally developed because carriers saw a need to communicate with each other and to the manufacturers. There was frustration among these groups because there was no consistent and easy way to interchange information. The ATA also wanted to study the types of information that are useful to fleet managers. Their reports have been models for many in-house and commercial computerized fleet information systems (FIS). The ATA VMRS standard code allows this communication because it describes the vehicle and the types of information fleet managers find useful.

The codes introduce a definitive description of all of the activities involved in managing a fleet. The ATA standardized data needed for comparisons include: equipment make, equipment model, specification, fleet ID number, operator, type of service. All repair data are collected: what was worked on (over 8000 detailed system codes), what parts were used, what was the failure mode (broken, bent, rusted, etc.), why was the repair initiated (breakdown, PM, scheduled), who worked on it, how long did the repair take, when was the work done, what was the cost for labor and parts, where did the repair take place.

The system also includes procedures, forms, and report format. The data are designed in a format to expedite computerization and *most computerized maintenance systems support VMRS to one degree or another.* It is important to investigate that while a system might be advertised as VMRS compatible, it may, in fact, only support some subset of the standard. Areas that vendors frequently cut corners on are system codes (8 digits), failure codes, and full inventory implementation.

A full VMRS implementation would be a very large system and would not be practical for most fleets unless there is significant use of bar codes and other methods of automated data entry. Consult the Glossary for the specific vocabulary of the ATA-VMRS.

ATA VMRS Repair Order

The American Trucking Association designed a repair order as part of their Vehicle Maintenance Reporting System. The

VMRS also includes equipment jackets, PM documents, unit birth certificates, and complete coding structures.

Figure 42-1 is a sample of the VMRS repair orders and the ATA explanation of the meaning of all of the fields. This is an excellent place to start. You have to decide how much detail you want to capture. Remember that details not captured on the repair order are lost.

Fuel Consumption

The fleet manager should know the answers to the following questions.

1. What are my total fuel purchases?
2. Where do I purchase my fuel?
3. What are the different prices by source?
4. How much do I use?
5. How much is missing?
6. Why don't I fuel in-house?
7. What is my fuel specification?

Much of the data for this section are derived from materials generated from the government's Voluntary Truck and Bus Fuel Economy Improvement Program (VTBFEIP). This program, put into place during the 1979 energy crisis, conducted research and disseminated information from manufacturers, users, and associated parties. We will first review the factors of fuel consumption.

Power

Fuel is consumed to generate power to turn the engine. Power is needed for four purposes.

1. It takes power to *overcome inertial resistance* to start the unit moving, or to accelerate the vehicle to needed speed. Power to overcome inertial resistance is determined by acceleration and gross mass. In addition to standard inertial resistance, there is grade resistance. The amount of power required for long constant grades is (this is a metric formula, expressed in metric tons, km/hr):

$$\text{HP Grade} = 0.03645 \times GM \times G \times V$$

where GM is gross mass, G is grade in %, and V is velocity.

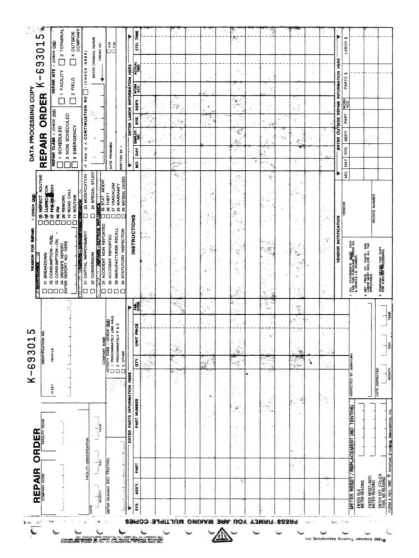

Figure 42-1. ATA VMRS Repair Order (first page).

2. Friction causes *rolling resistance.* Rolling resistance is the tendency for the unit to slow down unless engine power is applied. It is the internal resistance of the bearings, drive-train friction, and the visco-elastic resistance of the tires against the pavement. The vehicle mass, road surface, and design all impact rolling resistance.

3. As the unit moves faster, *aerodynamic drag resistance* (wind resistance) becomes the big factor. At 63 mph wind resistance consumes 50% of the available power. The standard square shape (contrast truck shapes to airplanes or speed boats) requires high amounts of fuel to merely part the air. The height, width, shape, and speed determine this resistance. Doubling the height or width (for the same shape) will double the wind resistance. Speed of the vehicle (or a head wind) has a major effect. Doubling the speed will increase the drag by 8.

4. *Accessories* include air conditioning, thermostatic fans, alternator, air compressor, power steering pump, etc. These items' use of power varies with engine speed (rather than road speed). The total of these loads varies from 3–4 HP under light conditions to a maximum of about 35 HP.

Total Power: The total of the four above factors is the minimum total power requirement. The margin between the power needed and the power available is apparent for hills and acceleration. Note that the margin decreases as speed increases. Items that will impact fuel consumption will impact power.

Ten Areas to Reduce Fuel Consumption (VTBFEIP Recommendations)

1. Steel belted radials reduce rolling resistance by 40%. Super singles decrease resistance additionally as well as provide mass (weight) improvements. Make sure tires are properly inflated.

2. Improved lubricants can decrease rolling resistance 2–3%. These include ester based lubes or ones with additives such as molybdenum, Teflon, or graphite. Run your own tests to separate the flashy ones from the ones that work. Additives have been shown to improve efficiency and reduce wear and tear. Only your testing in your environment will prove the claims.

3. Purchase smooth sided trailers with well rounded corners. Fuel consumption gains for farings is well established, except knowledge of use is important. Roof farings are only effective when used with vans. Performance is worse than before on dump trailers, flats, and bobtail. To maximize fuel efficiency, the trucks/tractors of the future sport smooth wheel shirts, panels between the tractor and trailer, and smooth curves on all corners. Smoothing and rounding will increase the efficiency of straight trucks and buses too.

4. Improve operational techniques. Underfill tanks in hot weather (for expansion). If possible, garage vehicles in cold weather to cut fuel on start-up. Improve routing to reduce mileage. If you are cube limited, doubles will provide better fuel gallons per ton mile. Running with full loads also improves efficiency and $/ton-mile.

5. Thermostatically controlled fans will reduce fuel use. In tests on 23 units, the fans ran less than 3% of total engine hours. This can save 20 HP on large engines.

6. Respecify engine to more nearly match HP actually required, including turbocharged and reduced fuel consumption engines. Consider re-powering before you are about to do any major work on the engine. Specify gear train to minimize RPM at road speed. In all cases, lightened components will reduce fuel consumption. Consider aluminum bodies to cut weight (and increase life).

7. Govern engines to limit maximum speed. Remember, wind resistance increases 8 times as speed increases twice. Also, govern engines to operate at maximum fuel efficiency which is usually also near the maximum torque RPM.

8. Keep vehicle maintained and, when replacing components, choose ones that will increase efficiency. Key areas include keeping air cleaners free, when replacing mufflers use low back pressure units (check for dents or twists that would increase back pressure), avoid a smoking diesel, use a torque wrench and don't estimate torque of heads, keep wheels and axles aligned (inspect frame for alignment), keep tire pressure right, change to disk wheels, adjust brakes.

9. Improved driving practices can save 1–10% of your fuel bill. These include observing speed limits, keeping RPM's down, maintain steady road speed, gradual acceleration, shutting off engine when not in use (3 minutes idling on startup and before shutdown), avoiding overfills, and shifting as little as possible.

10. If we add all of the ways fuel dollars are consumed (fuel used in vehicles, fuel pilfered, lost or never received, etc.), the potential for savings is dramatic. Remember that fuel consumes about 20% of the total fleet dollars.

Tire Analysis and Cost Reduction

Tires are the highest operating expense after fuel. In a tractor/trailer combination each mile wears 18 tires (18 tire-miles). With the cost of new radials in the $300 to $400 range, tire costs must be controlled.

Today you must factor in purchase cost, retread costs, number of safe caps, tire mileage per 1/32, theft, cost of carrying inventory, PM costs, costs of managing the rolling asset, and warranty recovery (policy adjustments).

There are several arguments raging about the best tires for the job. For most high-mileage over-the-road applications, the tubeless radial with recapping is the most economical choice. Other situations indicate other combinations. We will briefly review the arguments.

Radial:
   40% longer life

   5% increase in MPG

   increased puncture resistance

   improved traction

   higher load rating

Tube:
   nails tend to stay in tire

more repair for retreading

more damage to casings

Bias:
35% less expensive

increased sidewall strength

Tubeless:
quicker to mount, less storage

lower cost

less weight

less downtime

(We are indebted to Joe Pearce for his work on tire management techniques and his presentations at the ATA for a framework to view tire costs. Some material was also adapted from the Pepsi Cola Management Institute "Fleet Management Seminar.")

Vehicle and Equipment Specification

Vehicle specification has become a complex art where maintenance costs, fuel costs, ownership costs, and downtime costs have to be calculated and balanced. There are literally hundreds of thousands of combinations possible for trucks, buses, and tractors. Look at your own unique history and consider all four cost areas for each decision.

The fleet's primary job is to deliver the "goods" (we mean goods in the widest sense to include passengers, freight, school children, concrete, pallets, etc.) in the most efficient way possible. The job of the equipment specifier is to deliver the most efficient equipment for that purpose. While efficiency is hard to define over the entire life cycle, in most cases it is better to overspecify than underspecify.

Purchase price is not the only (or even the most important) consideration. There are many managers who believe that, since all trucks break down, "we might as well buy the cheapest." They then tie the hands of the fleet manager or hand them a negotiated contract (with trucks arriving later that afternoon). All levels of government have low bid purchasing provisions which make new equipment purchasing especially trying for the fleet manager responsible for maintenance.

The first step in specification is to ask a series of questions about your current and proposed operation.

1. What is your business (what product)? A specification for moving rolls of fiberglass insulation would be different than one moving steel, people, or explosives. Is your product time-sensitive like lettuce or auto parts for immediate assembly? Are there changes in product mix in the air?

2. What are the operating conditions? Are you operating in a major city, line-haul, or on construction sites? Consider the road conditions, terrain, climate, and type of traffic.

3. How long is the life cycle for similar types of equipment? Do you trade equipment on a four year cycle, when it is completely used up, or by a measure of utilization such as 500,000 miles?

4. What are your demand hours for the equipment? Do you need the equipment around the clock or just one shift? Is the equipment used year around?

5. Are there practical limitations to what our shop can handle? Consider both skills and physical limitations. The first diesels you add to a gasoline fleet will require some thought and training. You can't close the doors and repair a 48' trailer in a 45' bay (no matter how hard you try).

6. One of the most important items to consider in the specification cycle is driver productivity. How can you use the specification to increase productivity? A special design that saves time for drivers may be worthy of consideration.

7. Consider your special, unique history.

PM and Increased Availability

In a 250-unit fleet, if we can increase availability 3% (from 90% to 93%) that is equivalent to expanding the fleet by 7 vehicles. With low availability those 7 vehicles are already funded with depreciation, insurance, permits, and a license *even if they spend most of the year parked against the fence.*

With the increased availability, the units can be used for expansion, trade-ins, or analyzed and used to replace higher cost units. Several of the highest cost units in your fleet together could be used to fund a good new/newer unit. This constant winnowing of your fleet will help keep your costs in line.

Availability and Utilization Targets: 98% cars, light trucks; 94% medium duty trucks; and 90% heavy duty equipment.

The availability of units when your user department/group needs them is part of your mission. In many organizations units are kept as spares because needed units are down. In a fire-fighting-oriented atmosphere of a repair shop (not a maintenance facility), there is usually little idea of availability.

In conclusion, PM will decrease emergency breakdowns. Lower breakdowns mean that vehicles are more often available when the user group needs them. Higher availability also means that fewer spare units are required to maintain the same effective fleet size. Fewer vehicles means lowered ownership costs.

Remanufacture Versus Replacement

Many types of vehicles can be economically remanufactured. Some major fleets run captive remanufacture shops. The U.S. Postal Service has replaced the Jeep mail vans with a unit manufactured by Grumman. One of the advantages of the Grumman van (in addition to larger capacity) is an aluminum body. The postal service is figuring on a 15–20 year life for the aluminum body with 3–5 rebuilds (with new chassis).

Smaller fleets can also take advantage of this by ordering premium chassis in the first place to facilitate remanufacture. This is a totally economic decision. The cost of the remanufactured unit has to be lower than a new unit by a wide margin to make it pay.

Pepsi-Cola recommends to their bottlers that remanufacture usually costs 1/3 of new costs and they can expect an additional 5 years of useful life. They look at capital availability and costs. During times of high capital costs, remanufacture becomes very attractive. On beverage trailers and route truck bodies, remanufacture costs 1/5 of replacement.

Loss Prevention and Fleet Safety

In the longer term, loss prevention is the best method to cut insurance costs and provide a safe fleet for the operators and public. Loss prevention programs do for accidents what PM programs do for maintenance. These programs identify the risks and the contributors to accidents and manage them. Like PM programs, risk management is proactive. These programs anticipate problems and correct them before there is a loss.

Information for this section is derived in part from the National Highway Safety Council, Smith System (driver safety training organization), and materials provided to the Pepsi Bottlers for loss prevention. Steps to establish a loss prevention/safety program are as follows.

1. Set up a safety committee. The concept and organization of the safety committee is important. The mission of the safety committee is to determine whether an accident was preventable or non-preventable; to determine ways company policy can be modified to prevent similar accidents in the future; protect the operator's rights; and to involve more of the organization in the safety program.

The National Highway Safety Council defines a preventable accident as: ".. any accident involving a company vehicle which resulted in property damage or personal injury regardless of who was injured, what property was damaged to what extent, or where occurred, in which the driver in question failed to exercise every reasonable precaution to prevent the accident."

Committee members could include fleet management, safety, insurance people, drivers, and management. Usually the safety committee decides these issues on the data collected alone without the verbal testimony of the driver (only the driver's written report). In fact, Pepsi recommends that the committee not be given the name of the driver involved to try to insure a fairer hearing.

2. Calculate your fleet's accident rate. This formula will highlight the extent of the problem and how it has varied over the years. The National Highway Safety Council has a standardized formula to express the frequency of accidents (to establish rate and compare years to each other). Publish your current and historical rates with the average cost per incident.

$$\text{Accident rate} = \frac{\text{Number of Accidents} \times 1,000,000}{\text{Total Fleet Mileage}}$$

3. Gather all relevant information about the accident. This should include driver statement, witness statements, police report, medical reports on the condition of the driver (if this is allowed), insurance company reports, diagrams or photos, physical evidence, company investigation reports, and recent maintenance records.

4. Gather all cost data. This should include repair and downtime costs. Include non-financial costs, medical costs, disability estimated costs, and loss of time.

5. All accidents should be investigated. Salespeople, company administrators, and officers should be under the same review system.

43
Special Issues of Building Maintenance

Building maintenance is unique in the maintenance field because of the high asset value per maintenance worker. In most cases, the building maintenance person is covering more ground (except for a field service person), servicing more users, and is responsible for more raw assets. Another unique aspect is the lack of resources of all types. The building maintenance person operates on their own without large crews. The building maintenance field also has significantly more contracting out of maintenance activity than fleets or factories.

What is the appropriate level of maintenance and custodial service? More than any other type of maintenance, the building maintenance personnel must decide on and enforce some level of quality. This decision is related to who the customers are and to the expectations of the industry you are in. You have to decide what is the appropriate level of service and rehabilitation for your clientele, buildings, and location. This decision will influence the materials for maintenance, the frequency of inspections, and the way your organization responds to tenant/user complaints.

For example, the Four Seasons Hotel chain expects that a maintenance worker will be knocking on the door of any guest who calls the front desk for maintenance within 10 minutes. A nice rental complex with 200 units expects to address 98% of all tenant/user requests within 72 hours (emergencies in under 4 hours). Decide on your standard level of response. Measure yourself against it. The same hotel might polish brass railings daily or inspect and spot-clean restrooms hourly. You have to decide the appropriate level of service for your customers (either internal or external).

413

Your custodial department must answer the same question. There is an appropriate band of acceptable housekeeping frequency for each part of your building or facility. Ideally this acceptable level would be the result of a negotiation with management. The equation relates quality of cleaned surfaces and fixtures with frequency with dollar input.

Improve Your Building Maintenance Operation

Route maintenance is one of the best ideas to improve the public perception of the maintenance department and reduce costs. In a maintenance route, the building is divided into regions or sections. A route box (looks like a suggestion box) is installed in each section with short write-up forms or service request forms. A section leader or contact person is also designated.

Each section is assigned a time and day of the month (or week) for its route visit. For example, the reception area might be assigned the second and the fourth Thursday of the month at 9 am. That time and day should be repeatedly publicized. The dispatcher should remind any requestor of minor calls that comes in about the scheduled time and day or the route.

Be sure you can service the route at least 95% of the time. On the second and fourth Thursday mornings, the route person rolls into the section, picks up the route box requests, and checks in with the section leader. All minor work and local PM tasks are completed. Larger jobs are written up by the route person and turned in to the maintenance control point.

A successful route person should be personable with a friendly and professional manner, multi-skilled, able to work without close supervision, able to work with customers so they feel positive and satisfied.

The route person should fill out maintenance log sheets. This will help you sharpen your overall maintenance effort. A review of the log sheets will show you things that your route person needs to carry! The better equipped they are, the more likely they can make the repair efficiently.

The route person's cart should be intensively studied by both the maintenance personnel and management. Consider the phone company or gas company. Tremendous thought goes into how to outfit a service person's truck. Next time you have an opportunity, ask the telephone installer or gas repair person how their truck is set up and why. Apply the lessons to the route cart, van (or even 5-gallon bucket!).

The more often they have the needed part with them, the more money you save. When they can use items from stock, they take the best price rather than the local neighborhood hardware store price (or waste time with PO's and supply companies).

Management of Housekeeping

Housekeeping (or custodial management) is concerned with the management of three factors:

Factor	% of contribution
1. People	75%
2. Chemicals and supplies	15%
3. Machines	10%

As in all maintenance work, the people factor is by far the most time consuming and the most important. This is why knowledgeable housekeeping managers should always have a good people person (who can learn the chemistry and the machinery) as a supervisor rather than a pure technical person.

No discussion of custodial management is complete without a look at quality. In housekeeping, once the cleaning technique is satisfactory, quality becomes directly related to frequency. Frequency converts directly to dollars. A corridor that is cleaned daily will be cleaner than one cleaned weekly. Management has to be taught that reductions to the housekeeping budget will result in reductions in frequency which will translate to reductions in quality.

People

Since pay is usually low and status is also low, housekeeping departments experience high turnover with the attendant problems of inadequate training, low commitment, high security risk, low morale, and low reliability. Departments with people-savvy managers have ameliorated some of the problems mentioned above by improving morale. Their techniques include individual and team recognition, uniforms, advancement opportunities, using ideas from the crews, and good supervision, contests, use of relief/project crews, good organizational attitude toward house-keeping.

Special Considerations for Housekeeping Contracts

Proper monitoring of the housekeeping contract is critical to getting the full value being paid for. There are many ways to deceive the contractee organization, particularly where low bid forces someone to take the contract at too low a price. To protect yourself, don't jump on the lowest bid but also look at the reputation of the vendors. Look at them in action before awarding them the contract. Your contract should be a fixed input type for all inputs into the housekeeping equation. Build into the contract price a sufficient number of inspectors to monitor the inputs.

1. The contract is based on an agreed-upon number of custodians. These custodians have had an agreed-upon amount of training and will get an agreed-upon amount of ongoing training. The custodians should be checked by your security department (or through the local police for undisclosed criminal records). Reserve the right to exclude anyone unsuitable from your crew. Your inspector verifies the number of custodians daily and spot checks training documentation.

2. The calculations to determine a reasonable budget for a building are necessary to write the bid specifications. Of course, if the contractor has a better way to clean an area, you can listen and make the appropriate modifications.

3. The contract is based upon an agreed-upon number of supervisors (on site). These supervisors have an average of an agreed-upon amount of housekeeping supervisory experience. Your inspectors verify the amount and quality (where possible) of the supervision.

4. The crews use agreed-upon frequencies, supplies, chemicals, and techniques to clean your building. No substitutions are allowed without approval. Your inspector verifies that the supplies, chemicals, and techniques are approved and that dilution and use is correct.

5. Your contract documents should list the formats of any reports you expect from the contractor. Your contract and search for a contractor should follow the guidelines mentioned in other sections of this chapter.

6. The best term is usually around 3 years, with termination with cause clauses.

Manage Housekeeping with the Standing Work Order and the Maintenance Information System

The standing work order is defined as "a form used to manage jobs that are done routinely with known labor and materials." A single standing work order might be good for a week, month, or longer. This type of paperwork could be used to control routine housekeeping. Normally the manager of housekeeping would analyze the building and break it into logical job units. Each unit (a day's work or less) would be written onto the standing work order. The packet of custodial work orders would constitute the cleaning workload for that building or facility. The major advantages of this approach are as follows.

1. Consistent paperwork system from housekeeping to maintenance (and grounds too!).

2. This facilitates upward mobility for custodians to maintenance workers. They would already know the maintenance information system.

3. You might want the flexibility of assigning a custodian to minor maintenance work or assigning a maintenance worker to some cleaning project.

4. The maintenance information system collects cost data, frequency data, and material usage. These data might be useful for a more complete analysis of the cleaning function.

The drawbacks are mainly concerned with introducing paperwork into an area that has been run by word of mouth. There might be custodians who will have trouble with the reading and writing (this is also an issue in maintenance too). In most cases, the housekeeping staff will resent the control and the snooping that the paperwork allows.

Workload

One of the biggest gripes is uneven workload among custodians. The only fair way to evaluate jobs is by time. Each custodian should be expected to complete a reasonable day's work. This reasonable day is determined by an analysis of the areas and tasks required. This analysis uses techniques of work standards. Sources of information on standards include Ed Feldman's *Housekeeping Handbook for Institutions, Business and Industry.* Other sources of standards include GSA, *Housekeeping Handbook,* and the U.S. Navy *Engineering Performance Standards.*

Concepts that Can Aid in the Management of Housekeeping Personnel

Relief/Project Team

Of the several strategies for cleaning a building (area complete, crew or gang cleaning), the most advantageous one seems to be area complete plus projects. Project is defined as cleaning activity that takes place less than once a month. The area complete custodian is responsible for all cleaning in their area except project work.

The project team comes in to strip the floors, clean light fixtures, and perform other infrequent work. The project team also supplies relief people to cover absence. The main advantage is that the project work gets done without pushing anyone else's work behind schedule. The project team is a good training place for future lead people and supervisors. It also allows the concentration of heavy equipment and specialized know-how. It is also a good place for new people (as helpers) to see the whole facility and meet the crews.

Vertical Scheduling

High-rise buildings offer special challenges in cleaning. In the normal area complete model, a custodian would be assigned a floor (or subsection). In the evening, the entire building would be illuminated. This technique has disadvantages from security, public relations, and energy points of view.

Most organizations want to present the point of view that they are thrifty to their shareholders or the public. A fully lighted, vacant building at night doesn't convey this image. It will also cost as much as 15% more electricity to run this way. It is also difficult to provide security for an entire building at the same time.

Vertical scheduling was designed to solve these problems. Each custodian is given a section of the same floor. At an agreed-upon time (such as break) the entire crew goes down to the next level. Lights go out and security is restored on the completed floor. Energy savings, improvement to security, and PR improvement recommend this technique.

Janitor's Closet

Your janitor's closet is essential in a successful housekeeping campaign. Proper design, stocking, and inspection are elements in a good program. If you have inadequate closet spaces add hose bibs between the sinks in the restrooms at bucket height. Specifics on the design of the closets can be obtained from the *Housekeeping Handbook* mentioned in the Appendix. If the closets are disgusting upon immediate inspection, then your first project might be to clean them up and bring them up to spec.

Supply Custodians

Custodians are constantly out of their areas walking to central stock for small amounts of supplies. The solution is to appoint one custodian and make him/her responsible for weekly stocking the closets, ensuring two week's worth of supplies.

Chemicals and Supplies

Chemicals and supplies are a low percentage of the total housekeeping dollar. This is an area where purchasing high quality saves money and improves morale. It is generally accepted that for highest efficiency, get the best price on one of the higher quality lines of janitorial supplies.

Successful supply usage deals with the scope of chemicals used in housekeeping with attention to dilution, safe and effective use, disposal, and transport. Be sure you have easily available measuring devices and adequate hazard warning systems.

Look for:

1. The UL seal on floor wax (finish that passes the beanbag test).

2. One purchase cycle a year with monthly or just-in-time releases.

3. Few vendors as possible; purchase wax, stripper, sealer, spray buff material from the same vendor.

4. Five gallon packaging on wax and stripper; one gallon and pump spray on everything else.

5. Use Quaternary Ammonium germicidal detergent, except for special situations.

6. Carpet manufacturers sell carpet spot-cleaning kits. Learn the details of how to spot-clean your particular brand of carpets.

Avoid:

1. Scouring powder on restroom fixtures (will eventually scratch them and increase need for maintenance).

2. Poison label products (such as acid descalers with greater than 10% acid, because they increase your liability and your probability of a problem).

3. Routine acid descaling of toilet fixtures.

4. Eliminate ammonia and bleach from general supply (very dangerous to mix).

5. Unlabeled containers for anything.

6. Home cleaning agents; and do not allow workers to bring in their own favorite home chemicals for cleaning.

7. Do not allow custodians to run out of supplies.

8. No aerosols; use pistol grip sprayers instead.

Stocking and Purchasing Custodial Supplies and Materials

Low bid: Supplies constitute only a small percentage of the housekeeping budget. Quality supplies can reduce labor and preserve the surfaces to be cleaned. Many organizations purchase the cheapest materials. This is a false savings. The solution is to purchase the lowest price of the best quality material and train your crews in proper application.

Purchase to specs: Other organizations purchase to specifications. The problem is that the specifications do not always reflect the best material for the job (the manufacturers sometimes formulate to specs, but the material is not too good). The solution is to include a performance specification. Custodial chemicals change very often so that a preferred brand might change formulation before the contract year is over.

Packaging: Frequently in the search for the absolutely lowest cost, the organization buys supplies in 55-gallon drums. Many custodians have suffered smashed fingers, back injury, and smashed toes because they didn't have proper handling equipment. Also, product and time was wasted transferring products. Unlabeled jugs have led to custodian deaths when non-compatible chemicals were mixed. The solution is to purchase high-volume supplies in 5-gallon containers and everything else in 1-gallon containers. Investigation of portion-controlled packaging might be advantageous in some circumstances.

Dilution and mixing: Product like liquid hand soap was sometimes put out undiluted (which would clog up the dispensers). Custodians would figure that if 1 ounce per gallon was good, 2 ounces would be better. This would interfere with the cleaning and might damage the surface. The solution is to dilute or mix when the supplies are received into labeled containers marked ready to use. Many organizations have also gone to portion dispensing equipment in the custodian closets.

One of the most interesting trends is toward portion packs (some of them even have the packaging dissolvable). This solves one of the biggest problems in housekeeping by standardizing the solutions. Of course, you pay significantly for the privilege. These systems would work best where you have an inexperienced crew, small usages, satellite operations, route cleaning or smaller buildings or other similar situations. It is a very attractive alternative.

Salespeople: The sanitary supply business seems to attract hundreds of small vendors. Their salespeople can use up tremendous amounts of the housekeeping supervisor's time. The solution to this problem is to bid blanket contracts for supplies only once a year. Only see salespeople (except by appointment) for 30 or 60 days before the contract comes due. Blanket orders would also tend to lower your cost. The trend today is to concentrate your business among only a few vendors so that you are a larger customer for the distributor. As a larger customer you can get support that a smaller customer cannot.

Incompatibility of the chemicals: Particularly in the floor care area, the strippers are designed for the wax to be stripped. The solution is to always buy your stripper from the same vendor as the wax. This also encourages the trend toward fewer vendors, mentioned above.

Machines

Mechanization of housekeeping can provide significant return on investment to the organization. Introduction should be gradual, and adequate training should be provided. The rule in housekeeping (Feldman's Rule, from Ed Feldman) is to use the largest and most rugged machine that can do the job.

Many organizations try to save money by avoiding mechanization or keeping obsolete equipment. Housekeeping is usually the last area to mechanize.

1. Use automatic scrubbers instead of mops for large open areas.

2. Look into telescoping window washing equipment (available up to four stories).

3. Litter vacs are better than blowers because you don't have to deal with the same litter again.

4. Pack vacs now weigh under 10 pounds and are ideal where you are working in a congested area and cleaning ceiling fixtures.

5. Significant thought should go into the janitorial carts and the jobs your people routinely perform.

6. Be sure to budget for one machine per person/team. Sharing machines promotes downtime, prevents ownership, and provides *the feeling of responsibility for their "own" equipment.*

7. Payback on mechanization comes from reduced labor or increased quality.

Grounds Maintenance

Listed below are some possible tasks to consider based on time of year. (Recommended by Jeffrey Bourne, Chief of the Bureau of Parks, Howard Country, MD.)

WINTER:

Snow removal

Salting/cindering

Equipment overhaul (turf)

Safety related tree pruning

Refuse removal

SUMMER:

Grass cutting

Irrigation

Insecticide program

Growth regulators

Fungicide applications

Refuse removal

SPRING:

Equipment preparation

Turf seeding, fertilizer, treatment

Soil testing, weeding, begin edging

Turning over top soil, mulching

Pruning trees and shrubs

Refuse removal, clean debris

Test systems (irrigation, lighting)

Check for winter damage

FALL:

Hard surface repair

Snow equipment renewal

Winter prep of equipment

Spring bulb planting

Grass fertilizer/treatment

Tree replacement/pruning

Refuse and leaf removal

Protecting fragile trees and shrubs

Figures 43-1, 43-2, and 43-3 show examples of forms used.

GROUNDS SURVEY SHEET		Date:	Surveyer:

Site Name:

Size, description:

Skill	Fre-quency	Task
1	D	1. Pick up grounds: broken glass, bottles, litter, other
1	Q	2. Check condition of all sidewalks
1	W	3. Check condition of driveways, parking lots
1	W	4. Check sewer drains for back-up
1	W	5. Check condition of lawn and plants
1	D	6. Check cleanliness around dumpsters
1	M	7. Check fencing
1	M	8. Check condition stairs and hand railing
1	M	9. Check condition patio areas and benches

Comments:

Figure 43-1. Exterior task list.

#PMWO	PM WORK ORDER		Address 6535 LIMESTONE
Frequency: Quarterly	Tenant: John C. Cook	Keys: (BOX)	Phone: 535-0002

Priority: 4-PM	Reason for write-up: PM	Requested by: JL

	Description of PM Task:		1	(A) 2	3	4
1	Smell for gas leaks in basement, quick look for leaks		✓	✓	✓	✓
2	Test all smoke detectors	How many? 3	✓	✓✓	✓	✓
3	Check and tag fire extinguishers		✓	✓	✓	✓
4	Test and reset GFI's		✓	✓	✓	✓
5	Look for leaks and drips all sinks, under tub		✓	✓	(1)	✓
6	Flush toilet run sink check for slow drains		✓	✓	✓	✓
7	Pest spray		✓	✓	✓	✓
8	Look at general housekeeping		✓	✓	✓	✓
9	Remove debris from around heater, HW heater		✓	✓	R	✓
10	Take all meter readings		✓	✓	✓	✓

SKILL LEVEL	Unskilled	(Maint. person)	Licensed Trades	Engineer or other	Contractor:

Notes and problems:	MWU #
(1) Leak under tub-tighten trap nut 5Min	
(2) Tenant to remove debris in basement to other side	

Q	DATE	INIT	IN	OUT	TIME	MATERIAL	Q	DATE	INIT	IN	OUT	TIME	MATERIAL
1	3/1/92	JL	2PM	2^30	:30	—	3	9/20/92	JL	10AM	10^30	:30	
2	6/12/92	JL	9AM	10^00	1:	—	4	12/1/92	JL	10AM	10^30	:30	
					1:30							1:00	

(HRS(1:30)+(1:00))* CHGRT(25)+MAT'L 1(——)+MAT'L 2(——)=TOTAL(62.50)

Figure 43-2. Quarterly inspection, small building.

BUILDING EXTERIOR AND ROOF SURVEY SHEET		Date:		Surveyer:
Site Name:				
Materials, exterior			Roof:	

Skill	Fre-quency	Task
1	W	1. Check for and replace burnt out bulbs pole, wall, ground
1	M	2. Inspect all electrical hardware and connection boxes
1	W	3. Inspect roof for damage, leaks, clear broken glass
1	M	4. Inspect stack and stack supports
1	M	5. Inspect flashing, transitions, parapets
1	M	6. Check roof drains for clear drainage
1	Q	7. Check condition of paint
4	Q	8. Inspect mechanical house on roof
1	M	9. Check condition of antennae and wires
4	M	10. Check condition of walls

Comments:

Figure 43-3. Grounds task list, medium-sized facility.

Purchasing and Stocking Materials

A substantial percentage of your building maintenance budget is spent on materials. These materials range from toilets, to paint, to bug spray. It pays to have secure storage to take advantage of deep discounts. Any item that is to be stored should be about 10% less than the usual price for every 3 months of expected storage time. Many departments use their basement or garage for the materials, particularly when the basements have street-level entrances or elevators.

Many of the best wholesale-to-the-trade distributors are being put out of business by the home center superstores. They cannot compete, even for the wholesale trade, with the buying power and efficiency of these stores. The sacrifice is in inability to get sound trade advice from the counter people. The trend means that traditional sources of supply might have to be periodically looked at.

Standardization is another important strategy that the maintenance manager can take advantage of. Over time, you will know what brands work best and what brands are a waste of money. Even expensive brands make sense if they reduce overall costs and give your tenant/user better service. Also, standardization reduces the number of parts you need to carry.

Your best friend is the business-to-business yellow pages. Look up "Janitorial Supply," "Plumbing Distributors," "Electrical Distributors," "Industrial Equipment," and "Supplies, Tools." Establish accounts with the good vendors to the trades. These distributors usually have the best quality and selection. They have prices that are as good as a home center sale price. While the home center can sometimes beat the price of the distributer, you can't have a conversation with a knowledgeable counter person.

There are many mail order businesses that cater to the maintenance trade (big building complexes). We use Maintenance Warehouse and Johnson Supply (the addresses in the Appendix). Also look toward the industrial distributors. The biggest is W.W. Grainger (branches in most major cities).

Building Automation Systems

Building automation is the use of computers to control the functions of a building. The first systems were designed for energy efficiency and were basically smart thermostats. Low cost smart computerized controls and inexpensive sensors increased the capability of these systems by several orders of magnitude.

Some of the newer high-end systems feature graphic interfaces, and touch screens allow you to walk through a symbolic picture of the building. Most systems allow maintenance supervisors to dial in from home and check temperatures, valve positions, and air flow. Applications include integration of energy efficiency, comfort, security, building control, and maintenance schedules.

Building systems are rated by the number of control points. Each control point represents an input or output to the building control program. These systems monitor all the inputs and activate the outputs. Systems allow load shedding, night set-back, and remote control of functions.

Typical Control Points

1. *HVAC:* Outputs to compressors, furnaces, air handlers, dampers, zone control. Inputs from thermocouple in ducts, rooms, halls, walls. Inputs of humidity, fan status.

2. *Chilled water systems:* Outputs to pumps, valves. Inputs include water temperature (before and after chiller), outside air temperature, flow rate, valve position indicator.

3. *Lighting:* Outputs to light banks, sections. Inputs from room in use sensors, electric eye outside, timers that allow automatic shutdown and allowance for janitorial access.

4. *Security/fire system, access subsystems:* Outputs include alarms, fire pumps, printers, auto closing doors, outputs phone for fire/police, sirens, bells, security gates. Inputs from motion detectors, door switches, glass breakage detectors, card readers, sirens, smoke detectors, rate of rise detectors.

Ideas to Save Real Money in Building Maintenance

Management

1. All managers should sell the idea that the rent is determined by the expenses for the building. Rent raises are related to ex-

pense increases. A tenant/user suggestion system for saving money with small monetary (dollars off rent) or token rewards (dinner at McDonalds®) for suggestions accepted. Be sure to invite energy-saving suggestions. Publish winners to your other tenant/users.

2. Tenant/user orientation can save you money. Make up a list of the items you want to cover from mistakes in the past. Show them where the water shut-offs are, how the toilet works, how to change the battery and test a smoke detector, where the fuse box is, use of fire extinguisher, whom to call for what emergencies, the list of questions you will ask when they call for emergency service, etc. At the Naval Academy in Annapolis, the maintenance manager created a 1st lieutentant's desk book that featured all of the information needed.

3. The time to charge your tenant/user for damage is right after your monthly PM or survey, don't wait until they move out. It is easier to collect because you can document damage when it occurs. Get together with your crew or contractor and put together a schedule of reasonable charges for common types of damage. Fix it and collect it. Photograph all damage over $100.

4. Planned component replacement (called PCR) can sometimes be used to save headaches. A PCR program would be replacement of components on a planned basis rather than after breakdown. For example, write the date that you replace your hot water heater in magic marker on the heater. Replace it 10–12 years later, before failure.

5. Use your camera. Document your buildings for your files. Experiment with infrared film to see where heat is escaping. It is amazing what you can see when you take pictures every year (at the same time of year).

6. Be very careful, when doing construction in one part of a building, that you: put up plastic curtains to isolate the area from the rest of the building, and put down walk-off mats (8' long is best) for worker's feet. You can also turn off or turn down heat or A/C to this area.

Plumbing

7. In kitchens and restrooms, standardize to one major brand of faucet. Be sure to instruct the plumber to install fixture valves called stops on all fixture risers. Also, the best method of connection today is the stainless steel braided flexible supply. This allows a quick change of the faucet next time with only a wrench (no plumber needed).

8. In smaller buildings, Bernie Cleff, a manager/landlord in Philadelphia, recommends installation of shut-offs on both hot and cold water sides of hot water heaters and heating boilers. This allows removal and replacement of the unit without draining the system.

9. Tag all valves with what they turn off. Use plastic tags and freezer pen. Industrial distributors sell valve tags. If a valve gets stuck and won't turn off and there is no emergency, spray stem and bonnet nut with WD-40 and let it sit for a hour. You may have to repeat to free up the valve.

10. Outfit showers with low flow shower heads to reduce hot water. Insulate hot water pipes that run long distances. Better than that—relocate the hot water heaters nearer to where the hot water is used. Reduce the setting on the hot water heater to 120°F, higher normal settings might scald an elderly person or child, *and* waste energy.

11. Purchase low flow toilets. This reduces the almost 5 gallon flush to 1.6 gallons, saving 20 gallons a day for a family of 4.

12. It is usually more time consuming to find the correct washer for a faucet than the actual washer replacement. Next time you replace a washer, wire several of the same type to the shut-off valve for future replacement. If you need new washers in as little as 6 months, look into replacing the seats.

13. The most common reason that sinks get clogged is grease congealed in the trap. A quick fix is to direct a hair dryer on the trap, on hot setting. Flush the drain with hot water. If you have to disassemble the trap, consider putting grease on the threads of the large nuts. This will both seal the pipe and aid in later disassembly.

14. Water damage is very common behind the walls around the tub and shower. Install a plastic tub surround. The trick is to caulk all exposed edges. On the topic of caulk in restrooms—next time you stay in a hotel notice that they caulk all around the bathroom. Almost anything that could get wet is caulked. Use good quality silicone.

Carpet and Linoleum
15. If you decide to install carpet in-house, use these tips. The first key is to keep the carpet tight. Then lay the carpet with pile pointed toward the door. To reduce problems and make a neater job, tape seams on padding (padding is usually 6' wide) with duct tape before stapling it to floor. Install padding rough side down.

16. Dirt exclusion is the cheapest way to keep buildings clean. It costs $50 or more per pound to clean up dirt after it gets into your building. *Techniques:* Rough surface walk-off mats at least 8' long in first hall (remove and clean regularly), wooden or metal grate with catch pan in vestibule (best spacing is 1/10" between slats). (This tip is from Ed Feldman, a leading expert in building maintenance of Atlanta, Georgia.)

17. Allen Whiley, a maintenance manager for the Philadelphia School District, recommends installing resilient flooring (linoleum) over sheets of 30# roofing paper (felt). The felt does two important things. It contributes to the watertightness in a bath or kitchen, and it provides a soft base layer so that irregularities in the underlaying floor don't work their way through the floor covering. He says the job will last longer, protect the room below better, and feel better.

Painting
18. This tip is courtesy of Barry Romano, Director of Operations at the Albany Housing Authority. He is responsible for over 1000 rental units. He paints the woodwork and walls with the same off-white semi-gloss. Semi-gloss is harder to damage and can be washed. Woodwork and walls are cleaned with a solution

containing TSP (tri-sodium phosphate) such as Spic and Span professional with TSP (available from paint stores). Always prime wallboard with latex because it drys fast and seals both the paper covering of the wallboard and the joint compound.

19. Painting tips always save money. Some good ones: line your roller pans with plastic wrap or a kitchen garbage bag to speed clean-up time; put paint brushes in plastic bags to store for short time; use a bronze wire brush (barbecue type) to clean paint brushes, by combing like hair; clean the exterior with a rented pressure washer to speed work and improve quality; store cans of paint upside down and turn over every few months. Ideas courtesy of Practical Home Owner Publishing Co., New York.

20. More tips on painting are always useful for saving money. Paint stores sell a putty knife with a circular cutout to clean rollers. The cutout follows the contour of the roller to effectively clean it. Manager/landlords should consider investing in a spinner that spins brushes and rollers to clean them.

21. One of the most difficult problems is to paint textured ceilings and walls because they soak up so much paint and never look quite right. The trick is to paint the entire ceiling with a good alkyd sealer and a long-nap roller. Cover the whole floor with a tarp (or plastic since it might be a little messy). This tip and the last one are courtesy of *Popular Mechanics.*

Roofing and Chimney
22. Sean McGarry, a Philadelphia roofer and manager/landlord, gave us some tips to save money by preventing roofing problems. Be sure all gutters and downspouts are cleaned out regularly (particularly in the Autumn). Also clean off the entire roof of debris and keep people off!

23. Sean uses two products on his own buildings for quick fixes of minor roof problems. The two products are used together: Roof Cement (about $15/5 gal can), and Roof Fabric (about $4/4" roll). These items will fix all sorts of water problems in roofs, flashing, chimneys, and other areas. Repair by smearing the cement onto the problem area with a pointed trowel and putting the fabric over the cement. Cover the fabric with additional cement. Remember the old roofer's wisdom: "Water always flows downhill!"

24. John Foody, a public adjuster and manager/landlord in Philadelphia, has seen a lot of water damage. He recommends coating flat roofs with aluminized roof coat every three or four years. This is a good job for do-it-yourselfers.

Electric
25. Electric safety is most important for intermediate level handy-people. At this level, a person might be skilled enough to wire new circuits. If you are qualified, keep safety in mind. Keep a rubber mat by your fuse box and stand on it whenever you are working in or near the box. Always use only one hand when working hot (the other one is in your pocket). A supply of fuses and a flashlight can be lifesavers (literally). From *500 Terrific Ideas for Home Maintenance.*

26. The easiest way to save money with common area lighting is to reduce wattage. Watch out for special long life incandescent bulbs. They last longer but give off less light per watt. Restrict them to hard to reach areas where a fluorescent will not fit.

27. Forget about timers to turn hallway or outside lights on. The best technology today is the photo eye (remember to keep them clean).

28. If there are grates or covers over the ceiling fixtures, consider removing them if glare is not a problem. Dirt in the fixture reduces life (increases temperature) as well as light. Wipe down bulbs, reflectors, lenses, and grills on your annual inspection. Dirty reflectors cut light dramatically (they impact the light output twice; the light is reduced going into the reflector through the dirt and again on the way out).

29. The best invention for manager/landlords was the screw-in fluorescent light. These lamps can save big money in common areas. These lights use 1/5 the electricity and burn 15 times longer. The electricity savings for a 24 hour burn period is 0.75–

1 kW per day per bulb. In dollars the savings is $.06–.08 per day ($21.90–$29.20 per year per bulb). Also the cost of regular bulbs is $.50–.75 ea or $7.50–$11.25 plus the labor of 15 bulb changes. In this area the screw-in circle fluorescent bulbs are $6.99–$18.99.

30. What happens when the good fluorescent bulbs are stolen from the common areas? This is a particular problem in areas where there is access (the bulb can be reached without a ladder). The vendors came up with a fluorescent light that screws in but requires a tool to remove. They are more expensive at $25/ea in lots of 10 from Krell Lighting (201-391-7685) or your local wholesale lighting distributer. The ballast has a 7-year warranty. The bulbs can be changed separately.

Yards and Landscaping

31. Landscaping presents an opportunity for energy savings. Properly positioned trees will shade a property in the summer and insulate it in the winter from cold winds. Studies have demonstrated up to 10% reductions. Consider evergreens where there are no windows, trees that retain their lower limbs, or dense shrubs. Be sure to plant consistent with good maintenance and security guidelines.

32. Additional ideas on landscaping from Philip Christianson, a landscape expert in Atlanta, include: use a different schedule to cut rear and side lawns than you use for the front grass (such as 2 weeks on the rear, 1 week on the front). He suggests pruning hedges and shrubs deeply in early spring and let them grow out naturally (no trimming if you cut them back far enough in the first place).

33. Regarding flower beds, Philip advises to mulch to only 2"; use Round-Up instead of weeding (Round-Up leaves no residual effects in the soil); only plant evergreen shrubs like holly; don't try to put lawn over clay, instead mulch and put in a simple planting.

34. Save your trees from rabbits and other bark lovers by wrapping them with burlap up to about 3'. To nourish a root system, keep a circle around each tree free of grass and weeds and with loose dirt. Other tips include use saw dust from construction projects for mulch. From *500 Terrific Ideas for Home Maintenance.*

Weatherstripping, Windows

35. Make sure weatherstripping, caulk, window/door tightness is part of your inspection program. After a snow, the area on the building roof where snow melts first needs insulation/inspection.

36. Choose windows that are cleanable from the inside and where the screen is storable in the unit. We found that old sashes were rarely worth it to work on.

37. Awnings can make rooms more comfortable by blocking the summer sun (high in the sky) and allowing the winter sun in (low enough to get under the awning).

Locks

38. Standardize on locks.

39. Install brass plates that sandwich the door and protect from breakage.

Heating

40. When you pay for heat or hot water, look for high efficiency heaters. Install automatic dampers where you have older equipment. The automatic damper closes when the flame is off so hot combustion gases stay in the furnace heating the air or hot water.

41. Evan Lazar, a Philadelphia area manager/landlord, has a building where he has to pay the heat. He installed a large thermometer with $68°F$ marked off with a magic marker. The tenant/users are instructed to call only when the needle goes below the mark.

42. In smaller buildings, use set-back electronic thermostats that turn the heat down at night.

43. Where there are single thermostats for the heating system, consider: 1) putting several thermostats in series so that the heater only goes on when all the thermostats call for heat; 2) put the thermostat in a 4×4 electrical box with a blank cover out of the way so it will not be tampered with; 3) buy and install special thermostat that has only one electronic setting like 68°.

44. Of course, locate the thermostats away from drafts, heat vents, and windows.

45. Wash your radiators (or require your tenant/users to clean them). Most paint serves to insulate the radiator (less heat gets into the room). Put aluminum reflectors between the radiators and the wall. Insulate heat pipes or ducts in unheated areas.

44
Special Issues of Field Service

Field service is a unique branch of maintenance because of the distances involved and the fact that the customer is usually another organization. As a result of these two differences, management of field service is somewhat different than management of a conventional department.

In field service, being given the wrong part from your parts department and taking it to a job could be a catastrophe. In a typical factory or fleet, the wrong part will stimulate some irate comments and an extra walk or cart ride. Because of the distances, any planning mistake (wrong tools, wrong drawing, improper parts, lack of access like no keys) becomes a catastrophe. The other factor is that the technician usually cannot jump on another job while the problem is fixed by an overnight delivery.

The size of the territory impacts how the department should be looked at. In practice, a building services department approaches a field service department when the size and scope get too large to manage conventionally. Also, many traditional departments have a field service arm. This might include a construction company with a lubrication and minor repair truck that services equipment at all of the contract sites. It certainly includes a school district with a roving plumber.

In some cases, a traditional field service department becomes more like an in-house department. In the largest corporate or government data centers, the computer manufacturer or service company will typically have a technician or even a technical group on-site. These on-site technicians act like fixed base departments and set up a small inventory, keep a full set of tools and test equipment, and put in critical O & M files.

The area of field service is diverse. Included are copier or computer repair groups, fire alarm companies, contractors doing service calls and service contracts, punch press repair teams, and predictive maintenance companies—the scope is almost endless.

Parties (Who Does What)

First party maintenance is performed by the user's own maintenance department. First party maintenance is frequently a competitor against the second or third party service providers. In some cases, the OEM (second party) maintenance department encourages the user to provide maintenance themselves because the parts business is so lucrative. OEM's can sell the profitable parts without the hassle of maintaining a service operation.

Most service in major electronics is provided by the manufacturers (*second party* maintainers). Many companies such as IBM or Xerox grew powerful in their respective fields partially on the strength of their service departments. When copiers or mainframes were introduced (and for years afterward), all equipment was leased with full service. This guaranteed revenue built powerful departments. With the direct sale of equipment and the unbundling of services, customers were too leery of the technology to rely on anyone other than the manufacturers for service.

After a few years passed, service companies sprung up for specialized complex equipment. These *third party* service providers were usually ex-OEM service employees. They could charge ⅔ what the big companies could and make a tidy profit.

The problem was that few of these service providers had the critical mass necessary to set up and run economically in speciality areas such as disk drive repair/rebuild, board repair, and other specialities. This led to the birth of *fourth party* service providers. A fourth party provider was a behind-the-scenes support player for the third party service company. In the back of any computer service magazine, you can find disk drive repair, board repair, power supply repair, etc.

The last link is the salvage operations that take older equipment and strip it for parts. Similar to the automotive junk yard, there is now an active market for salvaged parts in computers and electronics.

The sources of parts follow the party system of service with some added twists. Finding inexpensive sources of parts in any field is a barrier to going into self-service. The OEM's are the primary source of parts. Parts make up a significant part of the profit picture. In fact, Ron Giuntini, in his article "Managing an OEM Parts Business" in January 1996 *AFSM*, says that parts should generate a bulk of an OEM's profit where the unit sales are adding less than 10% of the installed user base (most big companies with more mature product lines). This means high prices and monopolistic service.

There is an invisible source of parts that requires knowledge and significant expertise. It could result in very low prices. OEM's buy much of what goes into their products from vendors. In one case, a trailer manufacturer bought corner posts from a steel fabricator. It is frowned upon for the fabricator to solicit the users directly. Occasionally a large user will approach the original manufacturer of the part. In this case, they needed a few hundred corner posts to rebuild a batch of trailers. Bypassing the markup from the OEM can save the maintenance department a bundle if the user can buy in fairly large quantities. An additional benefit was the opportunity to improve the design to reduce future maintenance costs and effort levels.

In some established fields, the parts can be picked from catalogs or specification sheets. This would include parts such as power supplies, disk drives, bearings, seals, memory chips, limit switches, pulleys, gears, lamps, and other hardware. The OEM's engineering department designs from these catalogs to speed time to market and reduce their setup costs.

Fourth party repair and rebuild depots mentioned above are a large source of parts. Since they rebuild, they can offer limited warranties and good reliability. In some cases, they refurbish parts for the OEM themselves. The salvage companies are also good sources of parts if you are a sophisticated buyer. Only knowledgeable users or professional service providers should use this channel. In some cases on older equipment this might be your only source of supply.

Customer-Driven Maintenance

The major difference is how the customer is viewed. Most field service companies see the customer as the life blood of their business. While technicians complain about how the customers are this or that, it is easy to see that there is no work without customer calls and service contracts.

The criticality of the customer is true in fixed base maintenance departments but is not as obvious to the mechanics. It would be less likely to view a customer as an inconvenience in a field service department, especially if there hasn't been any work that day.

Much time and effort is spent in field service on improvements to the delivery of service to the customer. The rest of the maintenance field is just waking up to the need to give the customer service beyond expectations.

Field Service Benchmarks—Measuring Customer-Driven Maintenance

MTBF&V (Mean Time between Failure and Visit): The first important measure in field service is simply how long does it take to get a response. This measure goes well beyond time from call to time technician shows up at the door. It also includes the time it takes to realize there is a failure. The newest and most sophisticated assets alert the service company to an impending problem. This could cut the time MTBF&V to almost zero.

MTRV (Mean Time from Request to Visit): This is a subset of the above more inclusive measure. If you call in for copier service at 11 am, when do you expect a visit? What level of response would exceed your expectations?

FTFR (First Time Fix Rate): The second measure is how often does the technician fix the problem on the first visit. One of the most common ploys is when a service organization is backed against the wall with calls they send an unqualified person to "shut the customer up." This unsavory practice is immediately detected by an abominable FTFR.

MTTR (Mean Time to Repair): After the technician is on site, how long does it take for them to fix the problem. Consider a computer problem that occurred a few years ago. The major computer manufacturer responded within 4 hours. They assembled a team within 24 hours and worked around the clock for 4 days to fix the problem. The MTTR to that point had been 1 hour.

MTTT (Mean Time to Travel): While this is far less important to the customer than MTRV or MTBF&V (unless they are paying), it is an important internal tool. Excessive travel could mean a problem in communications, routing, or holes in the market (that could be filled in by sales effort).

There are many additional measures that a field service department would use as a business. The most common is revenue dollars per technician hour. This is a good general index of a service company. All of the costs could also be traced back to the hours. The most popular gross measure, Total Revenue per Year per Technician, is related.

Use of Statistics in Field Service

Statistical analysis for field service is more accurate and easier to calculate than for other types of maintenance (except fleet). The reason is that most field service organizations have many of the same model machines to service. They also have significant experience in these repairs. An article by Jeffrey Oelker in *AFSM Journal* demonstrates some applications of statistics and SPC to field service.

Service calls are divided up by model of equipment (not by problem). Travel time is subtracted so that you are looking at time on site (MTTR). Determine the mean and standard deviation for the MTTR. You can refer to formulas in the section on using statistics.

In the field of SPC, a process is either in control or it is not in control. An in-control process has a bulk of its measurements between an upper and lower control limit of ±2 standard deviations. Any process will vary within these control limits due to natural variations.

Figure 44-1 could be a chart of the average time it takes each of your technicians to repair a certain model pump, printer, or photo typesetter. The one technician well above the control limits would be looked at for training, techniques, or other source of variation.

The second use of statistics is to calculate the capability of your service process. You can determine how long it takes to ser-

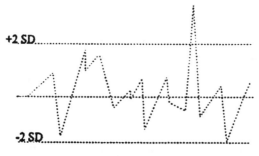

+2 SD

-2 SD

Figure 44-1

vice an average printer. You can further determine how often you meet that goal using the arithmetic of a normal distribution. Calculate the mean and standard deviation of the time to service a particular type of equipment. You can advertise that your technicians will fix that asset 95% of the time in less than 2 standard deviations plus the MTTR.

If the MTTR is 1 hour and the standard deviation is 20 minutes, then you can say to your customers that 95% of the time we will fix your printer in less than 1 hour and 40 minutes. You can give away the service call if you can afford to give up the revenue from 5% of your longest calls.

Your goals as a service manager are twofold: you examine the process for delivering this service so your variation gets smaller and smaller, and redesign the process to lower the MTTR.

Quick Test for Your Field Service Department

1. Call up your company and act like a customer. Try all the different numbers that your customers might have access to. Record all responses, time on the line, how you were treated. This test can be carried out by different members of your service team and discussed. The goal is to deliver the best service possible, treat the people respectfully, and solve their problem in a way that encourages them to come back.

2. Call a competitor and also act like a customer. How are you treated? Look at the same issues as above. If you are a member of an association, work with service providers in other parts of the industry or country. Test each other's systems and response.

3. Periodically call a lost customer. Do not focus on blame or even on getting the customer back. Your goal is to see why someone would leave and which parts of your process went wrong. This takes a strong ego not only to keep your mouth shut but to actively listen.

4. Spend time with the people who use the equipment that you service. Many great ideas flow from seeing how equipment is used. You could think up whole new ways to add value to your service offering. Ask the question over and over in different ways—how can we help you better? This is also a good exercise for your field service technicians. Discuss their findings in your team meetings.

5. Go to a customer's site and do something simple (like remove a wire) and have the customer call for service. Track every step of the process for review with your team.

6. Listen to your meetings with the field staff. Your field staff is alone out there. They are representatives of your organization and are frequently the only link between the customer and your own company. Are you providing the kinds of support that people this important need? Consider the educational support, emotional support, financial support that you provide. It should not surprise anyone that many customers defect when their favorite service rep moves to a competitor (or starts her/his own company).

Service Contract

One of the greatest inventions of the field service business is the service contract. It distills down to the essence what users want from a service company. Users want someone else (preferably an expert in the field) to be responsible for an asset. Users want to use—not care for—an asset.

The service contract is also a great source of revenue for service companies. If fact, whole industries live on their contracts, including copiers, computers, fire safety, elevators, HVAC systems, and building automation. Everyone wants the constant revenue stream of the service contract. We see automotive, tires, and other building services like lawn care, pest treatment, and windows, factory electronics, such as test gear or PLC networks, lighting, and others are offering service contracts.

The challenge was always to price the service high enough to make money, and low enough to be attractive to the customer. The more proprietary the equipment, the less likely the user can get a third party company to compete.

Information Superhighway, Here We Come

The Internet is changing the complexion of field service. You can review the sales of an individual Coke machine from the Internet. Copiers, computers, test equipment, CAT scanners, and other equipment are starting to have addresses on the Internet or direct dial access. Service contractors can call up the asset and get meter readings, operating parameters, and change settings. All it takes today is a telephone, a modem, and familiarity with the Internet communications protocol. Once the automated software is put into the machine's program and the machine has an account on the corporate Internet machine, the rest of the hook-up is just following a script.

This connectivity is changing the complexion of field service. The MTBF&V might change to zero if the asset calls field service when it detects an *impending* problem. The technician calls the asset back, and after a computer-to-human conversation, finds (s)he can work around the impending failure using on-board spares, degraded operation, redundancy, or extra capacity. (S)he can then schedule a new module for shipment and replacement by a semi-skilled on-site operator.

There will be whole new ways of providing service which are unknown today, that will change the face of the business.

One immediate benefit of the Internet has been the availability of up-to-date information for service from the OEM's. Using the World Wide Web (WWW) or an FTP or Telnet site, service providers can find out the latest technical changes and bulletins. Almost all sites have FAQ's (frequently asked questions) for novices. See the chapter on the Internet for additional information.

Expert Systems

Software is catching up with the enormous advances in hardware. All new automobiles have designed-in expert systems to aid service. With increasing complexity, there are fewer and fewer mechanics that understand the whole machine and all of the technology.

Expert systems are software versions of skilled and knowledgeable service people. In fact, some of the systems were a computerization of the logic and steps of the most expert technician solving a problem. The systems followed the logic step by step and took readings, made observations, or requested input (if the question could not be answered by a sensor or a reading). From this logic, a conclusion was reached, and either the machine reset itself to conform to the new directive or it requested an action from the mechanic or operator (change the O_2 sensor).

When expert systems are linked to the Internet, there is a powerful service possibility formed. The skilled technician can query the expert system and, in some cases, have a diagnostic conversation without leaving their office.

CD ROM and Other Powerful Hardware and Software

One of the biggest problems is carrying detailed documentation for a wide variety of models. For example, if you service personal computers you might face over 1000 different motherboards. Each has different characteristics. In that industry there is a CD available to service providers with specs from 9200 products from 1800 manufacturers including hard drives, motherboards, controllers, configuration, settings, jumpers, functionality.

It is now probable that a field service technician would be as likely to carry a notebook-sized computer as a big heavy tool case. A suitcase full of diagnostic gear is now software inside a 5-pound notebook computer.

Who Should Service?

Who is the best choice to service the customer? One issue of field service customers is which of the following is better.

1. Local distributors that service a variety of equipment and have varying levels of skill with the specific equipment that you own.

Advantages: Continuing business, a local rep would sell you other products so they have a stronger incentive to keep you happy. Local rep can be on-site quickly and fix most problems.

Easier to build a partnership. Cares more about the outcome. Might be less expensive.

Disadvantages: Might not be up to date on all revisions, service bulletins, field engineering modifications. Might not be certified on that particular equipment. Might not have extensive experience on your equipment. Is not qualified for deeper problems. Might not fix it the first time.

2. Factory or regional employee representatives who have to get on an airplane to provide service.

Advantages: Has probably seen your problem before. Is more likely to fix the problem right the first time. Is usually up to date with changes, service bulletins, and all other best practices. If equipment is on-line or dial-up, diagnosis can be done remotely, and the correct parts and tools brought to the job.

Disadvantages: May take too long to get service. Does not have time to build a relationship, let alone a partnership. Has less incentive to go beyond the usual barriers to provide great service.

APPENDIX
Resources

Magazines, Newsletters, and Books

1R. *AIPE Conference Proceedings* from 1995 Portland, OR.

2R. *Benchmarking* by Kathleen H.J. Leibfried and C.J. McNair. Published by Harper Business, 10 E. 53rd St, New York, NY 10022. One of the best books if you want to understand benchmarking.

3R. *The Book Of Five Rings* by Miyamoto Musashi. Published by Bantam Books, New York, NY. If you are interested in the Japanese view on business and business strategy, this is a translation of a 16th century book on swordsmanship used in Japan today to train new managers. We will need the resolve and willpower described in this book to effectively compete in the future.

4R. *Cleaning Management Magazine.* National Trade Publications, 13 Century Hill Drive, Latham, NY (518-783-1281). This is the leading magazine for the housekeeping professional and the publication of the CMI (Cleaning Management Institute). Each month is an educational experience. Take note of the cleaning calendar which lists training available.

5R. *The Complete Handbook of Maintenance Management* by John E. Heintzelman. Published by Prentice-Hall, New Jersey (TS192.H44). This is a very readable overview of the maintenance field.

6R. *Computerized Maintenance Management Systems* by Terry Wireman. Published by Industrial Press, 200 Madison Ave., New York, NY 10016. General rundown of the computer system vendors and system capabilities as of 1985. Some of the core items are still true today, and many of the vendors are still around.

447

7R. Deleading. Publication of the National Lead Abatement Council, published by Weil Communications, P.O. 535, Olney, MD 20832 (301-924-5490, Fax: 301-924-0265). This publication is the last work in lead abatement. Since the passing of new laws on lead abatement in 1992, this will be a bigger and bigger issue.

8R. Discipline of Market Leaders by Michael Treacy and Fred Wiersema. Published by Addison-Wesley Publishing Co., Reading, MA 01867 (800-238-9682). A very interesting read for maintenance management who are interested in seeing maintenance as a service business. Every business is great at something. What are you great at?

9R. Fleet Management by John Dolce. Published by McGraw-Hill. This is the best all-around fleet management book available. A recent new edition updates some of the older technical and computerization information.

10R. Juran on Leadership for Quality by J.M. Juran. Published by the Free Press Division of MacMillian, Inc., 866 Third Ave., New York, NY 10022. Juran is one of the three deans of the modern quality movement. It is essential to understand his body of work to be effective in quality issues.

11R. Handbook of Building Maintenance Management by Mel A. Shear. Published by Reston Publishing Co. (a Prentice-Hall Company), Reston, VA 22090 (TH3361.S45). This is a nuts and bolts book of building management. It combines ideas on the management of maintenance with the doing of maintenance. Excellent for maintenance departments that support the facility in addition to the equipment.

12R. Housekeeping Handbook for Institutions Business and Industry by Edwin Feldman. This excellent book (almost 500 pages, now quite old and still useful) is available from Ed himself (address in the People section of Resources). It is a complete handbook designed for the manager of the housekeeping department.

13R. How to Manage Maintenance by Joseph Johnstone and Kenneth Ward. Published by the American Management Association. The AMA is the largest professional education association that crosses functional boundaries. They have been conducting training in maintenance for several decades.

14R. *How to Slash Utility Bills* by Dr. John Studebaker. Published by Clemson University as a professional development seminar text. Using savings in utilities is one of the best ways to leverage maintenance activity.

15R. *If You Haven't Got Time to Do It Right, When Will You Find Time to Do It Over?* by Jeffery Mayer. Published by Simon and Schuster, New York. This short snappy book has excellent ideas and attitudes for the harried supervisor.

16R. *Maintenance Management* by Jay Butler. Available from Mr. Butler (see address in the People section). The author used a special version of Mr. Butler's book to teach maintenance management for several years. Much of my knowledge about the field comes from using the Butler text. It would be impossible to separate the knowledge gained from the text from knowledge gained independently. This book has been used to train 10,000 maintenance managers in the U.S., Canada, and the U.K.

17R. *Maintenance Management* by August Kallmeyer (out of print). One of the original seminar texts available for maintenance management. The author used this text to teach maintenance management for several years in the mid-1980's. Much of my knowledge about the field came from learning Mr. Kallmeyer's material and making it my own. Updates are still occasionally seen in seminars by its author.

18R. *Maintenance Management* by Don Nyman. Available from USC, 100 State Street, 4th Floor, Boston, MA 02110. This text is a summary of maintenance lore and information from Don's 30 years as a top level consultant. Excellent questionnaire to evaluate your department.

19R. *Maintenance Supervisor's Bulletin.* A twice-a-month newsletter with articles on all topics related to maintenance. Published by Bureau of Business Practice, Paramount Publishing Business Technical and Professional Group, 24 Rope Ferry Rd, Waterford, CT 06386-0001 (800-243-0876 X52).

20R. *Maintenance Supervisor's Standard Manual.* Published by Bureau of Business Practice, Prentice-Hall, Waterford, CT 06386. A good outline for a supervisor training (or self-training) effort. Brings up the concerns that all maintenance supervisors face.

21R. *Maintenance Technology* magazine. Published by Applied Technology Publications, 1300 S. Grove Ave., Barrington, IL 60010. This is an excellent trade publication that concentrates on the technology side of our businesses. (Free.)

22R. *Maintenance Time Management* by John Criswell. Published by The Fairmount Press, 700 Indian Trail, Lilburn, GA 30247 (404-925-9388). Good starter book on the systems of maintenance activity.

23R. *Managing Factory Maintenance* by Joel Levitt. Published by Industrial Press. Excellent overview of the specifics of factory maintenance management. Also presented as a live seminar in individual organizations.

24R. *One Page Management* by Riaz Khadem and Robert Lorber. Published by William Morrow, New York. This book is part of the One Minute Management library promised by Ken Blanchart and Spencer Johnson. All of these books have a common thread relating good time management to effective supervision. This book in particular speaks to the issue of paperwork and reporting (a major issue of supervision).

25R. *Out of the Crisis* by W.E. Deming. Published by MIT Press. This was the last major work before his passing. As such, it is a powerful summary of his life's work.

26R. *Plant Engineering Magazine*. Published by Cahners Publications, Cahners Plaza, 1350 E. Touchy Ave., P.O. 5080, Des Plaines, IL 60017-5080. PE is an excellent all-around journal for maintenance professionals. (Free.)

27R. *Project Management* by Project Management Mentors, located in San Francisco, CA. An excellent roundup of the technique and technology of managing projects.

28R. *Project Management* by John Reddish. Published by Advent Management Associates, Ltd. of West Chester, PA. This seminar introduces the basics of project management for all types of projects. Includes case studies.

29R. *Reliability Centered Maintenance RCMII* by John Moubray. Published by Industrial Press. A truly great book in understanding maintenance failure and how a department can work to get out of the traditional maintenance business. A "must read" for machine designers and serious students of maintenance.

30R. *Supplier Certification II: A Handbook for Achieving Excellence Through Continuous Improvement* by Peter Griece and Michael Gozzo. Published by PT Publications, Inc., 4360 North Lake Blvd., Palm Beach Gardens, FL 33410. Excellent primer in the field of purchasing. Your understanding best practices could help your organization's purchasing department serve you better.

31R. *Introduction to TPM* by Seiichi Nakajima. Published by Productivity Press, P.O. Box 3007, Cambridge, MA 02140 (617-497-5146). This book gives a complete overview of TPM. It and its companion volume below are essential reading to understand TPM.

32R. *Techno Trends* by Daniel Burris. Published by Harper Business, 10 E. 53rd St., New York, NY 10022. Technology has become the enabler of business strategies impossible 5 or 10 years ago.

33R. *Total Productive Maintenance*, An American Approach by Terry Wireman. Published by Industrial Press. This book explains the steps for TPM for an American organization. It has many excellent ideas for the organizations considering TPM. Terry is also the author of World Class Maintenance Management. For more information on the author, see the People section.

34R. *TPM Development Program* by Seiichi Nakajima. Published by Productivity Press, P.O. Box 3007, Cambridge, MA 02140. This book gives a complete systems design to the set-up and day-to-day working of a TPM system.

35R. *Uptime* by John Dixon Campbell. Published by Productivity Press. This is an excellent book to show upper management some of the issues of maintenance management in language that they can understand. See People section for more information on the author.

36R. *Why Buildings Fall Down* by Matthys Levy and Mario Salvadori. Catastrophe is a powerful teacher. Some organizations will only change after something goes very wrong. Levy and Salvadori's excellent book has many lessons for maintenance professionals.

People

1P. Jay Butler, 5 North Branch River Road, Somerville, NJ (201-725-3396) is an expert trainer and consultant in maintenance management. His special expertise is in computerization and application of predictive maintenance.

2P. John D. Campbell, Coopers and Lybrand Consulting, International Center of Excellence Maintenance Management, 145 King Street West, Toronto, Ont. M5H 1V8, Canada. Mr. Campbell is the author of *Uptime*. He is a widely traveled and knowledgeable consultant.

3P. Philip Christian is an expert in maintenance of grounds. His knowledge is used to reduce costs at universities, airports, and transit systems throughout the country. If you have a large exposure to grounds, seek him out.

4P. Edwin Feldman, P.O. Box 52729 Atlanta, GA 30355 (for list of books) has written several useful books about property maintenance and cleaning.

5P. Mark Goldstein, Ph.D. Dr. Goldstein is a revolutionary thinker in the field of maintenance and its connections to the corporate information systems. His many articles and seminars are shaping the cutting edge of the field.

6P. Ron Guilinti. Mr. Guilinti is an expert in maintenance stores and parts, and in the accounting for maintenance activity.

7P. Don Nyman, 67 Toppin Drive, Hilton Head Island, SC 29926 (803- 837-2264). Don is one of the leading experts in maintenance management, and leads seminars for Clemson University and others.

8P. Ron Turley, 801 Hopi Trail, Dewey, AZ 86327 (520-775-4897). One of the leading fleet consultants in the business.

9P. Terry Wireman. Terry is the resident guru in TPM for many organizations. He also teaches maintenance management courses for several universities.

Organizations that Publish Training Courses

1T. American Technical Publishers, 1155 West 175th St., Homewood, IL 60430 (800-323- 3471). A great source for trade training. This includes books for carpenters, electricians, and plumbers. Prices range from $10 to $50, with many of the offerings under $20.

2T. National Education Training Group, 1751 West Diehl Rd., Naperville, IL 60563-9099 (630-369-3000). Applied's catalog reads more like a phone book than a course catalog with over 2000 offerings. The two areas that caught my eye were hundreds of courses in shop disciplines, and hundreds more in computerization. Courses are available in interactive video, booklet, and in several languages.

3T. Baker Instrument Co., Box 587, Ft. Collins, CO 80522 (fax: 970-221-3013). Baker is a vendor of test instruments. They produce a seminar titled "Keep your motors running" concerning electrical testing for reliability and predictive maintenance.

4T. Clemson University, College of Commerce and Industry, Office of Professional Development, P.O. Drawer 912, Clemson SC 29633-0912. Clemson has a national program in maintenance management training with several excellent instructors. They visit almost every region on a periodic basis.

5T. Concordia, Center for Management Studies, 1550 Maisonneuve Blvd W, Suite 403, Montreal, P.Q., HSG1N2, Canada. Offers maintenance management seminars in both French and in English.

6T. Crisp Publications, 95 First St., Los Altos, CA 94022 (405-949-4888, fax 415-949-1610). The Crisp catalog offers a wide variety of books, cassettes, and videos. Topics run the gamut from time management, to business writing and 30 or 40 topics in between. Prices range from under $10 for books and cassettes to several hundred dollars for videos.

7T. Fisher-Rosemount Educational Services, 8301 Cameron Rd., Austin, TX 78754 (512-832-3111). Provider of distance training technology.

8T. One on One Computer Training, 2055 Army Trail Rd., Suite 100, Addison, FL 60101 (800-222-3547, fax: 630-628-0550). One of the many good training tools for computer expertise. Their audio training covers most major PC software. Includes a course guide. Prices range from $100 to $200 per course.

9T. General Physics Corp. Provides training services with over 250 courses in electrical, electronics, instrumentation, mechanical, and manufacturing. Contact Maintenance Technology (14T).

10T. Industrial Press, 200 Madison Ave., New York, NY 10016 (212-889-6330, fax: 212-545-8327). In addition to publishing the popular *Machinery's Handbook*, they publish several good maintenance management texts.

11T. Lion Technology Inc., P.O. Drawer 700, Lafayette, NJ 07848 (201-383-0800, fax: 201-579-6818). Lion is one of the leading firms in the area of training workshops on hazardous/toxic waste management; their popular 2-day workshop is listed at $600.

12T. Listen USA, 60 Arch St., Greenwich, CT 06830. Listen USA interviews top business managers on their theories of management. Cost $7.95 to $12.50. Available in bookstores.

13T. Maintenance Technology, 1300 South Grove Ave.,Suite 205, Barrington, IL 60010 (800-223-3423, fax: 847-304-8603). *Maintenance Technology* is one of the best maintenance magazines. They have developed a troubleshooting course complete with a hands-on troubleshooting platform. Their system includes workbooks and an administrators' guide.

14T. MW Corporation, 3150 Lexington Ave., Mohegan Lake, NY 10547 (914-528-0888). This organization specializes in up-to-date team-building seminars and tapes. If your manufacturing group is moving toward a continuous improvement process, see how maintenance can fit in.

15T. National Technology Transfer, P.O. Box 4558, Englewood, CO 80155-4558 (800-922-2820). This firm travels around the country and gives classes on important maintenance topics using unique fixtures with actual devices (hydraulic systems, PLC's, etc.) built in. It is an excellent training modality.

16T. Nightengale-Conant Corp., 7300 North Lehigh Ave., Niles, IL 60714 (800-323-5552). This is probably the best company in the business of self-improvement cassettes. The topics range from improving your selling skills to their monthly cassette magazine *Insight.*

17T. Pacific Luteran University, Center for Executive Development, Tacoma, WA 98447. PLU has an excellent continuing education department that sponsors a variety of maintenance courses in Seattle throughout the year.

18T. Perry Johnson, Inc., 3000 Town Center, Suite 2960, Southfield, MI 48075 (810-356-4410, Fax: 810-356-4230). This is one of the leaders in the area of ISO 9000 certification, consultation, and training.

19T. Ramsay Corp, Boyce Station Offices, 1050 Boyce Road, Pittsburgh, PA 15241-3907 (412-257-0732). This company designs and administers maintenance tests in all crafts.

We believe that it is essential to determine competency before training is prescribed.

20T. Rowan College, Management Institute, 201 Mullica Hill Road, Glassboro, NJ 08028. Rowan gives a certificate in maintenance management. They schedule courses in New Jersey twice a year.

21T. Springfield Resources, 902 Oak Lane Ave., Philadelphia, PA 19126-3336 (215-924-0270). This company specializes in maintenance management training. It markets several audio taped courses by the author in general maintenance, maintenance supervision, and industry-specific maintenance.

22T. TPC Training Systems, 750 Lake Cook Rd., Suite 250, Buffalo Grove, IL 60089 (847-808-4000). TPC is a craft training company. The courses consist of programmed learning texts, teacher's guide, and audio visuals. Some courses are available in Spanish.

23T. University of Alabama, Professional and Management Development Programs, Box 870388, Tuscaloosa, AL 35487-0388. UOA has one of the best conference centers in the South. They offer a wide range of courses for professional and management development.

Predictive Maintenance Equipment Vendors

Temperature (Infrared) Inspection

1V. Hughes Aircraft Company, Ind. Products Division, 6155 El Camino Real, Carlsbad, CA 92008.

2V. Kodak Film: Available on order from any commercial photography store. Ektachrome Infrared 135 (color negative film). Hi-speed Infrared 135 (Black and White film).

Ultrasonic Technology

3V. Panametrics, 221 Crescent St., Waltham, MA 02154 (617-899-2719). *Products:* Ultrasonic flaw detectors.

4V. U.E. Systems, 12 West Main St., Elmsford, NY 10523 (1-800-223-1325, in NY: 914-592-1220). *Product:* Handheld ultrasonic detectors.

Vibration Analysis

5V. Entek-IRD International, 1700 Edison Dr., Milford, OH 45150 (513-576-6151, fax: 513-576-6104). Products:

IRD is the "grand daddy" of the vibration field. Much of the original theoretical work was done by them after WWII. Excellent engineering reputation.

Visual Inspection

6V. Visual Inspection Technologies, 199 Highway 206, Flanders, NJ 07836 (201-927-0033). *Products:* This is typical of the visual inspection service companies.

Associations

1A. AMA (American Management Association), 135 West 50th St., New York, NY 10020. Seminars, books, and meetings about all facets of management.

2A. ATA (American Trucking Association), 2200 Mill Road, Alexandria, VA 22314. ATA is known for the VMRS (Vehicle Maintenance Reporting Standards), an excellent beginning to understand computerizing maintenance. Company membership is under $300/year (& up).

3A. Association of Energy Engineers, 4025 Pleasantdale Rd., Suite 420, Atlanta, GA 30340 (770-925-9558, fax: 404-381-9865), Order department: P.O. Box 1026, Lilburn, GA 30226. Texts available on all building systems that supply or consume (lighting, chillers, cooling towers, HVAC, pumps, EMS, etc.) energy. Other topics include energy audits, governmental policy, and engineering reference texts.

4A. AFE (Association for Facilities Engineering, formerly American Institute of Plant Engineers (AIPE)), 8180 Corporate Dr., Suite 305, Cincinnati, OH 45242 (513-489-2473). One of the more active associations for facilities engineering. They have extensive education opportunities. Their conferences are probably some of the best training values on the market today. Be sure to get a copy of the "Facilities Management Library," a list of 75 books for sale on building maintenance management.

5A. CMI (Cleaning Management Institute), 13 Century Hill Drive, Latham, NY 12110. A great source of information for the housekeeping professional. Branching into building maintenance support also.

6A. EMC (Equipment Maintenance Council), 113 Highland Lake Drive, Lewisville, TX 75067. Individuals can join for less than $100/year.

7A. Instrument Society of America. This association has developed interactive videos on systematic methods in troubleshooting. Each video contains 4–6 hours of instruction. Contact Maintenance Technology (14T).

8A. International Maintenance Institute, P.O. 266695, Houston, TX 77207. This organization is dedicated to furthering the goals of maintenance professionals. Dues are $25.

9A. PGMS (Professional Grounds Management Society), 120 Cockeysville Road, Suite 104, Hunt Valley, MD 21031 (Voice 410-584-9754, fax: 410-584-9756).

10A. Roofing Industry Education Institute, 14 Inverness Dr. East, Bldg. H, Suite 110, Englewood, CO 80112 (303-790-7200). This is one of many associations that offer training. Their courses include Reroofing and Retrofit (3-day, $595), and Roof Inspection, Diagnosis, and Repair (2-day $395).

11A. Society for Maintenance and Reliability Professionals, P.O. Box 3757, Barrington, IL 60011 (800-950-7354, fax: 312-527-6658). Network of professionals concerned about reliability.

Glossary

Accuracy: (From work sampling.) The tolerance limit of a study which falls within a desired confidence level, and a function of the number of readings. A larger number of readings will result in higher accuracy. Determine the accuracy that you require, and from that determine the number of observations.

Assembly: From ATA-VMRS. Part of the description of what work was done. The assembly defines the specific area worked upon. For example, air cleaner assembly or front brake assembly.

Asset: Either a machine, vehicle, building, or a system. It is the basic unit of maintenance.

Asset number: A unique number necessary to identify an asset. Useful in a manual system and essential in a CMMS. Same as fleet ID number and machine number.

Assignments and subcontracts: A purchase order cannot be assigned or the work subcontracted without written consent. You want to know who is actually doing the work (from terms on purchase order).

Autonomous maintenance: Routine maintenance and PM's are carried out by operators in independent groups. These groups, which may include maintenance workers, solve problems without management intervention. The maintenance department is only officially called for bigger problems that require more resources, technology, or downtime.

Backlog: All work available to be done. Backlog work has been approved, parts are either listed or bought, and everything is ready to go.

BNF equipment: Equipment left off of the PM system, left in the Bust 'n Fix mode (it busts and you fix—no PM at all).

459

Call back: Job where maintenance person is called back because the asset broke again or the job wasn't finished the first time. See *rework*.

Capital spares: Usually large, expensive, long lead time parts that are capitalized (not expenses) on the books and depreciated. They are protection against downtime.

Changes: (Terms on purchase order.) Must be agreed to in writing to be binding. Verbal changes are the subject of a large portion of lawsuits with vendors. This protects both organizations.

Charge-back: Maintenance work that is charged to the user. All work orders should be costed and billed back to the user's department. The maintenance budget is then included with the user budgets. Also called rebilling.

Charge rate: This is the rate in dollars that you charge for a mechanic's time. In addition to the direct wages, you add benefits and overhead (such as supervision, clerical support, shop tools, truck expenses, supplies). You might pay a tradesperson $15/hr and use a $35/hr (or greater) charge rate.

Class: (From ATA-VMRS.) Groups of *like* equipment in *like* service. Classes might include 12 bay straight beverage trucks in city delivery, UPS vans on rural routes, class 8 tractors, etc.

CM: See *corrective maintenance*.

Computerphobia: Irrational fear or dread of computers.

Confidence level: (From work sampling.) The percentage of confidence that your conclusions represent the true picture of what you are sampling. A confidence level of 95% means that 95% of the time your observations reflect reality and 5% you are off base.

Continuous improvement: In maintenance, reduction to the inputs (hours, materials, management time) to maintenance to provide a given level of maintenance service. Increases in the number of assets, or use of assets with fixed or decreasing inputs.

Core damage: When a normally rebuildable component is damaged so badly that it cannot be repaired.

Corrective maintenance (CM): Maintenance activity that restores an asset to a preserved condition. Normally initiated as a result of a scheduled inspection. See *planned work.*

Deferred maintenance: This is all of the work you know needs to be done that you choose not to do. You put it off, usually in hope of retiring the asset or getting authorization to do a major job that will include the deferred items.

Delivery: (From purchase order.) The promised date is important (time is of the essence). If a vendor fails to deliver in the specified time, you can cancel the order without any damages, or buy the product from another vendor and charge the first vendor the difference. For some items, delivery time is critical. This alerts the vendor that you can get damages from them if they miss deliveries (from terms on purchase order).

Design responsibility: (From purchase order.) The seller is totally responsible for the design of the product. The purchase order is a request to meet specifications, not a design of a product. If the design is defective, this places the blame on the vendor and protects your organization from certain liabilities (from terms on purchase order).

DIN work: "Do It Now" is non-emergency work that you have to do now. An example would be moving furniture in the executive wing.

Emergency work: Maintenance work requiring immediate response from the maintenance staff. Usually associated with some kind of danger, safety, damage, or major production problems.

Explicit task: PM where the task is completely described with pictures if necessary. Usually associated with TPM where non-maintenance people do maintenance tasks.

Failure code: Why did the part fail (broken, worn through, bent, etc.). From ATA-VMRS.

Feedback: (When used in the maintenance PM sense.) Information from your individual failure history is accounted for in the task list. The list is increased in depth or frequency when failure history is high, and decreased when it is low.

Fleet ID number (FID): Unique number assigned to each truck when it is purchased. FID must be used on all repair orders, fuel tickets, trouble tickets. Same as asset number. From ATA-VMRS.

FOB: (From purchase order.) (City, Shipping Point, or Delivered.) Free On Board (seller will load truck or rail car). The FOB point is important because of both the responsibility for the shipment and the freight charges. FOB Delivered keeps the vendor responsible for the shipment until it reaches your door. FOB Shipping Point or FOB Originating City makes you responsible for the shipment.

Force majeure: (From purchase order.) Either party can be excused from performance by Acts of God, fires, labor disputes, acts of government, and a whole list of possible problems.

Frequency of inspection: How often do you do the inspections? What criteria do you use to initiate the inspection? See *PM clock.*

FTFR: First Time Fix Rate. This measure is how often the technician fixes the problem on the first visit. One of the most common ploys is when a service organization is backed against the wall with calls, they send an unqualified person to "shut the customer up." This unsavory practice is immediately detected by an abominable FTFR.

Future benefit PM: PM task lists that are initiated by a breakdown rather than a usual schedule. The PM is done on a whole machine, assembly line, or process after a section or subsection breaks down. This is a popular method with manufacturing cells where the individual machines are closely coupled. When one machine breaks, then the whole cell's PM is initiated.

GLO: Generalized Learning Objective is the general items necessary to know to be successful in a job. Each job description would be made up of a series of GLO's.

Iatrogenic: Failures that are caused by your service person.

Idle: (From work sampling.) Person is not working, traveling, or waiting (not having lunch either). Drinking coffee is considered idle as well as wandering around the shop is idle.

Implicit task: PM task that relies on the skill, knowledge and experience of the mechanic. Task is not spelled out. Implicit tasks are dangerous where you cannot guarantee the competency of the inspectors.

In-bin work: Maintenance jobs not ready to release to the mechanic. You haven't approved, gotten money, parts are on order and not in, or other problem.

Inspection list: See *task list.*

Inspectors: The special crew or special role that has primary responsibility for PM's. Inspectors can be members of the maintenance department or can be members of any department (machine operators, drivers, security officers, custodians, etc.).

Interruptive (task): Any PM task which interrupts the normal operation of a machine, system, or asset.

Labor: Physical effort a person has to expend to repair, inspect, or deal with a problem. Expressed in hours, and can be divided by crafts or skills.

Life cycle: This denotes the stage in life of the asset. Three stages are recognized by the author: start-up, wealth, and breakdown.

Life cycle cost (LCC): A total of all costs through all of the life cycles. Costs should include PM, repair (labor and parts and supplies), downtime, energy, ownership, overhead. An adjustment can be made for the time value of money.

Log sheet: A document where you make log entry of all small jobs or short repairs.

Maintainability improvement: Also called maintenance improvement. Maintenance engineering activity that looks at the root cause of breakdowns and maintenance problems, and designs a repair that prevents breakdowns in the future. Also includes improvements to make the equipment more easily maintained.

Maintenance: The dictionary definition is "the act of holding or keeping in a preserved state." The dictionary doesn't say anything about repairs. It presumes that we are acting in such a way to avoid the failure by preserving the asset.

Maintenance prevention: Maintenance-free designs resulting from increased effectiveness in the initial design of the equipment.

Management: The act of controlling or coping with.

Meter master: Form designed to record meter readings. There is also space for the subtraction for usage calculations.

MSDS: Material Safety Data Sheets. These sheets should come with any chemicals that you purchase. They give the formal name of the chemical, describe its toxicity, and have warnings on use. One master copy should be kept in the maintenance technical library.

MTBF: Mean Time Between Failures. Important calculation to help set up PM schedules and to determine reliability of a system.

MTBF&V: Mean Time Between Failure and Visit. This measure is used in field service. It simply calculates how long it takes to get a response. This measure goes well beyond time from call to time technician shows up at the door. It also includes the time it takes to realize there is a failure.

MTL: See *technical library.*

MTRV: Mean Time from Request to Visit. This is a subset of the more inclusive measure MTBF&V. If you call in for copier service at 11 am, when do you expect a visit? What level of response would exceed your expectations?

MTTR: Mean Time to Repair. This calculation helps determine the cost of a typical failure. It also can be used to track skill level, training effectiveness and effectiveness of maintenance improvements.

MTTT: Mean Time to Travel. How long does it take to travel to the site of the breakdown? Can be a critical issue for field service and widely dispersed maintenance operations.

Non-interruptive task list: PM task list where all of the tasks can safely be done without interrupting production of the machine.

Non-scheduled work: Work that you didn't know about and plan for at least the day before. Falls into three categories: emergency, DIN, routine.

Observation: (Special definition from work sampling.) Note the activity going on at the instant you look. Do not prejudge or anticipate the next activity.

Occurrence, percentage of: (From work sampling.) To determine the percent of occurrence of an activity simply divide the total observations in the activity by the total observations for all activities. In a study with 1000 observations, if there were 200 observations of waiting, then the percentage of occurrence of waiting would be 20% (200/1000 = 0.20).

Parts: All of the supplies, machine parts, and materials to repair an asset, or a system in or around an asset.

PCR: Planned Component Replacement. Maintenance schedules component replacement to a schedule based on MTBF, downtime costs, and other factors. Technique for ultra-high reliability favored by the aircraft industry.

Pending work: Work that has been issued to a mechanic or contractor that is unfinished. It is important to complete all pending work.

Planned maintenance: See *scheduled work*.

PM: Preventive maintenance is a series of tasks that either 1) extend the life of an asset, or 2) detect that an asset has had critical wear and is going to fail or break down.

PM clock: The parameter that initiates the PM task list for scheduling. Usually buildings and assets in regular use days (for example, PM every 90 days). Assets used irregularly may use other production measures such as pieces, machine hours, or cycles.

PM frequency: How often the PM task list will be done. The frequency is driven by the PM clock. See frequency of inspection.

Predictive maintenance: Maintenance techniques that inspect an asset to predict if a failure will occur. For example, an infrared survey might be done of an electrical distribution system looking for hot spots (which would be likely to fail). In industry, predictive maintenance is usually associated with advanced technology such as infrared or vibration analysis.

Price warranty: (From terms on purchase order.) The seller warranties that the price charged is the best price charged to any customer for this volume. Price shown on the P.O. is complete and there are no hidden charges. This keeps the vendor from charging you a higher price than anyone else buying that quantity. Hidden charges have to be unhidden to be acceptable.

Priority: The relative importance of the job. A safety problem would come before an energy improvement job.

Proactive: Action before a stimulus (antonym is reactive). A proactive maintenance department acts before a breakdown.

Probability, the law of: (From work sampling.) We use the laws of probability as the mathematics to "prove" the rightness of work sampling. By applying the formulas, we can prove that if you take so many observations randomly, then your results will conform to the actual situation with a certain accuracy and confidence level.

Randomly selected time: (From work sampling.) A day has 480 minutes. A randomly selected time means that each minute has an equal chance (or probability) of being selected. This is true each time a selection is made (after a selection of a specific time is made, that time has an equal chance of being selected again).

Reason for write-up: (Also called reason for repair). Why the work order was initiated. Reasons could include PM activity, capital improvements, breakdown, vandalism, and any others needed in that industry.

Repair order: Primary repair document. All work except DIN and Emergency must have a repair order issued before the work is started. Repair order required for any parts. DIN, emergency repair order to be written up when job is complete or when parts are required. Same document as work order. From ATA-VMRS.

Repair site: Location of the repair—in the terminal, on the road, in the shop, etc. From ATA-VMRS.

Rework: All work that has to be done over. Rework is bad and indicates a problem in materials, skills, or scope of the original job. See *call back.*

RM: Replacement/rehabilitation/remodel maintenance. All activity designed to bring an asset back into good shape, upgrade an asset to current technology, or make an asset more efficient/productive.

Root cause (root cause analysis): The root cause is the underlying cause of a problem. For example, you can snake out an old cast or galvanized sewer line every month and never be confident that it will stay open. The root cause is the hardened buildup inside the pipes which necessitates pipe replacement. Analysis would study the slow drainage problem and figure out what was wrong and also estimate the cost of leaving it in place. Some problems (not usually this type of example) should not be fixed. Root cause analysis will show this.

Route maintenance: Mechanic has an established route through your facility to fix all the little problems reported to them. The route mechanic is usually very well equipped so they can deal with most small problems. Route maintenance and PM activity are sometimes combined.

Routine work: Work that is done on a routine basis where the work and material content is well known and understood. Example is daily line start-ups.

Scheduled work: Work that is written up by an inspector and known about at least 1 day in advance. The scheduler will put the work into the schedule to be done. Sometimes the inspector finds work that must be done immediately which becomes emergency or DIN. Same as planned maintenance or corrective maintenance.

Short repairs: Repairs that a PM or route person can do in less than 30 minutes with the tools and materials that they carry.

SLO: Specific Learning Objective is the detailed knowledge, skill, or attitude necessary to be able to do a job.

SM: Seasonal Maintenance. All maintenance activities that are related to time of year or time in business cycle. Cleaning roof drains of leaves after the autumn would be a seasonal demand. A swimming pool chemical company might have some November activities to prepare for the next season.

String based PM: Usually simple PM tasks that are strung together on several machines. Examples of string PM's would include lubrication, filter change, or vibration routes.

Survey: A formal look around. All of the aspects of the facility are recorded and defined. The survey will look at every machine, room, and throughout the grounds. The surveyor will note anything that looks like it needs work.

SWO: Standing Work Order. Work order for routine work . A standing work order will stay open for a week, month, or more. The SWO for daily furnace inspection might stay open for a whole month.

System code: Describes the general area worked upon. For example, an engine or cab system. Full description of all system codes on back of repair order. From ATA-VMRS.

Talking: (From work sampling.) In conversation with another employee or supervisor. It is not necessary to know the content of the conversation or whether job-related information is being exchanged.

Task: One line on a task list (see below) that gives the inspector specific instruction to do one thing.

Task list: Directions to the inspector about what to look for during that inspection. Tasks could be inspect, clean, tighten, adjust, lubricate, replace, etc.

Technical library: (Maintenance Technical Library—MTL.) The repository of all maintenance information including (but only limited by your creativity and space) maintenance manuals, drawings, old notes on the asset, repair history, vendor catalogs, MSDS, PM information, engineering books, shop manuals, etc.

Termination: (From purchase order.) For stock items you can cancel the order and only pay for items shipped. For a custom item you will pay for only the work accomplished. Certain overhead and profit might be paid after negotiation. Termination before completion of an order is always a sticky affair. This limits your liability. (From terms on purchase order.)

Terotechnology: "A combination of management, financial, engineering and other practices applied to physical assets in pursuit of economic life-cycle costs (LCC). Its practice is concerned with specification and design for reliability and maintainability of plant machinery, equipment, buildings, and structures with their installation, commissioning, maintenance, modification, and replacement and with feedback of information on design, performance and costs." (From the definition endorsed by the British Standards Institute.)

TPM: Total Productive Maintenance. A maintenance system set up to eliminate all of the barriers to production. It uses autonomous maintenance teams to carry out most maintenance activity.

Travel: (From work sampling.) Transporting him/herself, tools, materials to and from a job, such as a road call or work undertaken in the yard; or transporting the unit to the mechanic. The amount of time is the travel time.

UM: User Maintenance. This is any maintenance request primarily driven by a user. It includes breakdown, routine requests, and DIN jobs.

Unable to locate: (From work sampling.) Cannot find person at assigned job. Do not look.

Unit: The asset that the task list is written for in a PM system. The unit can be a machine, a system, or even a component of a large machine.

Utilization: Measure of use (mileage, hours, tons, yards, feet, etc.).

Waiting: (From work sampling.) Person is observed waiting for materials, tools, another crew member, instructions, next job; includes "looking for" type activities. Sometimes this activity is hard to distinguish from idle.

Warranties: Any products bought by P.O. are warranted by the vendor, are fit for the intended purpose, and should be free from defects. Seller to fix, replace defective items, or refund all money including shipping charges. Products should do what they are supposed to do—and if not, you can return them. (From terms on purchase order.)

Work order: Written authorization to proceed with a repair or other activity to preserve a building.

Work request: Formal request to have work done. Can be filled out by the inspector during an inspection on a write-up form or by a maintenance user. Work requests are usually time/date stamped.

Index

ABC (activity based costing), 166

ADA (American's with Disabilities Act), 377

ANSI ASQCQ90, 146

access to equipment, 222, 396, 398

accounting issues , 28, 164, 175, 354–356

asset number (also ID number), 50, 224, 313, 463

asset inventory, 264–265

autonomous maintenance see TPM, 463

average rate of return (also see ROI), 30

BNF (bust'n'fix) equipment, 21, 23, 225, 464

backlog, 179, 321, 464

bar code, 188, 192

bath tub curve, 38

benchmarks, 28, 66–76, 107, 154, 440–441

Belanger, Tom, 284

big ticket item, 345

bolting, 21, 131, 210, 235, 238–239

breakdown phase, 41, 232

budget items, 45, 56, 87–95

building automation, 427

building maintenance, 42–47, 60, 413–436

building maintenance, saving money, 428–436

Butler, Jay, 29, 181, 243, 263, 265, 344, 451, 454

CD ROM, 320, 384, 445

CM (corrective maintenance), 72, 90, 211, 214, 465

CMMS (Computerized Maintenance Management System), 51, 187, 262, 298–334, 395

CMMS costs, 304–306

CMMS, evaluation and shopping for, 309–312, 320–324

CMMS, installation (also see PM, install), 299, 310

CMMS return on investment (ROI), 301–304

call back, , 68, 76, 464

capital spares see insurance policy stock

Campbell, John, D., 24, 119, 454, 455

certified operators, 24, 39, 379

charge-back, 173–175, 194, 356, 464

charge rate, 46, 169–173, 464

chemical analysis (see oil analysis)

Christianson, Philip, 434, 455

class of equipment, 287, 464

cleaning, 21, 131, 210, 236–238

communications, 110–111, 125, 157, 193, 194, 300, 364, 372

competition 6, 397

component numbering, 265–266

component rebuilding, 332, 410

computerphobia., 307

condition based PM, 221

consequence of failure, 5, 10–12, 155, 162

consignment inventory, 334

contracting see outsourcing

continuous improvement (in maintenance): 2, 141, 325, 465

corrective maintenance (see CM)

cost of acquisition, 171, 342, 344

cost of possession , 171

cost control, 6, 194, 355

cost cutting, reduction, 6, 21, 69, 158, 268–269, 354

cost saving ideas see building maintenance, saving money

credit, 164

cross training see multi-skilling

customer service, 106–111

DOT (Department of Transportation), 5, 205, 215

Damiani, A.S. Migs, 363

daily transactions (also see work order), 312–314

data audit, 192, 224, 313

data entry, 189–193

data needed for analysis (also see work order sections), 300

date wanted, 201

debit, 164

deferred maintenance, 465

Deming, W.E., 140–145, 453,

deterioration, 37–41, 42, 247

disaster planning, 269–272, 361

distance training (also distance learning), 385, 388

Dolce, John, 172, 450

downsize, 113, 398

downtime, 6, 36, 75, 134, 207, 395–396

downtime costs, 102, 133, 198, 202, 325

EDI (Electronic Data Interchange), 335

EPA, 5, 205, 215

EOQ (economical order quantity), 337, 340, 341–344

E-mail, 110, 177, 350

environment, 104

equipment effectiveness, 130–135

estimating, 287–291

experiments, 303

expert systems, 444–445

explicit PM task, 132, 230, 466

FAQ (frequently asked questions), 346
FTFR (First Time Fix Rate) , 440, 467
factory maintenance, 60, 129–136, 157, 394–400
failure analysis see root failure analysis
failure experience, 217
failure modes, 161
feedback (when used in the maintenance PM sense), 217, 303, 466
Feldman, Ed, 128, 136, 268, 431, 451, 455
field service, 437–446
flat rates, 288–289
fleet maintenance, 59, 130, 401–412
fleet fuel consumption, 403–407
fleet tire analysis, 407–408
fleet vehicle specification, 408–409
frequency of inspection, 216, 217, 226–227
front end filter, , 182–183
full service leasing, 23, 26, 157
function (of maintenance dept), 1, 108, 120, 139
functional failure, 160
future benefit PM, 220–221

Goldstein, Mark, PhD, 5, 113, 171, 342, 352, 394, 455
grounds maintenance, 422–424
guaranteed maintainability, 57, 58, 136–138, 140
Guintini, Ron, 168, 439, 455

historical work standards, 289
housekeeping, 415–422
housekeeping chemicals and supplies, 419–422
housekeeping contracts, 416–417
housekeeping mechanization, 422
housekeeping work order, 417

ID numbers (see asset number)
ISO 900X, 137, 146–152, 262, 379
ISO 14000, 152
iatragentic, 155, 467
implicit PM task, 230, 467
in-bin work, 178
incident report , 188
infantile mortality, 37–39
infrared scanning, 255–258
insourcing, 129–135, 399
inspectors see PM inspector
inspection list see task list
insurance policy parts model, 326, 355
Internet, 346–353, 444
interruptive (task) , 223, 468
inventory control (also see stock room and parts), 336–345
inventory control, elements of, 337
inventory control, inadequate, 337
inventory reduction technique, 345–346
inventory types, 336

JIT (Just In Time), 135, 398

janitor's closet, 419
job control center see work control center
job tracking, 295
Jones, Edwin, 114

latent defect analysis, 40
leadership, 141, 363–374
legends, 363
life cycle cost (LCC), 23, 96–105, 409, 468
lock-out/tag out, 198, 201
logistical support for maintenance strategy (also see chapters 28–40), 20
log sheet, 197,
loss prevention, 411–412
lubrication, 21, 131, 156, 239–241
lubrication, automated systems, 241–242

MSDS sheets (Material Safety Data Sheets), 50, 469
MRP (Materials Requirements Planning), 312, 394–395
MTBF (Mean Time Between Failures), 75, 155, 245–246, 276, 322
MTBF&V (Mean Time Between Failure and Visit), 75, 273, 440
MTR (see MTBF&V)
MTTR: (Mean Time To Repair), 75, 155, 276, 440
MTRV (Mean Time from Request to Visit), 440
MTTT (Mean Time to Travel), 441

maintenance costs (categories), 100
maintenance economic analysis, 29–34, 42–47,
maintenance engineering, 41, 155, 157, 227, 399
maintenance history, 321
maintenance improvement., 25, 41, 157, 398–399
maintenance inventory (see inventory, parts, stock room), 321, 322
maintenance planner, 281
maintenance planning, 53, 54, 193, 196, 278–281, 365, 371
maintenance planning tasks, 280–281
maintenance prevention, 131
maintenance ratio, 43
maintenance request, 180–184
maintenance scheduling, 292–297, 321
maintenance scheduling problems, 296–297
maintenance scheduler, 294–296
maintenance scrap book, 263
maintenance technical library, 261–263
masterfiles, 314–318
McGarry, Sean, 432
Mianus River bridge, 10
motivation, 368–374
Moubray, John, 160, 215, 232, 243, 247, 453
multi-skilling, 297, 380

NUCREC, 205
non-interruptive task list, 223, 469

non-scheduled work, 202–204

Nyman, Don, 73, 75, 217, 452, 455

OSHA, 5, 9, 205
oil analysis, 241, 250
operating costs, 100
other functions carried out by maintenance, 108, 120, 139
out source, 23, 24, 120, 121–128, 322, 334, 400, 413
overhead costs, 46, 87, 88, 101, 169–170, 173–173
ownership, 371
ownership costs, 98–99

PCR (Planned Component Replacement), 23, 157, 242–246, 276
PM (Preventive Maintenance), 24, 52, 130, 153, 156, 209–260, 276, 470
PM, cost analysis, 40, 41, 52–53, 90, 214, 225, 232, 323
PM clock, 216–219, 226, 470
PM frequency, 216, 470
PM inspector, 224, 227–228
PM install, 209, 223–227
PM task force, 223
PM tasklist development (also see tasklist), 213, 228–230, 231–246, 303
paperless work order, 189
parts, 55, 71, 157, 167, 195, 296, 303, 325–335, 470
parts catalog, 328
parts, usage, 327–329
parts, cross reference (also called interchange), 329

parts, vendors, 330–332
partnership with vendors, 23, 241, 332–333, 358
pattern, maintenance (also see chapters 4, 4–11), 10–19, 20
patterns, company, 20–26
payback, 31, 34
pending work, 179
perceived level of service, 7
Peterson, David, 68, 70, 75, 76
physical inventory of parts, 266
planned discard (also see PCR), 243
planned maintenance, 72, 177
planned overhaul (rebuild), 243
planning see maintenance planning
predictive maintenance, 153, 210, 226, 323, 471
productivity, 68, 77–86, 103
priority, 198, 204
present value, 32
proactive, 21
process of maintenance, 1, 140
project management, 194, 282–286
purchasing issues in maintenance, 140, 149, 356–360
purchasing blanket orders, 328, 334
purchasing, improve the relationship, 359
quality, 133, 139–145, 146–152, 413, 442

RAV (Replacement Asset Value), 43, 44, 47

RCM (Reliability Centered Maintenance), 23, 40, 41, 158, 160, 225, 232

RIME (system for assigning priority), 205

RM (Rehabilitation Maintenance), 91, 93, 94

random deterioration, 37

random time table, 84

reason for write-up (also called reason for repair), 49, 198, 202, 471

reengineering, 112–120, 153–163, 399

relief team (in housekeeping), 418

remanufacture see component rebuild

repair order see work order

reorder point, 337, 341–345

repair history file, 211

resource leveling, 285

return on investment (ROI), 29, 155

rework, 50, 472

root cause (and root cause analysis, root failure analysis), 155, 195, 207, 472

route maintenance, 414–415, 472

routine work, 204, 472

SM (Seasonal Maintenance), 344

SWO (Standing Work Order), 197

safety, 7, 104, 194, 360–361, 411–412

scrap, defects, 133, 139

scheduling see maintenance scheduling

scheduled work, 211, 473

selling maintenance improvements, 27

service contracts, 121–128, 443–444

service values, 108, 139

short repairs, 211, 473

short sightedness, 15, 143

slotting, 288

specifications, 358

speed losses, 133

social demands, 91, 93–94

stakeholders, 35,

standing work order, 197

start-up phase, 39

start-up counter measures, 39

statistics, 151, 273–277, 441

strategy, maintenance, 20, 22

strategic planning

strategies, alternative

stock-out, 340–341

stock room (also see inventory, parts), 153, 167, 336–345

stock room computerization, 339–341

stock room layout, 338

string based PM , 199, 220, 324

supervision, 366–368, 381

survey, 204, 262

TLC (Tighten, Lubricate, Clean), 21, 23, 40, 130–135, 156, 209, 235–241

TPM (Total Productive Maintenance), 23, 130–135, 156, 453, 454, 474 task list (also see PM), 156, 189, 209, 474

technical library see maintenance technical library

tools, 264

top management's views, 63–65

traffic, 360

training, 58, 141, 151, 195, 211, 370, 375–393

training, adults guidelines, 390–393

training, job analysis, 377

training, organization, 378–379

training, sources, 382–386

tribology, 240

UM (user maintenance), 90, 475

ultimate customer service, 107–109, 258–259

ultrasonic inspection and detection, 258–259

un-funded maintenance liability, 213

unions, 113, 365, 381, 386

unit based PM, 219

VMRS (Vehicle Maintenance Reporting Standard), 401–403, 460

vertical scheduling (in housekeeping), 418

Veteran's Stadium, Philadelphia, 15

vibration analysis, 253

visual inspection, 10, 259–260

warranty recovery, 196, 302

wealth phase and counter measures, 40

Williamson, Robert, , 115

Wireman, Terry, 454,455

work control center, 175

work order, 49, 186–208, 475

work order design, 187

work order header, body, 199–208

work request (also see maintenance request) 475

work sampling, 80–86

work standards, 227

write-up form see work order

year 2000 bug, 318

zero-base budget, 89–95